International Encyclopaedia of ENERGY RESOURCES

International Encyclopaedia of ENERGY RESOURCES

— *Volume 3* —

— *Edited By* —
Dr. M.P. Singh
Head of the Department of Forest Sciences
Birsa Agricultural University
Ranchi - 834 006 (Jharkhand)

2005
DAYA PUBLISHING HOUSE
Delhi - 110 035

ISBN 81-7035-357-2 (SET)
ISBN 81-7035-358-0 (Vol. 1)
ISBN 81-7035-359-9 (Vol. 2)
ISBN 81-7035-360-2 (Vol. 3)
ISBN 81-7035-361-0 (Vol. 4)

Published by : **Daya Publishing House**
1123/74, Deva Ram Park
Tri Nagar, Delhi - 110 035
Phone: 27383999
Fax: (011) 23244987
e-mail: dayabooks@vsnl.com
website: www.dayabooks.com

Showroom : 4762-63/23, Ansari Road, Darya Ganj
New Delhi - 110 002
Phone: 23245578, 23244987

Laser Typesetting : **Classic Computer Services**
Delhi - 110 035

Printed at : **Chawla Offset Printers**
Delhi - 110 052

PRINTED IN INDIA

Preface

This book is for people who still think that there are energy problems ahead, and who are not satisfied with an energy future dominated by coal and nuclear power.

Fortunately, the time has come when one can examine other alternatives seriously–in particular, energy conservation and renewable energy–without being relegated at once to the outer fringes. Southern California Edison, Pacific Gas and Electric, and numerous other progressive utilities, having embraced these alternatives, are a bit too formidable to be drummed out of the corps. The Office of Technology Assessment, the Gas Research Institute, and the Harvard Business School likewise are not thought of as outposts of a lingering counterculture.

Still, not enough people in the energy supply industries, in government, or in private agencies concerned with energy policy are aware of the full range of possibilities in energy conservation and in the numerous renewable energy technologies. For them, this book is an attempt to lay out, in an ordered fashion, the basis in fact, logic, and judgment, for an energy -efficient, renewable energy future.

One cluster of views about energy may be characterised as "supply oriented." The response to the oil shortage was quite expectedly framed in terms of increasing energy supply, especially in forms other than oil. The basic premises motivating supply-side arguments included the following assumptions:

Energy demand will grow with output. A healthy and growing economy requires a growing supply of energy. In early formulations the rate of energy supply growth was about equal to the growth rate of the economy.

While this volume does provide a great deal of information about energy sources, its more appropriate use is as a stimulus for discussion. Towards this end the book often raise more questions than they answer. The complexity of the ethical dilemmas posed by energy sources is such that we are still trying to identify whai questions should be asked. The volume recognises that the first step in making the types of ethical decisions which are increasingly demanded of us is searching for the right questions. The book vary greatly in both length and orientation. This will make it useable by a wide range of groups–from students in the classroom, to professionals such as scientists, medical doctors, business people, lawyers, and ethicists, as well as environmental and other socially active groups. It is hoped that this volume will contribute in a better way to a society.

Dr. M.P. Singh

Contents

74

South Africa

The Republic of South Africa, a major coal producer and exporter, has a highly developed synthetic fuel industry and small reserves of oil. South Africa is the continent's largest energy consumer (primarily coal and petroleum) and Africa's second largest energy producer (mainly coal) behind Algeria. South Africa is a prominent member of the Organization of African Unity (OAU), the Southern Africa Development Community (SADC), and the Southern Africa Customs Union (SACU).

General Background

South Africa's economic growth averaged only 1.8 per cent from 1980-2000, but has picked up in the last three years; real gross domestic product (GDP) grew at 2.5 per cent in 1999, 3.1 per cent in 2000 and 2.3 per cent in 2001. Real GDP is predicted to grow by 2.3 per cent in 2002. In the 1980's, South Africa experienced double-digit inflation, but the government's fiscal policies have helped to reduce inflation significantly. Inflation was 5.3 per cent in 2000, and 6.3 per cent in 2001. Despite these positive trends, Foreign direct investment remains below expectations. Fixed investment, which accounted for 27.5 per cent of GDP in 1981, accounted for only 14.9 per cent of GDP in 2000. South Africa's currency, the rand, lost 6.5 per cent of its value against the dollar during the first eight months of 2001. Following the terrorist attacks on the United States, the rand lost nearly 10 per cent of its value against the dollar, the euro, and the U.K. pound. The falling currency, when combined with higher prices for imported oil, has prompted fears of increased inflation in the country.

Coal

Coal is the primary fuel produced and consumed in South Africa and is one of the country's largest sources of foreign exchange. The country's coal reserves are mainly bituminous, with relatively high ash content (about 45 per cent) and low sulfur content (about 1 per cent). South Africa's recoverable coal reserves are estimated at 61 billion short tons (6 per cent of the world reserves). The South African government plans to conduct a national study to determine the extent of the country's

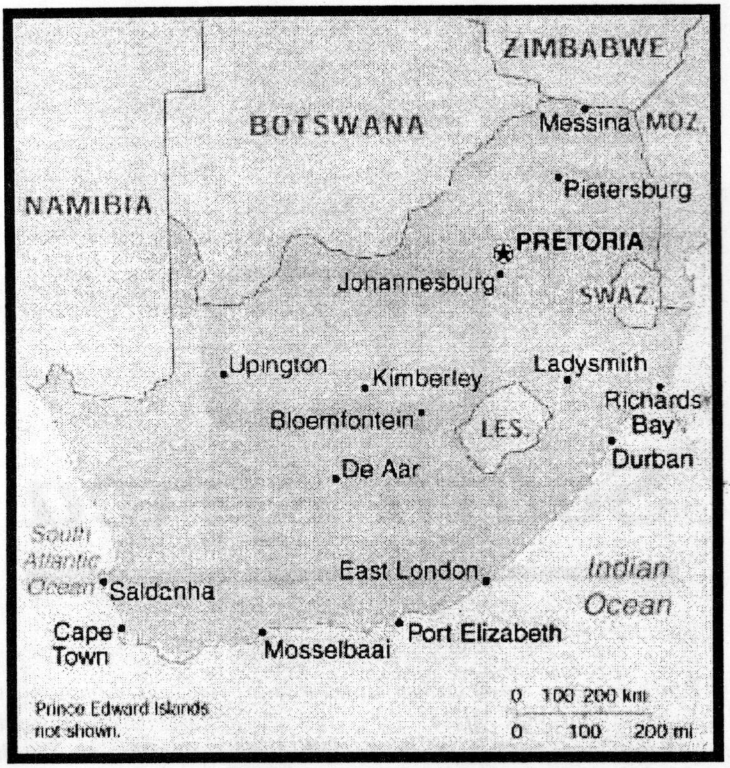

remaining coal resources. The study will also survey the discard (waste) coal from current and former coal operations in the country. As much as 1 billion tons of discard coal is on the surface in South Africa.

Production

According to data from the South Africa's Department of Mineral and Energy Affairs (DME) and Chamber of Mines, coal production in 2000 was 247.2 million short tons (mmst) or 224.3 million metric tons. In 1999, South Africa was the world's sixth largest coal producer. BHP Billington subsidiary, Ingwe Coal (Ingwe), Anglo American's coal division, Anglo Coal (Anglo), and Swiss-based Glencore's Enex Resources (Enex), are the largest coal producing companies in South Africa. Coal production in South Africa is concentrated in a few regions, with the Mpumalanga Province accounting for 80.8 per cent of the national total. The Northern Province produced 10.3 per cent, the Free State 7.5 per cent and KwaZulu-Natal produced the remaining 1.4 per cent. Elitheni Coal is studying the possibility of developing coal deposits in the Emalahleni district of the Eastern Cape Province, where coal was first mined from 1857 to 1917. Mining was discontinued as larger reserves with better quality coals were found in Mpumalanga and KwaZulu-Natal. There is currently no market for the Emalahleni coal, a low-grade smokeless anthracite. Elitheni Coal and Marubeni Europower of London, a subsidiary of Japan's Marubeni Corporation, have formed a joint-venture with the objective of establishing boiler operations in industrial areas to supply steam.

As the South African mining industry is still predominantly White-controlled, emphasis has been placed on stimulating Black-owned companies in the coal industry, as well as throughout the energy sector as a whole. In compliance with South African law and the objectives stated in the government's Energy Policy White Paper, large corporations have made room–often through the sale of existing assets–in the South African energy sector for emerging Black-owned firms. These Black-owned firms are commonly referred to as "empowerment" groups by industry, the government and the South African media. In November 2000, Anglo and Ingwe sold assets for $222 million to the black empowerment group Eyesizwe Coal (Eyesizwe), creating South Africa's fourth largest coal mining company. Anglo (11 per cent) and Ingwe (9 per cent) will retain equity interest in Eyesizwe and have undertaken to hold their shares for at least two years. Eyesizwe anticipates production of 19.9 mmst per year, of which nearly three-quarters is under contract to the state-owned electric utility, Eskom.

In November 2001, Eyesizwe announced that it had initiated talks with Kumba Resources (Kumba) about establishing a strategic partnership in the South African coal sector. The deal would create the second largest provider of coal to South Africa's domestic market. Kumba, the newest major coal producer in South Africa, was formed following the split of domestic firm Iscor's mining and steel assets. Eyesizwe stated in January 2002 that it hoped to finalize its deal with Kumba in the first half of 2002. In January 2002, Kumba announced, that its main empowerment partner, Tiso Consortium, had exercised its option to acquire 5 per cent of the issued share capital of Kumba. In March 2002, Anglo announced it had purchased a 9.6 per cent share of Kumba, with an option to acquire an additional 10.5 per cent.

In November 2001 BHP Billington announced that it was closing its Rietspruit colliery (coal mine) six months ahead of schedule. Reitspruit's output was reduced by 40 per cent in December 2001, and mining operations was planned to end in May 2002. 150 workers, less than a fourth of the normal operating personnel, will be retained to rehabilitate the facility.

Exports

South Africa is the world's second largest net exporter (defined as production less consumption) of coal. In 1999, South African net coal exports were 85.2 mmst behind Australia's 178.3 mmst of net exports. In 2000, the DME reported that South Africa exported 77.1 mmst. The top recipients of South African coal in 2000 were Spain, the Netherlands, France, Israel, India, Germany, Italy, the United Kingdom, Portugal, South Korea, Taiwan, Belgium, Morocco, Denmark, Japan, Brazil and Turkey. These countries received more than 90 per cent of all South African exports.

The vast majority (96 per cent in 2000) of South African coal exports are shipped through the Richards Bay Coal Terminal (RBCT), the world's largest coal export facility. Only shareholding members of the RBCT company can use the export facility. Current shareholders include: Ingwe, Anglo, Enex, Tesa (Total Exploration South Africa), Sasol, Kangra and Eyesizwe. Ingwe, Anglo and Enex combined own 86 per cent of the RBCT. RBCT has the capacity to export 79.4 mmst of coal annually.

Other port facilities utilized for coal exports are Durban and Matola Coal Terminal (MCT) located in Maputo, Mozambique. Durban had coal exports of 2.1 mmst in 2000. Kumba currently exports its coking coal through the Durban facilities. 1.2 mmst of South African coal was exported through MCT in 2000. Exports through MCT could increase in the future as infrastructure is upgraded and added. The rail link between South Africa and Maputo is undergoing a $13.8 million upgrade, and there are plans to dredge the harbor to accommodate larger vessels. MCT management anticipates the facility to have the capacity to export 5.5 mmst of coal by 2006.

To help facilitate the participation of small independent and empowerment companies in the coal export sector, a new coal export facility was planned. The South Dunes Coal Terminal (SDCT), was to be located at Richards Bay, but apart from the RBCT. The SDCT was to have a capacity to export 13.2 mmst of coal annually. SDCT partners include: Kumba, Eskom Enterprises and Golang Coal. In June 2001, the RBCT exporters and the SDCT partners reached an agreement to expand the RBCT. The expansion will increase the annual export capacity of RBCT by 11 mmst, and the SDCT firms entitlement will be 7.2 mmst per year. the expansion of the RBCT is considered the most cost-effective method of expanding South Africa's coal export capability as no new rail infrastructure is needed. Ingwe and Anglo, the two largest shareholders in the RBCT, will forego their participation in the expansion of the export facility. In March 2002, the SDCT firms secured $41 million (R475 million) in financing to help expand the RBCT. Total cost of the expansion is $52 million (R600 million), with the remaining $11 million (R125 million) to be financed by RBCT shareholders. The RBCT expansion is expected to be completed by mid-2004.

Consumption

South Africa's domestic consumption of coal was 163.0 mmst in 1999. The DME reported domestic coal sales of 170.5 mmst in 2000. Electric power generation and the synthetic fuel industry account for approximately 90 per cent (source DME) of South Africa's coal consumption. Other major coal consuming sectors include: the non-synfuels industrial sector (3.3 per cent of local consumption), metallurgical industries (3.7 per cent), and the merchant & domestic sectors (2.5 per cent). Iscor's steel plants are the main consumers of domestic coking coal in South Africa. Three-fourths of the requirements are met by Kumba's mining operations in South Africa. Iscor also utilizes imports of coking coal and supplies from other local producers to satisfy the remainder of its requirements.

Synthetic Fules

South Africa has a highly developed synthetic fuels industry, which takes advantage of the country's abundant coal resources and offshore natural gas and condensate production in Mossel Bay. The two major players are Sasol (coal-to-oil/chemicals) and Mossgas (natural gas-to-petroleum products). Sasol has the capacity to produce 150,000 barrels per day (bbl/d), and Mossgas 45,000 bbl/d.

Sasol is the world's largest manufacturer of oil from coal, with coal liquefaction plants located at Secunda (oil) and Sasolburg (petrochemicals). Started by the government in the 1950s to help reduce South Africa's dependence on imported oil, the company was privatized in 1979. In 1996, Sasol began an upgrade and expansion program at its Secunda facilities to reduce costs and to help it remain competitive. The project was completed in 2001 following the installation of the ninth new synthetic fuel reactor.

In early 2000, Sasol launched a study of the feasibility of replacing coal with natural gas as the synthetic-fuel feedstock, utilizing natural gas reserves in neighboring Mozambique. Sasol estimates that the switch to natural gas will reduce investment expenditures in its coal mining operations and the high costs of compliance with environmental regulations associated with coal. The project is expected to start delivering natural gas to South Africa during the first half of 2004. The project consists of several major components including the development of the Pande and Temane gas fields in Mozambique. The 536-mile (865-kilometer) transport pipeline will run from the Mozambican fields to Secunda. Sasol will convert its existing pipeline-gas network to natural gas, and supply natural gas to industries in South Africa, including its own facilities. Sasol will switch its Sasolburg plants from coal to gas feedstock and utilize natural gas at Secunda to supplement coal-based growth there.

The pipeline will be owned by a joint venture between Sasol, SAG, and the government of Mozambique. The parties have made provision for the future inclusion of Black empowerment shareholders as well as privatization initiatives. Sasol has signed memoranda of understanding with two empowerment groups for local gas distribution companies in Mpumalanga and KwaZulu-Natal. On May 3, 2002, President Chissano of Mozambique and President Mbeki of South Africa, officially launched the project at a ceremony held at the Temane gas field in the Inhambane Province of Mozambique.

Mossgas began production in 1993 and remains state-owned. The company is part of the Central Energy Fund (CEF) group of companies through which the SAG's interest in the liquid fuel industry is owned, developed and managed commercially. The DME announced in 1999, in line with the recommendations of the Energy White, the merger of the CEF commercial interests into a national oil company. Mossgas, along with Soekor Exploration & Production (Soekor), and elements of the Strategic Fuel Fund (SFF) will be joined to create the Petroleum Oil and Gas Corporation of South Africa (PetroSA). The merger process is expected to be completed in 2002.

The Mossgas plant receives feedstocks of natural gas and condensate from gas fields in Mossel Bay through a pair of 56-mile (91-kilometer) pipelines. The facility also has the ability to process up to 8,000 bbl/d of imported condensate. The onshore plant is situated approximately 8 miles (13 kilometers) west of Mossel Bay. Mossgas converts the gas into a variety of liquid fuels including motor gasoline, distillates, kerosene, alcohols and LPG.

Oil and Natural Gas

South Africa has recently begun to develop and exploit its reserves of conventional oil. The country imports crude oil primarily from the Middle East, with Saudi Arabia and Iran as its chief suppliers. South Africa has been trying to diversify its sources of imported crude and to reduce its dependence on oil imports from Iran, which previously dominated South African oil imports. Nigeria is now the third largest supplier of imported oil to South Africa.

Exploration

PetroSA (formerly Soekor) plans to concentrate its exploration efforts on South Africa's western and southern coasts. Several discoveries have been made on Block 9, which is located within the Bredasdorp Basin. The FA, EM and EBF natural gas fields currently supply feedstocks to the Mossgas synfuel facility. Oribi, the first significant and commercially viable oil discovery on Block 9 was made in 1990. Two other significant finds, Oryx and Sable have been made on the block. PetroSA, and its Sable field partner Pioneer Natural Resources (Pioneer), announced the results of their Boomslang discovery in February 2001. The find, located on the southern portion of Block 9, tested at a combined rate of 3,120 bbl/d of oil, 26 million cubic feet of natural gas per day (Mmcf/d), and 300 bbl/d of condensate. PetroSA and Pioneer plan to drill additional wells (appraisal) on Boomslang, as well as on the EBB discovery. EBB, discovered in 1991, originally tested at 46 Mmcf/d of natural gas and 1,830 bbl/d of condensate. Pioneer also holds the rights to offshore Block 7.

Two natural gas discoveries are located on Block 11A, which lies east of Block 9. PetroSA made the Ga-A find in 1969. The discovery had a combined flow rate of 24 Mmcf/d from two reservoirs.

The Ga-Q field was discovered in 1983, and it had an initial test flow rate of 11.4 Mmcf/d. Additional appraisal drilling is planned on Block 11A.

An offshore natural gas discovery was made in March 2000 close to South Africa's border with Namibia. Located in Block 2, off South Africa's western coast, the find is reportedly part of the same reservoir, which extends to the Kudu prospect (operated by Shell) off the coast of Namibia. Kudu's potential reserves are estimated at 20 trillion cubic feet (tcf). Denver-based Forest Oil Corporation (Forest) reported that the find, AK-1, flowed at a rate of 52.8 Mmcf/d of natural gas and 342 bbl/d of condensate. Forest's initial estimates placed recoverable reserves at 200 billion cubic feet (bcf) of natural gas. Forest drilled three appraisal wells in 2000-2001, of which two were successful. One appraisal well flowed at 71.4 Mmcf/d of natural gas and 1,376 bbl/d of condensate, while the other had flows of 53 Mmcf/d and 182 bbl/d respectively. Forest revised its estimate of total reserves for the discovery, renamed the Ibhubezi field, to 2.5 (tcf). Forest stated that production, which may be channeled towards regional electricity production, is not expected to begin prior to 2004. Forest, and its partner, Denver-based Anschutz Overseas Corporation, control 70 per cent and 30 per cent respectively of South Africa's Blocks 1 and 2. In September 2000, Forest and Anschutz offered a 10 per cent share in both blocks to South African firm, Mvelaphanda Holdings (MVP), in accordance with SAG's directive encouraging Black participation in the upstream sector.

South Africa's recent offshore success is sparking interest in further developments. Sasol holds the rights to Blocks 3A and 4A, which are adjacent to the West Coast and south of Forest's holdings. In February 2002, Colorado-based Global Energy Holdings (GEH) announced the formal approval of its prospecting agreement for Block 3B/4B from the DME and the South African Agency for Promotion of Petroleum Exploration and Exploitation (Petroleum Agency SA). The seven and one-half year agreement for Block 3B/4B covers 7.1 million acres (29,000 square kilometers) offshore western South Africa in waters ranging in depth from 900 feet to 4,000 feet (300 to 1,200 meters). In January 2002, Petroleum Geo-Services (PGS) and Petroleum Agency SA announced a joint cooperation agreement to promote deepwater exploration acreage in South Africa. The area, Block 2B and acreage west of Blocks 5 and 6, contains 39.5 million acres (160,000 square kilometers). PGS will shoot and market 2D seismic data, with the survey commencing in the first half of 2002.

Production

The Oribi oil field began production from a floating production, storage and offloading vessel (FPSO) in May 1997, South Africa's first conventional oil production. The field currently produces a light oil (API 42°) at the rate of 10,000 bbl/d with 15 Mmcf/d of associated gas, which is flared. The Oryx oil field lies 3.7 miles (6 kilometers) from the Oribi field and was tied back to the Oribi's FPSO production facility. Oryx began production in May 2000, and currently produces at 12,000 bbl/d. The Oryx reservoir is similar in type and age to Oribi. Combined, the Oribi and Oryx fields have to date produced over 23 million barrels of oil.

PetroSA and Pioneer announced plans for the development of the Sable field in June 2001. The field will be developed with six subsea wells tied back to a FPSO. The FPSO will have the capacity to process 60,000 bbl/d of oil, re-inject 80 Mmcf/d of natural gas and recover natural gas liquids. Production of 40,000 bbl/d is expected to begin in the first quarter of 2003. Total recoverable oil reserves are estimated to be 25 million barrels. Associated gas, which will be re-injected to improve liquids recovery, may be recovered at a later date as part of a planned natural gas development project.

The FA natural gas field currently produces at a rate of 194 Mmcf/d gas and 9,500 bbl/d of condensate. The FA production platform is one of the largest single structures ever constructed in South Africa. Nine production wells have been drilled from the platform. Four production wells on the FAR and FAH satellite gas fields are linked to the platform by subsea systems. The production wells on the EM and EBF gas fields are connected to the FA platform by a 32-mile (52-kilometer) pipeline, 18 inch diameter pipeline that has been designed for the future tie-in of other gas fields in the area.

Refining and Downstream Oil Activities

South Africa has the second largest refining capacity in Africa. South Africa's total refining capacity (excluding synthetic fuel plants) of 468,547 bbl/d is surpassed only by Egypt's. Its refined products are both sold in the local market and exported, mainly to other parts of Southern Africa, but also into both the Indian and Atlantic basin markets. Refinery problems in mid-2001 forced South Africa to reduce fuel exports by half to several SADC-member countries. Botswana, Lesotho, Namibia and Swaziland import nearly all of their oil products from South Africa. Zimbabwe, due to a shortage of foreign exchange, also had been relying heavily on South African fuel imports. The Natref refinery is undergoing a $123-million capacity expansion project. When completed in 2002, refining capacity will increase by nearly 17,000 bbl/d, and include the ability to produce low-sulfur diesel.

Multinational companies, including BP, Shell, Caltex (ChevronTexaco), and TotalFinaElf are major participants in South Africa's downstream petroleum markets. In July 2001, it was announced that Black empowerment group Thebe Investment Corporation purchased a 25 per cent share of Shell's South African downstream retail and marketing business. Shell, with its Sapref refinery (Durban) partner BP, signed an agreement with Black empowerment firm Southern Tankers to transport oil from the refinery to other South African locations. Southern Tankers will cover all of the refinery's coastal shipping requirements.

Several local firms also are involved in South Africa's downstream including the Black-owned firms, Naledi Petroleum and Afric Oil. Proposed amendments to South Africa's Petroleum Products Act would allow synthetic fuel producers Sasol and Mossgas to enter the retail market. In August 2001, it was announced that Sasol had signed an agreement with Petromoc, Mozambique's state-owned oil marketing and distribution firm, to market petroleum products to Mozambican service stations and commercial customers. The joint venture also plans to develop service stations in Mozambique.

The first privatization in South Africa's gas distribution sector was completed in August 2000, when a consortium led by the U.S.-based Cinergy and the Black empowerment group, Egoli Empowerment Holdings, was chosen to purchase Johannesburg's Metro Gas Company. Renamed Egoli Gas, the consortium announced in September 2000 the signing of a 20-year contract with Sasol Gas. In terms of the supply agreement, Sasol Gas will provide the Johannesburg area with 2.5 million cubic feet per year, with an option to increase the supply up to 7 million cubic feet per year.

Other projects to use natural gas are planned for South Africa. Negotiations are continuing between the SAG and Shell, operator of Namibia's offshore Kudu gas field. Initial plans call for the gas to piped from the Kudu field to Cape Town, where it will supply fuel for a 1,200-2,000 megawatt(MW) power station. The SAG would like for Shell to commit gas supplies to other buyers, possibly extending the pipeline to the Mossgas synfuel facilities at Mossel Bay. Sasol and Belgian-based Tractebel have signed a memorandum of understanding (MOU) for the development of gas-fired co-generation in South Africa.

Electricity

Parastatal company Eskom, one of the largest utilities in the world, generates nearly all (approximately 95 per cent) of South Africa's electricity. Eskom's generating capacity (39,870 MW), which is primarily coal-fired (35,627 MW), also includes one nuclear power station at Koeberg (1,840 MW), two gas turbine facilities (342 MW), six conventional hydroelectric plants (661 MW), and two hydroelectric pumped-storage stations (1,400 MW). Eskom also has 3,556 MW of mothballed coal-fired generation, and is currently constructing 1,426 MW of additional coal-fired capacity at its Majuba plant. South African municipalities own and operate 2,436 MW of generating capacity, of which the majority (1,932 MW) is coal-fired. An additional 836 MW of generating capacity is privately held. In December 2001, U.S.-based AES completed its purchase of a 600-MW coal-fired power plant. The Kelvin power plant was purchased from the Johannesburg Metropolitan Municipality for the price of $23 million.

South Africa's National Electricity Regulator (NER)is responsible for the licensing of electricity generators, transmitters and distributors in the country. NER is also overseeing the restructuring of South Africa's electricity supply industry (ESI) in accordance with existing legislation and the Energy Policy White Paper. The legislation and regulation is crucial to the government's continuing electrification program. The NER licensed Eskom as the National Transmitter for South Africa. The transmission license provides for: non-discriminatory access by generators being dispatched centrally, offering a transmission service to parties who are in a position to take supply directly off the transmission system, central dispatch of power stations participating in the National Power Pool, and organizing the exports and imports of electricity to South Africa. A transmission license was issued to a private company, Montraco, to provide a specific transmission service from the National Transmission System to specific supply points in Mozambique and Swaziland.

In addition to serving the domestic market, Eskom also exports power to Botswana, Lesotho, Mozambique, Namibia, Swaziland and Zimbabwe. South Africa and the other members of the SADC signed a memorandum of understanding in August 1995 to establish the Southern African Power Pool (SAPP). Eskom is expanding its involvement across the continent. In October 2000, Nigeria's National Electric Power Authority (NEPA) signed a partnership agreement with Eskom to help improve electricity supply. Eskom will help develop NEPA's repair capabilities, execute transmission line projects, and participate in rehabilitating, operate and transfer (ROT) schemes for the running of Nigeria's power stations. Uganda, in September 2001, chose Eskom as one of two entities to bid for two concessions in its power sector. In November 2001, Zimbabwe's state-owned utility, Zimbabwe Electricity Supply Authority (ZESA), awarded Eskom a contract to assist in the management of its main power station. ZESA's Hwange station provides 40 per cent of the country's power. Eskom purchased a 51 per cent interest in Zambia's Lusemfwa Hydropower Company in December 2001.

Environment

In 2000, 74.4 per cent of total energy consumption in South Africa was from coal consumption. This reliance on coal, a highly carbon-intensive fossil fuel, has negative environmental impacts. For example, electricity generation from coal combustion in the industrial sector is the prime contributor to air pollution. Mining, itself, adversely effects the environment through the disruption of ecosystems and the pollution of groundwater.

South Africa is a comparatively energy and carbon intensive country with respect to other African countries, as well as relative to many "developed" nations. In 2000, energy intensity in South Africa was 27.2 thousand Btu per $1995 GDP, while carbon intensity was 0.62 metric tons of carbon per thousand $1995 GDP. Per capita energy consumption and carbon emissions, on the other hand, are high in comparison with the rest of Africa, but relatively low when compared to many "developed" nations. In 2000, South Africa's per capita energy consumption was 106.3 million Btu, while per capita energy-related carbon emissions registered 2.4 million metric tons of carbon. The most widely used renewable energy source in South Africa is fuelwood, which meets the daily energy needs of more than one-third of the country's population. Deforestation attributed to increased fuelwood consumption by a growing population has prompted interest in developing other renewable energy sources, particularly solar, which could play an important role in supplying power to isolated rural areas not currently connected to the electric power grid.

Country Overview

Independence: May 31, 1910 (from United Kingdom)

Population (2001E): 43.6 million

Location/Size: Southern Africa / 1.2 million square kilometers (471,445 square miles, nearly twice the size of Texas)

Major Cities: Pretoria (capital), Johannesburg, Durban, Cape Town

Languages (official): Afrikaans, English, Ndebele, Northern Sotho, Southern Sotho, Swazi, Tsonga, Tswana, Venda, Xhosa, Zulu

Ethnic Groups: Black (75.2 per cent), White (13.6 per cent), mixed race (Colored) (8.6 per cent), Asian (2.6 per cent)

Religions: Christian (Dutch reformed, other Protestant and Catholic denominations) 60 per cent, Jewish, Hindu, Muslim, and traditional African religions

Defense (1999E): Army, 54,300; Navy, 8,000; Air Force, 11,140; Medical Corps, 6,000

Economic Overview

Currency: 1 rand (R) = 100 cents

Market Exchange Rate (06/22/02): US$1 = 10.26 R

Nominal Gross Domestic Product (2001E): $114.7 billion

Real GDP Growth Rate (2000E): 3.1 per cent *(2001E):* 2.3 per cent *(2002F):* 2.3 per cent

Inflation Rate (consumer prices) (2000E): 5.3 per cent *(2001E):* 6.3 per cent *(2002F):* 5.9 per cent

Merchandise Exports (2001E): $33.2 billion

Merchandise Imports (2001E): $27.3 billion

Trade Balance (2001E): $5.9 billion

Major Exports (2001): Gold, diamonds, other metals and minerals, machinery and equipment

Major Imports (2001): Machinery, foodstuffs and equipment, chemicals, petroleum products, scientific instruments

Major Trading Partners: Japan, Germany, Italy, United Kingdom, United States

Current Account Balance (2001E): $1.7 billion

Energy Overview

Proven Oil Reserves (1/1/02E): 29.4 million barrels

Oil Production (2001E): 224,000 barrels per day (bbl/d), of which about 186,000 bbl/d is synthetic

Oil Consumption (2001E): 482,000 bbl/d

Net Oil Imports (2001E): 258,000 bbl/d

Crude Refining Capacity (1/1/02E): 468,547 bbl/d

Natural Gas Production/Consumption (2000E): 49.4 billion cubic feet

Recoverable Coal Reserves (2000E): 61 billion short tons

Coal Production (2000E): 247.2 million short tons (mmst)

Coal Consumption (2000E): 170.5 mmst (domestic sales)

Coal Exports (2000E): 77.1 mmst

Electricity Generation Capacity (1/1/00E): 43.1 gigawatts

Electricity Generation (2000E): 194.4 billion kilowatthours (bkwh), of which 180.0 bkwh was thermal, 13.0 bkwh nuclear and 1.3 bkwh hydroelectric

Electricity Consumption (2000E): 181.5 bkwh

Environmental Overview

Minister of Environmental Affairs and Tourism: Mohammed Valli Moosa

Total Energy Consumption (2000E): 4.6 quadrillion Btu* (1.17 per cent of world total energy consumption)

Energy-Related Carbon Emissions (2000E): 105.9 million metric tons of carbon (1.6 per cent of world carbon emissions)

Per Capita Energy Consumption (2000E): 106.3 million Btu (vs. U.S. value of 351.0 million Btu)

Per Capita Carbon Emissions (2000E): 2.4 metric tons of carbon (vs. U.S. value of 5.6 metric tons of carbon)

Energy Intensity (2000E): 27,218 Btu/$1995 (vs. U.S. value of 10,918 Btu/$1995)

Carbon Intensity (2000E): 0.62 metric tons of carbon/thousand $1995 (vs. U.S. value of 0.17 metric tons/thousand $1995)

Sectoral Share of Energy Consumption (1998E): Industrial (57.1 per cent), Transportation (17.9 per cent), Residential (15.6 per cent), Commercial (9.4 per cent)

Sectoral Share of Carbon Emissions (1998E): Industrial (65.4 per cent), Transportation (17.5 per cent), Residential (11. per cent), Commercial (6.1 per cent)

Fuel Share of Energy Consumption (2000E): Coal (74.4 per cent), Oil (21.1 per cent), Natural Gas (1.1 per cent)

Fuel Share of Carbon Emissions (2000E): Coal (81.4 per cent), Oil (17.9 per cent), Natural Gas (0.8 per cent)

Renewable Energy Consumption (1998E): 386 trillion Btu* (0 per cent increase from 1997)

Number of People per Motor Vehicle (1998): 7 (vs. U.S. value of 1.3)

Status in Climate Change Negotiations: Non-Annex I country under the United Nations Framework Convention on Climate Change (ratified August 29th, 1997). Not a signatory to the Kyoto Protocol.

Major Environmental Issues: Lack of important arterial rivers or lakes requires extensive water conservation and control measures; growth in water usage threatens to outpace supply; pollution of rivers from agricultural runoff and urban discharge; air pollution resulting in acid rain; soil erosion; desertification.

Major International Environmental Agreements: A party to the Antarctic-Environmental Protocol, Antarctic Treaty, Biodiversity, Climate Change, Desertification, Endangered Species, Hazardous Wastes, Law of the Sea, Marine Dumping, Marine Life Conservation, Nuclear Test Ban, Ozone Layer Protection, Ship Pollution, Wetlands and Whaling.

The total energy consumption statistic includes petroleum, dry natural gas, coal, net hydro, nuclear, geothermal, solar and wind electric power. The renewable energy consumption statistic is based on International Energy Agency (IEA) data and includes hydropower, solar, wind, tide, geothermal, solid biomass and animal products, biomass gas and liquids, industrial and municipal wastes.

Energy Industries

Organization: Ingwe, Anglo, Enex, Eyesizwe, Kumba–major private coal producers; *Eskom*–parastatal electric power company; *Sasol*–coal-to-liquid synthetic fuels & chemicals group (privatized in 1979); *Petronet*–petroleum pipelines and tank farm; *Petroleum Oil and Gas Corporation of South Africa (PetroSA) whose subsidiaries include: Mossgas*–gas-to-synthetic fuels plant; *Soekor (the Southern Oil Exploration Corporation)*- oil and gas exploration/development; & *Strategic Fuels Fund*–strategic oil storage facility and state oil imports;

Major Coal Fields: Waterberg, Witbank, Highveld

Major Coal Ports: Richards Bay Coal Terminal, Durban

Oil Refineries (1/1/02E Capacity): Shell/BP–Durban (172,000 bbl/d); Caltex–Cape Town (105,000 bbl/d); Engen–Durban (104,000 bbl/d); National Petroleum–Sasolburg (87,547 bbl/d)

75

South Asia: Regional Overview

South Asia is important to world energy markets because it contains 1.3 billion people–more than one-fifth of the world's population–and is experiencing rapid energy demand growth. South Asia also is a major, and growing, contributor to global emissions of carbon dioxide.

General Background

The South Asian region (Bangladesh, Bhutan, India, the Maldives, Nepal, Pakistan, and Sri Lanka) has a huge population (more than one-fifth of the world total) which is growing rapidly. At the same time, and despite rapid economic growth during the 1990s, the region has among the lowest per capita incomes in the world. For 2002, Pakistan is expected to experience a growth rate in real Gross Domestic Product (GDP) of 4.8 per cent, with India at 5.5 per cent. India is by far the largest South Asian country, in terms of population, GDP, land area, energy consumption, energy production, and carbon emissions. After India, Pakistan and Bangladesh are the next largest South Asian countries in these categories.

Overall, South Asia is in a period of transition as it strives to implement effective economic, political, social, and legal structures to support sustained growth. The International Monetary Fund (IMF) and the World Bank have arranged several billion dollars worth of assistance to the region, with the IMF prescribing such measures as cuts in subsidies (including energy subsidies), deregulation, anti-poverty efforts, and increased privatization.

Energy

Economic and population growth in South Asia have resulted in rapid increases in energy consumption in recent years–well above the rate seen in the OECD. The Energy Information

Administration (EIA) estimates that South Asia's primary energy consumption (NOTE: EIA energy statistics include only "commercial" energy sources, but not animal waste, wood, or other biomass, which accounts for more than half of the South Asia region's total final energy consumption) increased nearly 58 per cent between 1991 and 2000. In 2000, South Asia accounted for approximately 3.9 per cent of world commercial energy consumption, up from 2.8 per cent in 1991. Despite rapid growth in energy demand, however, South Asia continues to average amongst the lowest levels of *per capita* energy consumption in the world, but among the highest in terms of energy consumption *per unit of GDP*.

Table 75.1: Economic and Demographic Indicators for South Asian Countries

	Gross Domestic Product (GDP)				Population
	2001E (Billions of U.S. $–PPP*)	Real GDP Growth Rate		Per Capita GDP, 2000E (U.S. $–PPP)	2000E (Millions)
		200 Estimate	2002 Projection		
Bangladesh	$203	N.A	N.A.	$1,570	131
Bhutan	$2.3	N.A.	N.A	$1,100	2
India	$2,200	5.5 per cent	5.9 per cent	$2,200	1,029
Maldives	$0.6	N.A.	N.A.	$2,000	0.3
Nepal	$33.7	N.A.	N.A.	$1,360	25.3
Pakistan	$282	3.4 per cent	4.8 per cent	$2,000	145
Sri Lanka	$62.7	N.A	N.A.	$3,270	19.4

Unit. PPP = Purchasing Power Parity exchange rates

Not counting "non-commercial" sources of energy like animal waste, wood, and other biomass, South Asia's *commercial* energy mix in 2000 was 44 per cent coal, 34 per cent petroleum, 13 per cent natural gas, 7 per cent hydroelectricity, 1 per cent nuclear and 0.1 per cent "other." There are significant variations within the region. Bangladesh's energy mix, for instance, is dominated by natural gas (68 per cent in 2000), while India relies heavily (53 per cent) on coal. Sri Lanka is overwhelmingly dependent on petroleum (78 per cent in 2000), while Pakistan relies on natural gas (45 per cent in 2000), petroleum (41 per cent), and hydroelectricity (12 per cent), and the Maldives is 100 per cent dependent on petroleum. The Himalayan countries of Nepal and Bhutan have high shares of hydroelectric power in their energy consumption mix. In recent years, natural gas has been growing in importance as a source of energy in South Asia, especially for use in power generation, fertilizer and petrochemical production. However, the problems encountered by some developers of proposed liquefied natural gas (LNG) projects, and continuing tensions with Pakistan, which effectively blocks potential natural gas pipeline routes from the Persian Gulf and Central Asia, may present obstacles to the future enlargement of natural gas as a share of India's fuel mix.

The major energy issues facing South Asian nations today are keeping up with rapidly rising energy demand and beginning to promote cross-border energy trade. Already, most of South Asia is grappling with energy shortfalls, usually in the form of frequent, costly, and widespread electricity outages. Given this situation, and in particular its potential economic and political ramifications, improving the supply of energy in general, and electricity in particular, is a major concern among regional governments. In order to accomplish this goal, South Asia thus is faced with the challenges of diversifying traditional energy supply sources (and expanding the use of indigenous energy resources), promoting additional foreign investment for energy infrastructure development, improving energy efficiency, reforming and privatizing energy sectors, and expanding regional energy trade and investment.

Oil

South Asia contains reserves of only 5.2 billion barrels of oil, around 0.5 per cent of the world total. In 2000, the region consumed around 2.5 million barrels per day (bbl/d) of oil, and produced 0.80 million bbl/d, making South Asia a net oil importer of around 1.7 million bbl/d. The vast majority (around 740,000 bbl/d) of South Asia's oil production comes from India, including its offshore Bombay High field (which accounts for about one-third of total Indian oil output). Most of the remainder (around 58,000 bbl/d) of South Asia's oil production comes mainly from Pakistan. Most South Asian crude oil imports come from the Middle East, and this is likely to remain the case for years to come. South Asia's oil imports are expected to grow sharply as production remains about flat while demand soars. Oil imports are projected to more than double by 2020. The vast majority of this oil is expected to come from the Middle East (located close to South Asia, with more than two-thirds of world oil reserves), with only small volumes coming from other areas.

Growing demand for transportation fuels and increased industrial power demand have been major factors behind the growth in South Asian oil consumption in recent years. Between 1991 and 2000, South Asian oil consumption–led by India–grew by about 47 per cent, and EIA's *International Energy Outlook 2002* projects India's oil consumption will grow another 25 per cent by 2005, reaching 2.5 million bbl/d (up from 2.2 million bbl/d in 2000). Sri Lanka imports all of its crude oil, which is used largely for electricity generation and transportation, and has refining capacity of 50,000 bbl/d. In recent years, Sri Lanka has increased its oil imports in an effort to diversify away from overreliance on hydroelectricity, which varies depending upon rainfall amounts. Between 1991 and 2000, Sri Lankan oil consumption roughly doubled.

Table 75.2: Energy Consumption and Carbon Dioxide Emissions in South Asian Countries, 2000

	Commercial Energy Consumption[1]								CO_2
	Total Quadrillion Btu)	*Petro-leum*	*Natural Gas*	*Coal*	*Nuclear*	*Hydro-electric*	*Other Electric**	*Net Electric Imports*	*Emissions (Million metric tons)*
Bangladesh	0.50	30%	68%	1%	0%	1%	0%		7.7
Bhutan	0.01	24%	0%	21%	0%	55%	0%		0.1
India	12.67	32%	7%	53%	1%	6%	0.2%		253.3
Maldives	0.004	100%	0%	0%	0%	0%	0%		0.1
Nepal	0.06	57%	0%	14%	0%	24%	5%		0.8
Pakistan	1.91	41%	42%	5%	0%	12%	0%		29.5
Sri Lanka	0.18	78%	0%	0%	0%	22%	0%		2.6
Total	**15.28**	**34%**	**13%**	**44%**	**1%**	**7%**	**0.2%**		**294.1**

[1]Note: Does NOT include such "non-commercial" energy sources as animal waste, wood, and other biomass, which account for more than half of South Asia's total final energy consumption.

[2]Other includes consumption of wind electric power for India and net imports of electricity for India and Nepal. Other does NOT include biomass or other "noncommercial" sources of energy.

[3] Includes carbon dioxide emissions from the consumption of petroleum, natural gas, and coal, and from the flaring of natural gas. Tons of carbon can be converted to tons of carbon dioxide gas by multiplying by 3.667.

Note: percentages may not add to 100 per cent because of independent rounding.

Table 75.3: Energy Supply Indicators–South Asian Countries

	Fossil Fuel Proved Reserves			*Fossil Fuel Production, 2001*			*Electric Generating Capacity,*	*Crude Oil Refining Capacity,*
	Crude Oil, 1/1/02 (Million Barrels)	*Dry Natural Gas, 1/1/02 (Trillion Cubic Feet)*	*Coal 12/31/98 (Million Kilo-watts)*	*Petroleum 2001 (Thousand Barrels/ Day)*	*Dry Natural Gas 2000 (Trillion Cubic Feet)*	*Coal (Million Short Tons)*	*1/1/00 (Million Kilowatts)*	*1/1/0 (Thousand barrels/day)*
Bangladesh	57	10.6	0	4	0.34	0.06	3.6	33
Bhutan	0	0	0	0	0	0	0.36	0
India	4,840	22.9	93.0	736	0.79	345.0	108.1	2,134
Maldives	0	0	0	0	0	0	0.03	0
Nepal	0	0	0.002	0	0	0.01	0.39	0
Pakistan	298	25.1	2.5	57	0.86	3.8	17.1	239
Sri Lanka	0	0	0	0	0	0	1.61	50
Total	4,195	58.6	95.5	797	1.99	348.9	131.2	2,456

Refining

The construction of adequate refining capacity to keep up with growing oil demand is of great economic importance to South Asia. Each South Asian country is proceeding with refinery construction plans. The largest recent project, the Reliance Industries refinery at Jamnagar in India, began operation in late summer 1999, and has a capacity of 540,000 bbl/d. The 100,000 bbl/d "Pak-Arab" refinery in Pakistan came online in late 2000, helping alleviate the country's refined products dependence. Petronet India, a company created in early 1998 as part of an agreement among India's three government-owned refineries (IOC, Hindustan Petroleum, and Bharat Petroleum) is building product pipelines that will add about 500,000 bbl/d to current pipeline capacity of about 325,000 bbl/d (all operated by IOC). Completion of these projects will shift the main transportation mode for petroleum products from rail to pipeline.

Natural Gas

At the beginning of 2002, South Asia's proven natural gas reserves totalled around 58.6 trillion cubic feet (Tcf), or about 1 per cent of the world total, with potentially larger resources suspected but unproven. The region consumed and produced around 1.99 Tcf of natural gas per year in 2000. Around 43 per cent of this was accounted for by Pakistan, 40 per cent by India, and the remaining 17 per cent by Bangladesh. If long-term projections of rapidly increased gas demand for South Asia are correct, the region will require either significant increases in production, imports, or most likely both. Gas imports to the region would require construction of infrastructure–either pipelines or liquefied natural gas (LNG) facilities. Without such infrastructure, natural gas supplies could be constrained, and increased reliance on other fuels could result. This appears more likely in light of the financial issues which have delayed the completion of the Dabhol LNG and power project in India, and renewed tensions between India and Pakistan, which effectively block potential pipeline routes from large natural gas reserves in the Persian Gulf and Central Asia to India.

Natural gas usage has increased rapidly in South Asia, growing about 72 per cent between 1991 and 2000. Natural gas is seen in the region as playing an important part in supplying new power plants in the region, plus as a means of diversifying away from expensive oil imports. An obstacle to the expansion of natural gas usage in South Asia is the region's inadequate domestic gas infrastructure. Indian proposals to import, for instance, will require that supporting infrastructure be in place before such imports can proceed. Cross-border gas pipelines also would hinge upon the successful construction of domestic gas pipeline systems first.

Natural gas is Bangladesh's main source of commercial energy. Bangladesh currently has estimated proved natural gas reserves of around 16.3 trillion cubic feet (Tcf) in approximately 20 fields (mainly onshore), although gas reserves estimates vary widely, with foreign energy companies such as Shell and Unocal, for instance, believing that Bangladesh's gas reserves actually might be much higher than the official figure. The US Geological Survey recently estimated that Bangladesh contains an additional 32.1 Tcf in additional "undiscovered reserves." If the higher estimates prove to be correct, Bangladesh could at some point become a major gas producer (as well as supplier to the vast potential market in neighboring India). Gas exports are controversial within Bangladesh, however, with many people feeling that Bangladeshi gas resources first should be used for domestic purposes (*i.e.*, electric power generation, fertilizer production, transportation), and also that the size of the country's gas reserves remains highly uncertain, particularly in relation to future domestic demand projections.

Indian consumption of natural gas has risen faster than any other fuel in recent years, and now accounts for around 7 per cent of the country's energy demand. From nearly 0.8 Tcf in 2000, Indian gas demand is projected in the *International Energy Outlook 2002* to reach 1.2 Tcf in 2005 and 1.6 Tcf in 2010, a significant reduction from previous forecasts, but still a large increase. Increased use of natural gas in power generation will account for most of the increase, as the Indian government is encouraging the construction of gas-fired electric power plants in coastal areas where they can be easily supplied with LNG by sea. While India's Foreign Investment Promotion Board (FIPB) has approved 12 prospective LNG import terminal projects, several have been delayed or cancelled outright between 2000 and 2002, and only the Dabhol project has had any construction activity, which is now halted due to financial disputes.

Pakistan has 25.1 Tcf of proven gas reserves, and currently produces nearly 0.86 Tcf of natural gas per year, all of which is consumed domestically. Pakistan plans to make gas the "fuel of choice" for future electric power generation projects. This will necessitate a sharp rise in production of natural gas.

Several import schemes also have been under discussion in recent years. In mid-2000, Pakistan's government stated that it would permit a gas pipeline linking Iran's massive gas reserves to rival India to cross its territory. Pakistan would earn transit fees for Iranian gas supplied to India and also be able to purchase gas from the pipeline itself. While Iran and Pakistan have shown great interest in the project, India has been reluctant to move forward as long as political and military tensions with Pakistan over Kashmir persist. Another possibility is that Pakistan could eventually be linked into the Dolphin Project, a scheme to supply gas from Qatar's North Dome gasfield to the United Arab Emirates and Oman, via a subsea pipeline from Oman. Even though Pakistan has signed a preliminary agreement to eventually purchase gas from Qatar, it seems increasingly unlikely that Pakistan will be included in the project in the near-term, due to it financial weakness and uncertainty about whether there will be sufficient demand growth. A third possible gas pipeline would link gas-rich Turkmenistan with Dalautabad in central Pakistan via Afghanistan. Unocal had been the main foreign backer of the plan until August 1998, when it withdrew from the project after the U.S. strikes against terrorist training camps associated with Osama bin Laden in Afghanistan. None of these proposed projects, however, seems to have made any real progress over the last couple of years.

Bhutan, the Maldives, Nepal, and Sri Lanka do not currently produce or consume any natural gas.

Coal

South Asia contains 86 billion short tons of coal, or around 15 per cent of the world total. Currently, coal accounts for 44 per cent of South Asia's energy consumption. Nearly all of this is produced and consumed by India, the only South Asian country with significant coal reserves, and the world's third largest coal producer (after the United States and China). Indian coal generally is of poor quality–*i.e.*, low in calorific content and high in ash–and primarily is located far from major consuming centers. Power generation accounts for about 70 per cent of India's total coal consumption, followed by steel and other industries. India's coal consumption is expected to increase to 546 million short tons (Mmst) by 2020, up 48 per cent from 369 Mmst in 2000. Largely for this reason, South Asia's carbon emissions are expected to increase sharply in coming years.

Coal currently plays a relatively minor role in Pakistan's energy mix (5 per cent in 2000), but the discovery of large volumes of low ash, low sulfur lignite in the Tharparkar (Thar) Desert in Sindh province will have a positive impact on consumption levels. Bangladesh and Sri Lanka have small coal reserves, and consume almost no coal, though Bangladesh has current plans to begin commercial coal production and build a coal-fired power plant with Chinese assistance.

Biomass (Non-Commercial Fuels)

As is the case in many developing countries and regions, South Asia continues to rely heavily on biomass (*i.e.*, animal waste, wood, etc.) for its energy needs. As of 1995, for instance, biomass accounted for 56 per cent of the region's final energy consumption, and 46 per cent of its primary energy use (according to the International Energy Agency–IEA). Also according to the IEA, around 20 per cent -30 per cent of South Asia's biomass use is animal waste, with another 20 per cent -30 per cent made up of agricultural residues, and only small amounts of charcoal. Biomass generally is burned directly using traditional, low-efficiency equipment. Biomass is consumed mainly in rural areas of South Asia. The IEA has projected South Asian biomass use to be approximately flat through 2020.

Electricity

In 2000, South Asia generated 633 billion kilowatt hours (Bkwh) of electricity. Of this, around 81 per cent was from conventional thermal power plants, 17 per cent from hydroelectric plants, and 2 per cent from nuclear, and less than 1 per cent for "other renewables" (like wind and solar). Also in 2000, India accounted for the vast majority (86 per cent) of regional electricity generation, followed by Pakistan (10 per cent), Bangladesh (2 per cent), Sri Lanka (1 per cent), Nepal, Bhutan, and the Maldives (1 per cent total). Regional electricity generation is expected to increase significantly in coming years. While natural gas use had been expected to increase at the expense of coal's share, this is much less certain given the delays nad problems in supplies of imported natural gas to India. Nuclear and hydro are expected to increase their shares, with oil maintaining an approximately constant share. Non-hydroelectric "renewable" capacity (*i.e.*, wind, solar, ocean, biomass, geothermal) is small at present, but is increasing, with wind power considered the most promising.

Electricity demand in South Asia is currently outstripping generating capacity, except in Pakistan, and the region as a whole is characterized by chronic electricity shortages. The main reasons for this situation are: shortfalls in building new power plants; low plant load factors due to aging generators and poor maintenance of equipment at existing plants (plus low-quality coal in many cases); and losses of power due to poor-quality transmission lines and theft. South Asia's rapidly rising electricity demand has heightened the need for additional investment by independent power producers (IPPs). However, bureaucratic obstacles and underdeveloped regulatory policies guiding such investment have led to construction delays as well as to foreign investor disillusionment, and most large IPP projects in the region have been delayed or cancelled over the past two years. Electricity rates are widely subsidized in South Asia, and state electricity companies are faced with the challenge of paying IPPs their asking price for power while simultaneously providing low rates for electricity to their customers, and losing a substantial percentage to theft. Meanwhile, the IMF and the World Bank have encouraged liberalization of South Asian power sectors, including reduction of subsidies. Meeting future electricity demand promises to pose a major challenge for South Asia.

India, which accounts for about four-fifths of South Asian electricity generation, is facing serious power supply problems (current generation is about 30 per cent below demand, according to the Indian government). As a consequence, India is faced with the need to invest heavily in new electric generating capacity. Although about 80 per cent of India's population has access to electricity, power outages and brownouts are common. India has forecast total additions of 111,500 MW will be needed by 2007, but it looks unlikely that this target will be met. Overall, Indian power demand is projected in the *International Energy Annual 2002* to increase to 923 billion kilowatthours (Bkwh) in 2020, about 81 per cent above the 510 Bkwh consumed in 2000, but a substantial downward revision to the previous forecast. India generates around 83 per cent of its electricity from conventional thermal power plants, around 14 per cent from hydroelectric plants (located mainly in the north and northeast of the country), and the rest from nuclear plants.

As of 2000, Pakistan had 17 gigawatts (GW) of installed electric generating capacity. Thermal plants (oil, gas, and coal) make up about 71 per cent of this capacity, with hydroelectricity making up 28 per cent and nuclear plants 1 per cent. Pakistan's total power generating capacity has increased rapidly in recent years, due largely to foreign investment in the mid-1990s, which has now decreased due to payment problems at those projects. Rotating blackouts ("load shedding") are, however, necessary at times in some areas. Transmission losses are about 30 per cent, due to poor quality infrastructure and a significant amount of power theft. Seasonal reductions affect the availability of hydropower. With much of the Pakistan's rural areas yet to receive electric power, and less than half of the population connected to the national grid, significant demand growth is expected in the long term, though in the short term, Pakistan has some excess generation capacity.

In Bangladesh, only around 18 per cent of the population (25 per cent in urban areas and 10 per cent in rural areas) has access to electricity, and per capita commercial energy consumption is among the lowest in the world. With power demand growing rapidly (a 60 per cent increase from 1991 to 2000), Bangladesh's Power System Master Plan (PSMP) foresees a doubling of required electric generating capacity by 2005 at a cost of $4.4 billion. On top of this, Bangladesh likely will need to replace 30 per cent -40 per cent of existing generating capacity. Power shortages can have serious social consequences, as demonstrated on April 10, 1999, when violent clashes took place in Dhaka between police and people protesting inadequate power supplies and demanding better service. Bangladesh generates its electricity mainly at thermanl power plants (93 per cent), but also has some hydroelectric dams (7 per cent). Discussions have been underway among Southern Asian nations for development of a regional electricity grid. Such a grid could lead to increased efficiencies and reduced power generation and transmission costs. Nepal and Bhutan have substantial untapped hydroelectricity potential. This power could be consumed in those two countries and also exported to India, Pakistan, and Bangladesh. In March 1999, it was reported that India's Power Grid Corporation had completed a feasibility study on possible exchange of 150 MW of power between Bangladesh and India. Interconnection points would be Ishwardi, Bangladesh-Farakka, India and Shahjibazar, Bangladesh-Kurnarghat, India.

Sri Lanka is almost totally reliant on hydropower for its electricity, making it vulnerable to fluctuations in rainfall. The Sri Lankan government is trying to attract foreign investors to build independent thermal power plants, but has yet to clarify its regulatory policies. Net electricity consumption in Sri Lanka doubled between 1991 and 2000.

Like Sri Lanka, Nepal relies almost exclusively on hydroelectricity for its power needs. Nepal also has large untapped hydroelectric potential, which could be developed both for domestic consumption (only about 15 per cent of Nepal's population currently has access to electricity) as well as for export (the main potential export market being India). Renewable power sources, including micro-hydro, and solar energy, is gaining popularity in Nepal, particularly in remote regions of the country. In recent news, the 144 MW Kaligandaki "A" hydroelectric dam began generating electricity in March 2002. Australia's SMEC International signed a contract in January 2002 for the development of the West Seti hrdroelectric dam, which will take five years to develop and have an installed capacity of 750 MW. Most of the power from the West Seti project will be exported to India.

Bhutan also has the potential to become a major hydropower exporter. Some estimates have put the country's potential hydropower resources as high as 30 GW. The main large project currently under development is the Tala hydropower project, which is to be operational by 2004-2005. It will have an installed capacity of 1,020 MW, and produce power for export to northern India.

76

South China Sea Region

The South China Sea region is the world's second busiest international sea lane. More than half of the world's supertanker traffic passes through the region's waters. In addition, the South China Sea region contains oil and gas resources strategically located near large energy-consuming countries.

General Background

The South China Sea encompasses a portion of the Pacific Ocean stretching roughly from Singapore and the Strait of Malacca in the southwest, to the Strait of Taiwan (between Taiwan and China) in the northeast (see the footnote for a more precise definition). The area includes more than 200 small islands, rocks, and reefs, with the majority located in the Paracel and Spratly Island chains. Many of these islands are partially submerged islets, rocks, and reefs that are little more than shipping hazards not suitable for habitation; the total land area of the Spratly Islands is less than 3 square miles. The islands are important, however, for strategic and political reasons, because ownership claims to them are used to bolster claims to the surrounding sea and its resources.

The South China Sea is rich in natural resources such as oil and natural gas. These resources have garnered attention throughout the Asia-Pacific region. Asia's economic growth rates have been among the highest in the world, and this economic growth will be accompanied by an increasing demand for energy. Between now and 2025, oil consumption in developing Asian countries is expected to rise by 3.0 per cent annually on average, with more than one-third of this increase coming from China alone. If this growth rate is maintained, oil demand for these nations will increase from about 14.5 million barrels per day in 2000 to nearly 29.8 million barrels per day by 2025.

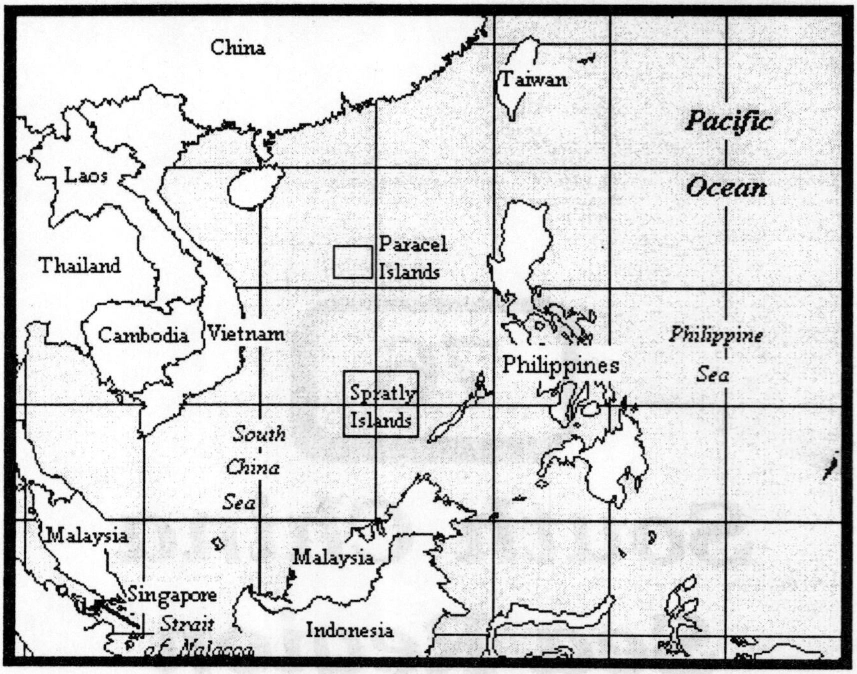

Much of this additional demand will need to be imported from the Middle East and Africa. Excluding cargoes bound for South Asia, most of this volume would need to pass through the strategic Strait of Malacca into the South China Sea. Countries in the Asia-Pacific region depend on seaborne trade to fuel their economic growth, and this has led to the sea's transformation into one of the world's busiest shipping lanes. Over half of the world's merchant fleet (by tonnage) sails through the South China Sea every year. The economic potential and geopolitical importance of the South China Sea region has resulted in jockeying between the surrounding nations to claim this sea and its resources for themselves.

South China Sea Territorial Issues

Competing territorial claims over the South China Sea and its resources are numerous, with the most contentious revolving around the Paracel Islands and Spratly Islands. However, ownership of virtually all of the South China Sea is contested. The disputed areas often involve oil and natural gas resources:

1. Indonesia's ownership of the natural gas-rich fields offshore of the Natuna Islands was undisputed until China released an official map with unclear maritime boundaries indicating that Chinese-claimed waters in the South China Sea may extend into the waters around the Natuna Islands. Indonesia responded by choosing the Natuna Islands region as the site of its largest military exercises to date in 1996. Since then, however, drilling in the natural gas fields has proceeded, and China has not voiced a specific objection to their development.
2. The Philippines' Malampaya and Camago natural gas and condensate fields are in Chinese-claimed waters. China has not, however, voiced a specific objection to the development of these fields.

3. Many of Malaysia's natural gas fields located offshore Sarawak also fall under the Chinese claim, but as with the Philippine gas fields, China has not specifically objected to their development. In July 2002, a new oil discovery by Murphy Oil (working under a contruct with state-owned Petronas) about 100 miles offshore from Sabah on island of Borneo rekindled interest in a latent dispute between Malaysia and Brunei over offshore rights. Brunei had asserted a 200-mile exclusive economic zone (EEZ) off its coastline in 2000. Negotiations between the two governments to resolve the issue are continuing.
4. Vietnam and China have resolved their dispute over areas in the Gulf of Tonkin to the south of China's Guangdong province. An agreement signed in December 2000 delineated the boundary between their EEZs, opening the way for oil and gas exploration.
5. Maritime boundaries in the natural gas-rich Gulf of Thailand portion of the South China Sea have not all been clearly defined. Several companies have signed exploration agreements but have been unable to drill in a disputed zone between Cambodia and Thailand. Overlapping claims between Thailand and Vietnam were settled on August 8, 1997, and cooperative agreements for exploration and development were signed for the Malaysia-Thai and Malaysia-Vietnam Joint Development Areas (the latter effective June 4, 1993).

Most of these claims are historical, but they are also based upon internationally accepted principles extending territorial claims offshore onto a country's continental shelf, as well as the 1982 United Nations Convention on the Law of the Sea.

UN Law of the Sea

The 1982 convention created a number of guidelines concerning the status of islands, the continental shelf, enclosed seas, and territorial limits. Among the most relevant to the South China Sea are:

1. Article 3, which establishes that "every state has the right to establish the breadth of its territorial sea up to a limit not exceeding 12 nautical miles";
2. Articles 55–75 define the concept of an Exclusive Economic Zone (EEZ), which is an area up to 200 nautical miles beyond and adjacent to the territorial sea. The EEZ gives coastal states "sovereign rights for the purpose of exploring and exploiting, conserving and managing the natural resources, whether living or non-living, of the waters superjacent to" (above) "the seabed and of the seabed and its subsoil.".
3. Articles 76 defines the continental shelf of a nation, which "comprises the seabed and subsoil of the submarine areas that extend beyond its territorial sea throughout the natural prolongation of its land territory to the outer edge of the continental margin, or to a distance of 200 nautical miles.". This is important because Article 77 allows every nation to exercise "over the continental shelf sovereign rights for the purpose of exploring it and exploiting its natural resources".
4. Article 121, which states that rocks that cannot sustain human habitation or economic life of their own shall have no exclusive economic zone or continental shelf.

The establishment of the EEZ created the potential for overlapping claims in semi-enclosed seas such as the South China Sea. These claims could be extended by any nation which could establish a settlement on the islands in the region. South China Sea claimants have established outposts on the islands (mostly military) in order to conform with Article 121 in pressing their claims. The Law of the Sea Convention states that countries with overlapping claims must resolve them by good faith negotiation. The use of the Joint Development Area principle, followed in the Gulf of Thailand, is one model that has been successfully used by South China Sea claimants.

Regional Conflict and Resolution

All of the Spratly Islands claimants have occupied some of the islands, and/or stationed troops and built fortified structures on the reefs. Brunei, which does not claim any of the Spratly Islands, has not occupied any of them, but has declared an Exclusive Economic Zone that includes Louisa Reef.

Military skirmishes have occurred numerous times over the past two decades. The most serious occurred in 1974, when China invaded and captured the Paracel Islands from Vietnam, and in 1988, when the Chinese and Vietnamese navies clashed at Johnson Reef in the Spratly Islands, sinking several Vietnamese boats and killing over 70 sailors.

Indonesia has taken the leading role in diplomatic initiatives and cooperative agreements to resolve South China Sea issues, particularly through the ASEAN (Association of Southeast Asian Nations) forum, which has called for the peaceful arbitration of territorial claims. ASEAN includes all South China Sea nations except for China and Taiwan, and has held a number of working groups with China and Taiwan on related issues that have the potential to foster the cooperation and friendship needed to resolve the more contentious issues in the region. Indonesia hosted the first of these workshops in 1990. These issues have also been discussed at the larger ASEAN Regional Forum (ARF), held in conjunction with the ASEAN Post Ministerial Conference, which draws together 22 countries which are involved in the security of the Asia Pacific region, including all ASEAN members.

ASEAN ministers agreed in 1996 that there should be a regional code of conduct for the South China Sea to permit activities such as scientific research and efforts to combat piracy and drug trafficking without invoking the contentious issue of sovereignty. At the ASEAN Summit in November 1999, ASEAN members put forth a general code of conduct for resolving disputes which had been drafted by the Philippines and Vietnam.

Any such agreements would need to involve non-ASEAN members such as China and Taiwan in order to be comprehensive. China, which is a member of the ARF, has argued in the past that the resolution of territorial disputes should be a bilateral issue. However, other ARF members, such as the United States, have argued that all ARF members had an interest in issues affecting the peace and stability of the region, and that the ARF forum was appropriate for discussing these issues. Views on this issue are varied:

1. China has begun a dialogue with ASEAN on the idea of a "code of conduct" governing actions by claimants, but progress has been slow. In general, ASEAN members have pushed for specific committments to refrain from additional occupation of reefs or new construction, which China has favored a more vague committment to refrain from actions which would "complicate the situation." In November 2002, China and the 10 members of ASEAN signed a Joint Declaration on the Conduct of the Parties, which pledged to "undertake to resolve their territorial and jurisdictional disputes by peaceful means" without "resorting to the threat or use of force."
2. Malaysian Foreign Minister Syed Hamid bin Syed Jaafar Albar stated that it was his belief that ASEAN nations had agreed that the territorial disputes were an ASEAN issue, and should not be resolved in other international forums.

3. Vietnam has had bilateral working groups with China to resolve disputed boundaries in the Gulf of Tonkin (referred to as the Beibu Wan by China, the Vinh Bac Bo by Vietnam) and the Spratlys, as well as land boundaries. The Gulf of Tonkin dispute was resolved in an agreement concluded in December 2000. Vietnam has wanted to include the dispute over the Paracel Islands in any "code of conduct," but the idea is not supported by other ASEAN members because the Paracels are disputed only between Vietnam and China.
4. Malaysia and Brunei have held talks in 2003 on their conflicting EEZ claims, but have not yet reached an agreement. There have been incidents in 2003 in which naval vessels from Malaysia and Brunei have acted (without the actual use of force) to prevent exploration vessels from working in the disputed area.

Oil

The focus of most attention regarding the South China Sea's resources has been on hydrocarbons in general, and on oil in particular. Oil deposits have been found in most of the littoral (adjacent) countries of the South China Sea. The South China Sea region has proven oil reserves estimated at about 7.0 billion barrels, and oil production in the region is currently around 2.5 million barrels per day. Malaysian production accounts for almost one-half of the region's total. Total South China Sea production has increased gradually over the past few years, primarily as additional production from China, Malaysia and Vietnam has come online.

The fact that surrounding areas are rich in oil deposits has led to speculation that the Spratly Islands could be an untapped oil-bearing province located near some of the world's largest future energy consuming countries. Speculation that the Spratly Islands could have great strategic value has fueled disputes over ownership. In fact, there is little evidence outside of Chinese claims to support the view that the region contains extensive oil resources. Because of a lack of exploratory drilling, there are no proven oil reserve estimates for the Spratly or Paracel Islands, and no commercial oil or gas has been discovered there.

Resource estimates for this region that have been reported in the Chinese press or attributed to Chinese officials vary greatly. Optimistic Chinese estimates of the South China Sea region's oil potential, however, have helped encourage interest in the area, with one report suggesting that the Spratly Islands region could become another Persian Gulf. One of the more moderate Chinese estimates suggested that potential oil resources (not proved reserves) of the Spratly and Paracel Islands could be as high as 105 billion barrels of oil, and another suggested that the total for the South China Sea could be as high as 213 billion barrels. A common rule-of-thumb for such frontier areas as the Spratly Islands is that perhaps 10 per cent of the potential resources can be economically recovered. Using this rule, these Chinese estimates imply potential production levels for the Spratly Islands of 1.4-1.9 million barrels per day (at reserve/production ratios of 15 and 20). The highest Chinese reserves estimate implies production levels that are twice as high as this.

China's optimistic view of the South China Sea's hydrocarbon potential is not shared by most non-Chinese analysts. A 1993/1994 estimate by the U.S. Geological Survey, for example, estimated the sum total of discovered reserves and undiscovered resources in the offshore basins of the South China Sea at 28 billion barrels. Using the same rule-of-thumb, these reserves could yield a peak oil production level for the Spratly Islands of 137,000-183,000 barrels per day, the same order of magnitude as current production levels in Brunei or Vietnam.

Natural Gas

Though sometimes overlooked, natural gas might be the most abundant hydrocarbon resource in the South China Sea. Most of the hydrocarbon fields explored in the South China Sea regions of Brunei, Indonesia, Malaysia, Thailand, Vietnam, and the Philippines contain natural gas, not oil. Estimates by the U.S. Geological Survey and others indicate that about 60 per cent -70 per cent of the region's hydrocarbon resources are gas.

At the same time, natural gas usage among developing Asian countries is expected to rise by about 4.5 per cent annually on average through 2025–faster than any other fuel–with almost half of this increase coming from China. If this growth rate is maintained, demand will exceed 21 trillion cubic feet (Tcf) per year–nearly triple current consumption levels–by 2025. Natural gas consumption could increase even faster if additional infrastructure is built. Proposals have been made to link the gas producing and consuming regions of the Pacific Rim region of Asia by pipeline, with the South China Sea geographically central to these regions.

Malaysia is not only the biggest oil producer in the region, it is also the dominant natural gas producer as well, and until recently has been the primary source of growth in regional gas production. The development of natural gas resources outside of Malaysia has been hampered by the lack of infrastructure. Despite this constraint, natural gas exploration activity elsewhere in the region has been increasing. Much of this new activity had occurred in the Gulf of Thailand, offshore China, in Indonesia around the Natuna Islands, and in Vietnam in the Nam Con Son basin southeast of Vietnam.

As with oil, estimates of the South China Sea's natural gas resources vary widely. One Chinese report estimates that there are 225 billion barrels oil equivalent of hydrocarbons in the Spratly Islands alone. If 70 per cent of these hydrocarbons are gas as some studies suggest, total gas resources (as opposed to proved reserves) would be almost 900 Tcf. If the rule of thumb for frontier areas were applied to these resource levels, the Chinese estimates would imply potential production levels for the Spratly Islands of almost 1.8-2.2 Tcf annually (at common natural gas reserve/production ratios in the region of 40-50). The entire South China Sea has been estimated by the Chinese to contain more than 2,000 Tcf of natural gas resources. As with oil, China's optimistic view of the South China Sea's natural gas potential is not shared by most non-Chinese analysts.

Liquefied Natural Gas (LNG)

The bulk of the world's LNG trade passes through the South China Sea, and LNG shipments through the Sea to Northeast Asian Markets constituted well over half of the world's LNG trade in 2001. Japan is by far the world's largest consumer of LNG, with shipments to South Korea (the world's second largest consumer of LNG) and Taiwan (the world's fifth largest consumer of LNG) accounting for most of the remaining shipments through the Sea.

Shipping

More than half of the world's annual merchant fleet tonnage passes through the Straits of Malacca, Sunda, and Lombok, with the majority continuing on into the South China Sea. Oil flows through the Strait of Malacca leading into the South China Sea are three times greater than through the Suez Canal/Sumed Pipeline, and fifteen times greater than oil flows through the Panama Canal. Virtually all shipping that passes through the Malacca and Sunda Straits must pass near the Spratly Islands. The other major shipping lane in the region uses the Lombok and Makassar Straits, and continues into the Philippine Sea. Except for north-south traffic from Australia, it is not used as extensively as the Strait of Malacca and the South China Sea, since for most voyages it represents a longer voyage by several hundred miles.

Shipping (by tonnage) in the South China Sea is dominated by raw materials en route to East Asian countries. Tonnage via Malacca and the Spratly Islands is dominated by liquid bulk such as crude oil and liquefied natural gas (LNG), with dry bulk (mostly coal and iron ore) in second place. Nearly two-thirds of the tonnage passing through the Strait of Malacca, and half of the volume passing the Spratly Islands, is crude oil from the Persian Gulf. Oil flows through the Strait of Malacca were 10.3 million barrels per day in 2002, and rising Asian oil demand could almost double these flows over the next two decades.

Northeast Asian nations are heavily dependent upon energy shipments through the South China Sea. More than 80 per cent of the crude oil supplies for Japan, South Korea, and Taiwan flow through the Sea from the Middle East, Africa, and South China Sea nations such as Indonesia and Malaysia. LNG (above) and coal from Indonesia, South Africa, and Vietnam are also shipped via this route. As a result, about two-thirds of South Korean energy supplies, and almost 60 per cent of Japan and Taiwan's energy supplies flow through the Sea.

Piracy

The large volume of shipping in the South China Sea/Strait of Malacca littoral has created opportunities for attacks on merchant shipping. Oil product tankers have been among the ships attacked, with 1999 targets including the Thai tanker MV Tenyu off the coast of Malaysia, the Singapore-owned tanker Petro Ranger sailing from Ho Chi Minh City to Singapore, the Indonesian tanker MT Atlanta in the Riau Straits off Sumatra, and the Honduran tanker MT 1 off the eastern coast of Malaysia. The three littoral States of the Malacca Strait (Indonesia, Malaysia and Singapore) implemented a coordinated patrol and other counter-measures in the region in 1992.

77

South Korea

The Republic of Korea (South Korea) is important to world energy markets as the fourth largest oil importer, and the second largest importer of Liquefied Natural Gas (LNG).

General Background

After making a strong recovery in 1999-2000 from the effects of the Asian financial crisis in 1997-1998, South Korea's economy was negatively affected by the global economic slowdown of 2001-2002, but has begun to recover in 2002. Growth in real gross domestic product (GDP) is projected at 5.8 per cent for 2002, up from 3.3 per cent in 2001 but down from the 9.2 per cent achieved in 2000. The recovery has been fuelled by domestic demand, even though export growth has been somewhat weak. Increased government spending, largely on infrastructure projects, has been a major contributor to increased domestic demand.

In the wake of the Asian financial crisis, South Korea has began an economic reform program designed to address some of the conditions which made its economy vulnerable. Most importantly, the South Korean government has begun to break the hold of the chaebols (large, multi-industry conglomerates) over the financial sector. The lack of an "arms length" business relationship between borrowers and lenders had led to many South Korean financial institutions having a very large ratio of non-performing loans. While there is no intention of forcing the chaebols to divest their financial subsidiaries, the government has increased regulation to prevent chaebols from abritrarily channeling money into other subsidiaries. Chaebols also have been pressed to spin off their non-core businesses and to rationalize their corporate structures. To stimulate domestic demand, the South Korean government under President Kim Dae-jung enacted a package of tax cuts directed at lower and middle-income workers.

The South Korean government has plans to privatize several large state-owned enterprises (SOEs), including the state electricity utility, Korean Electric Power Corporation (KEPCO) and natural

gas monopoly Korea Gas Company (KOGAS). The privatization program has moved at a slower pace than originally planned, due in part to strong opposition from labor unions to some of the privatizations and delays in passing implementing legislation.

South Korea recently joined the International Energy Agency (IEA). Its membership became effective in April 2001, upon its fulfillment of the requirements for membership, and was formally ratified by the South Korean government in March 2002.

Oil

With no domestic oil reserves, South Korea must import all of its crude oil. Oil makes up the largest share of South Korea's total energy consumption, though its share has been declining in recent years. Petroleum accounted for 56 per cent of primary energy consumption in 2000. South Korea consumed more than 2.1 million barrels a day (bbl/d) of oil in 2000, down from a high of nearly 2.3 million bbl/d in 1997, all of which was imported. South Korea is the seventh largest oil consumer and fourth largest crude oil importer in the world.

South Korea's total reliance on oil imports has led to a policy of securing and diversifying the country's oil supply. South Korea has both a short-term and a long-term approach to fulfilling its oil needs. In the short-term, it has developed a strategic petroleum reserve, which is managed by the state-owned Korea National Oil Corporation (KNOC). Strategic stocks are roughly equivalent to a 90-day supply. The period of "import cover" was expanded from 60 days in early 2001, in part to meet the requirements for entry into the IEA. This reserve serves as a safety net against supply disruptions.

In the long term, KNOC is pursuing equity stakes in oil and gas exploration around the world. KNOC has 18 overseas exploration and production projects in 13 countries. This includes 4 producing fields in Yemen, Argentina, Peru, and the North Sea, and 4 fields under development in Yemen, Venezuela, Libya, and Vietnam. KNOC also is exploring domestic blocks offshore from South Korea. KNOC reported a new oil find in August 2001 at the Vung Tau site offshore from Vietnam, which is expected to be developed and in production by 2003. Recoverable reserves at Vung Tau are estimated at 420 million barrels. The South Korean government has stated that it plans for KNOC to provide for 10 per cent of the country's oil needs by 2010.

The South Korean refining industry was strongly affected by the country's economic crisis in 1997-1998, especially because it already suffered from significant overcapacity before the downturn in demand. In September 1998, South Korea's four downstream oil companies raised the retail price of gasoline and diesel oil following a government tax hike. In October 1998, the South Korean government, under financial pressure, decided to fully deregulate the refining industry, accelerating this decision from the original January 1999 deadline in order to attract badly needed foreign investment. Foreign backing has proved critical in maintaining cash flows and preserving the creditworthiness of the refining industry.

Several corporate consolidations and selloffs occurred as a result. In September 1998, Hanwha's 270,000-bbl/d refinery in Inchon was taken over by Hyundai Oil Refinery Company, giving Hyundai the country's third largest refining capacity (after SK Corporation and LG-Caltex) with 580,000 bbl/d. In October 1999, Hyundai completed the sale of a 50 per cent interest in its refining operation to the Abu Dhabi International Petroleum Investment Corporation, which was intended to reduce the company's highly leveraged debt-to-equity ratio. Ssangyong Group sold its 28.4 per cent stake in Ssangyong Oil Refining Corporation to its majority shareholder, Saudi Aramco, in 2000. The firm's name was changed to S-Oil.

Despite the consolidation in South Korea's refining sector, it has yet to fully recover from the effects of the Asian financial crisis and the shock of the 1998 deregulation. Profit margins have remained very weak through 2002.

Natural Gas

South Korea currently relies on imported liquefied natural gas (LNG) to meet its entire demand for natural gas, though a project currently under development will make the country a minor natural gas producer by early 2003. Imports of LNG began in 1986, after the founding of the state-owned monopoly LNG importer Korea Gas Company (Kogas). South Korea currently gets most of its LNG from Indonesia, Malaysia, and Qatar, with smaller volumes from Brunei and Oman. The supplies from Qatar began in August 1999, under a contract with Qatar's new Ras Laffan LNG (RasGas) venture. The first shipment of Omani LNG was loaded in April 2000. In 2000, natural gas comprised around 10 per cent of South Korea's primary energy consumption. South Korea is the second largest importer of LNG worldwide, importing 669 billion cubic feet (Bcf) of LNG in 2000. South Korea's annual LNG imports increased by 140 per cent between 1993 and 1997, but fell by 9 per cent in 1998 due to the effects of the Asian financial crisis. In 1999, however, South Korea's LNG imports grew by 22 per cent, followed by 10 per cent growth in 2000, as its economy recovered and new supplies from Qatar came onstream. South Korean natural gas demand is split almost evenly between the electricity sector and the residential heating sector.

Despite the temporary downturn, Kogas is planning to push ahead with projects for the expansion of LNG receiving terminals. South Korea is increasing capacity at its existing terminals (Pyongtaek and Inchon). Also, Mitsubishi Corporation of Japan and Pohang Iron and Steel Corporation signed a letter of intent in October 1998 to build an LNG receiving terminal in South Korea at Kwangyang. Construction of the facility started in June 2002, and current plans call for it to be completed in the first half of 2005.

The South Korean government announced in 1999 that it intends to privatize Kogas. An initial public offering of 33 per cent of Kogas equity was carried out in December 1999. Privatization plans initially stalled, however, due to questions about the structure of the companies, which would result if Kogas were split for privatization, and opposition from labor unions representing Kogas employees. Current plans call for the Kogas privatization to take place in 2003, but legislation necessary to put the process in motion has not been passed by the South Korean legislature. The uncertainty over the future structure of the industry has led to delays in Kogas concluding agreements for new LNG supplies, even though additional volumes of LNG beyond current contracts are expected to be needed by 2004.

In addition to LNG imports, South Korea will have a small amount of domestic natural gas production starting in 2003. KNOC's $320 million Donghae-1 development project is developing a natural gas deposit offshore from Ulchin in southeastern South Korea estimated to contain 200 Bcf of reserves. Donghae-1 is a relatively minor development, however, and will satisfy only about 2 per cent of South Korea's gas demand once it comes onstream.

Meanwhile, South Korea also is exploring the possibility of a gas pipeline from the Kovykta natural gas deposit in the Irkutsk region of Eastern Siberia. The pipeline would supply China as well as South Korea, and might run through North Korea. The project as currently envisioned would supply about 1 Bcf/d to South Korea, and a larger volume to China, possibly beginning around the end of the decade. The two Koreas agreed in September 2001 to conduct a joint feasibility study of the pipeline project, which has not yet concluded. Since the only overland route the pipeline could take would cross North Korea, renewed tensions on the Korean peninsula have called South Korea's participation in the project into question.

Coal

Coal supplies about 20 per cent of South Korea's total energy requirements. Most of this coal is imported, since the only indigenous coal resources consist of low-quality anthracite used in home heating and small boilers. Bituminous coal supplies (steam coal for power plants and industrial boilers and metallurgical coal for steelmaking) come mainly from Australia and China, with the United States also among the suppliers. State power company KEPCO has invested in several Australian coal mines. China has become a significant supplier of coal to South Korea in the last two years, as its coal export volumes have increased, displacing some of the volume from Australia.

Electricity

South Korea uses a combination of thermal (oil, gas, and coal), nuclear, and hydroelectric capacity to meet its demand for electric power. Total power generation capacity was 50 gigawatts (GW) as of the beginning of 2000. The South Korean government estimated in May 2002 that its electricity demand will rise at an average annual rate of 3.4 per cent per year through 2015.

In September 1998, KEPCO officially dedicated its Ulchin Number 3 nuclear reactor and launched the construction of Ulchin Nuclear Power Plants Numbers 5 and 6. Ulchin Number 3 has a generating capacity of 1 GW and is the first nuclear power plant built completely with South Korean technology from design to construction. The Number 4 Ulchin nuclear plant was completed in late 1999, and Numbers 5 and 6 are targeted to be completed in 2004 and 2005.

The South Korean government is moving ahead with plans to break up and privatize KEPCO. The South Korean government plans to split KEPCO into separate generation, transmission, and distribution units. Progress has been slow, however, due to opposition from labor unions. In early 2001, KEPCO split its power generation holdings into six separate subsidiaries, in a preliminary move to facilitate a split into competing companies. Five of the six operate thermal and hydroelectric facilities and are of roughly equal size in terms of installed generating capacity–between 7 and 8 GW. The sixth is comprised of all of KEPCO's nuclear plants, which will be kept together in one corporation under government ownership. The privatization plan has been controversial, with unions fearing layoffs by new management and some politicians opposing foreign ownership. Current plans call for the first of the five generating units, Korea Southeast Power, to be sold off by January 2003. The other four thermal generation companies are to be sold by 2005.

While most of South Korea's generating capacity is still controlled by KEPCO, a few independent power producers (IPPs) exist. LG Power, owned by the LG Group conglomerate, operates a 540-megawatt (MW) independent power plant at Bugok near Asan Bay. The facility began operation in April 2001. LG Power purchased the existing Anyang and Puchon plants in June 2000, with a combined capacity of 950 MW, from KEPCO after a competitive tender. Tractebel is also investing in a new 519-MW IPP plant in Yulchon in partnership with Hyundai. In another significant development, South Korea's original IPP, Hanwha Energy was spun off from its chaebol parent company in June 2000, in a deal in which El Paso Energy acquired a 50 per cent stake. Hanwha Energy operates a 1,800-MW plant at Inchon.

While South Korea is not a party to the Kyoto Protocol on greenhouse gas emissions, its future plans emphasize the development of more nuclear power plants to reduce growth in carbon emissions. A dozen additional nuclear plants are planned before 2015.

Environment

South Korea's rapid growth has resulted in significant environmental side effects, as industrial emissions from factories have caused serious acid rain problems. Increased car ownership also has led to a corresponding rise in carbon emissions from the country's transportation sector, contributing to South Korea's air pollution problems. Transboundary pollution is a major concern in the region, leading environmental ministers from South Korea, China and Japan to establish a joint commission to attempt to tackle the problem.

Although slowed by the Asian financial crisis, South Korea's energy consumption has rebounded and continues to increase as the country's economy grows. In 2000, South Korea consumed 7.9 quadrillion Btu's of energy, up 7.8 per cent over 1999 and 15.9 per cent higher than in 1998. South Korea's energy intensity (energy consumption per $1995 of GDP) increased from increased from 11,054 Btu per $/1995 in 1980 to 12,759 Btu per $1995 in 2000, putting the country on par with countries such as Thailand and the Philippines. The country's per capita energy use has increased with its economic growth, from 88.1 million Btu per person in 1990 to 166.7 million Btu per person in 2000, but South Korea's per capita carbon emissions unfortunately have increased at a similar pace.

Thus far, there has been little emphasis on the development of renewable energy resources in South Korea, but that is changing as the 1997-1998 financial crisis focused attention on South Korea's dependence on imported oil to meet domestic energy demand. One of the country's stated goals in its

'National Vision for Environmental Policies in the 21st Century' is the promotion of green development schemes such as increased usage of photovoltaic power and fuel cells. South Korea's environmental outlook depends on its ability to shift its energy supply mix to cleaner-burning fuels and de-link the increase in carbon emissions from economic growth.

Country Overview

Independence: August 15, 1945

Population (7/01E): 47.9 million

Location/Size: Eastern Asia/(98,480 square kilometers 38,000 square miles), about the size of Indiana

Major Cities: Seoul (capital), Pusan, Taegu, Inchon, Kwangju

Language: Korean (English widely taught)

Ethnic Groups: Korean, with a small Chinese minority

Religions: Christianity, 49 per cent; Buddhism, 47 per cent; Confucianism, 3 per cent; Other, 1 per cent

Defense (8/98): Army, 548,000; Navy, 60,000; Air Force, 52,000 (plus 35,910 U.S. troops)

Economic Overview

Currency: Won (W)

Exchange Rate (8/13/02): US$1 = 1,201 Won

Real GDP Growth Rate (2001E): 3.3 per cent *(2002E)* 5.8 per cent

Inflation Rate (consumer prices)(2001E): 4.1 per cent *(2002E)* 3.1 per cent

Unemployment Rate (5/02E): 3.7 per cent

Current Account Balance (2002E): $4.6 billion

Merchandise Exports (2002E): $164.7 billion

Merchandise Imports (2002E): $152.8 billion

Merchandise Trade Balance (2002E): $11.9 billion

Major Exports: Electronics, textiles, ships, automobiles, steel, computers, footwear

Major Imports: Crude oil, food, machinery and transportation equipment, chemicals and chemical products, base metals and articles.

Top Trading Partners: U.S., Japan, China, Germany

Energy Overview

Oil Consumption (2001E): 2.1 million barrels per day (bbl/d); all imported

Crude Oil Refining Capacity (1/1/02): 2.6 million bbl/d

Natural Gas Consumption (2000E): 669 billion cubic feet (bcf)–all imported LNG

Recoverable Coal Reserves (2000E): 86 million short tons (Mmst)

Coal Production (2000E): 4.6 Mmst

Coal Consumption (2000E): 71.7 Mmst

Electric Generation Capacity (1/1/00E): 50.0 gigawatts

Electricity Generation (2000E): 273.2 billion kilowatthours

Environmental Overview

Total Energy Consumption (2000E): 7.9 quadrillion Btu* (2.0 per cent of world total energy consumption)

Energy-Related Carbon Emissions (2000E): 115.3 million metric tons of carbon (1.8 per cent of world carbon emissions)

Per Capita Energy Consumption (2000E): 166.7 million Btu (vs. U.S. value of 351.1 million Btu)

Per Capita Carbon Emissions (2000E): 2.4 metric tons of carbon (vs. U.S. value of 5.6 metric tons of carbon)

Energy Intensity (2000E): 12,759 Btu/$1995 (vs U.S. value of 10,919 Btu/$1995)**

Carbon Intensity (2000E): 0.19 metric tons of carbon/thousand $1995 (vs U.S. value of 0.17 metric tons/thousand $1995)

Sectoral Share of Energy Consumption (1998E): Industrial (55.7 per cent), Transportation (20.2 per cent), Commercial (15.1 per cent), Residential (9.0 per cent)

Sectoral Share of Carbon Emissions (1998E): Industrial (49.8 per cent), Transportation (24.0 per cent), Commercial (17.1 per cent), Residential (9.1 per cent)

Fuel Share of Energy Consumption (2000E): Oil (58.3 per cent), Coal (20.4 per cent), Natural Gas (9.5 per cent)

Fuel Share of Carbon Emissions (2000E): Oil (56.0 per cent), Coal (34.6 per cent), Natural Gas (9.4 per cent)

Renewable Energy Consumption (1998E): 177 trillion Btu* (9 per cent decrease from 1997)

Number of People per Motor Vehicle (1998): 4.4 (vs. U.S. value of 1.3)

Status in Climate Change Negotiations: Non-Annex I country under the United Nations Framework Convention on Climate Change (ratified December 14th, 1993). Signatory to the Kyoto Protocol (signed September 25th, 1998–not yet ratified).

Major Environmental Issues: Air pollution in large cities; water pollution from the discharge of sewage and industrial effluents; drift net fishing.

Major International Environmental Agreements: A party to the Antarctic-Environmental Protocol, Antarctic Treaty, Biodiversity, Climate Change, Endangered Species, Environmental Modification, Hazardous Wastes, Law of the Sea, Nuclear Test Ban, Ozone Layer Protection, Ship Pollution, Tropical Timber 83, Tropical Timber 94, Wetlands and Whaling. Has signed, but not ratified, Desertification.

The total energy consumption statistic includes petroleum, dry natural gas, coal, net hydro, nuclear, geothermal, solar, wind, wood and waste electric power. The renewable energy consumption statistic is based on International Energy Agency (IEA) data and includes hydropower, solar, wind, tide, geothermal, solid biomass and animal products, biomass gas and liquids, industrial and municipal wastes.

Energy Industries

State Energy Companies: Korea National Oil Corporation(KNOC); Daehan Oil Pipeline Corporation (DOPCO); Korea Electric Power Company (KEPCO); Korea Gas Corporation (KOGAS)

Major Oil Companies (Private): SK Corporation; LG-Caltex; S-Oil (formerly Ssangyong Oil); Hyundai Oil

Major Refineries (1/1/02 Capacity): Ulsan (817,000 bbl/d); Onsan (520,000 bbl/d); Yocheon (633,600 bbl/d); Daesan (310,000 bbl/d); Inchon (270,000 bbl/d)

Major Ports: Pusan, Inchon, Kunsan, Mokpo, Ulsan

LNG Regasification Terminals: Pyongtaek, Inchon

78

Southeastern Europe

The countries of Southeastern Europe–including Romania, Bulgaria, and Moldova–occupy a strategic location on the west side of the Black Sea, exporting electricity through much of the Balkan Peninsula and transporting Russian natural gas to Western Europe and Turkey. Southeastern Europe also is a potentially significant transit region for Caspian oil exports to Europe.

General Background

The countries of southeastern Europe–here including Romania, Bulgaria, and Moldova–share a troubled history in addition to their geographical location. Since the Eastern European revolutions of 1989 and the fall of the Soviet Union in 1991, the three countries have been independent democracies, but each has had significant problems in transitioning from a centrally-planned economic system to a market-based economy. While Bulgaria and Romania avoided the warfare and bloodshed that devastated the Balkans region in the 1990s, they were both significantly affected by the economic embargo placed on Yugoslavia, suffering several billion dollars' worth of losses due to disrupted trade, transport, and investment.

Moldova, although relatively less affected economically by the wars in the former Yugoslavia, suffered through a civil war of its own in the 1990s. Fighting broke out shortly after the country received its independence, paralyzing the country's already stagnant economy. Russian settlers and Moldovans on the industrialized left bank of the Dnistr River set up the secessionist Trans-Dnistrian Republic as the conflict stalemated. Moldova's economy has crept along as fighting has subsided, but there is no formal resolution to the conflict in sight and Western investment, which is desperately needed, is nearly nonexistent.

Unlike in central Europe and in the Baltic countries, the process of shedding the totalitarian past has proceeded slowly in southeastern Europe. Political reform did not match the sweeping changes elsewhere in the former Eastern bloc, and as a result, former Communist leaders were able

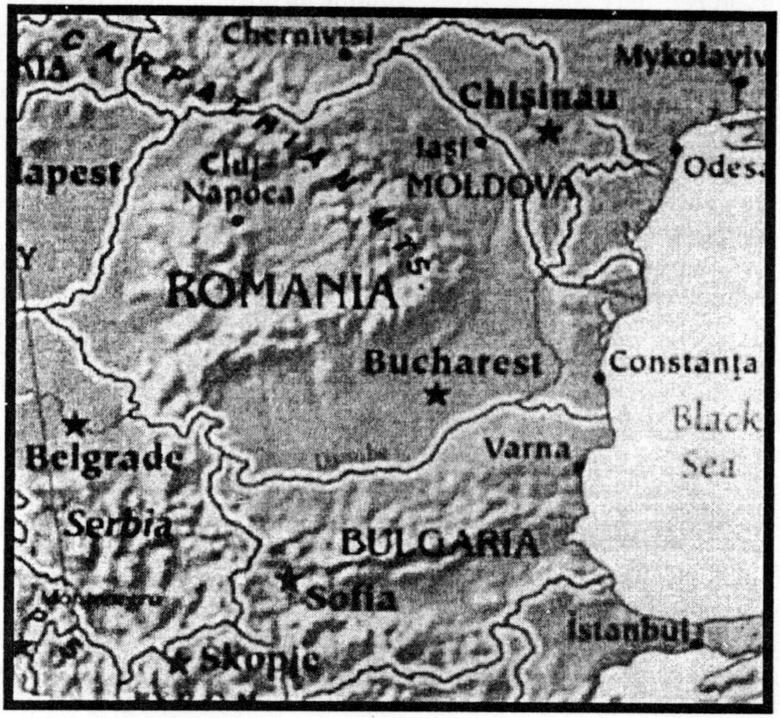

to hold on to the administrative controls of government. As a result, economic and structural reform in southeastern Europe was delayed. Although the pace of reform has picked up, the transition to democracy and a market-based economy in Romania, Bulgaria, and Moldova has lagged behind other parts of Europe.

Table 78.1: Economic and Demographic Indicators for Southeastern Europe

Country	*Gross Domestic Product (Nominal GDP), 2001E (Billions of U.S. $)*	*Real GDP Growth Rate, 2001 Estimate*	*Real GDP Growth Rate, 2002 Projection*	*Per Capita GDP, 2001E (Market Exchange Rate)*	*Population 2002E (Millions)*
Bulgaria	$13.7	4.0%	3.8%	$1,727	7.9
Romania	$39.6	5.3%	5.0%	$1,768	22.4
Moldova	$1.5*	6.1%	5.9%	$408*	4.3
Total/weighted average	**$54.8**	**5.0%**	**4.7%**	**$1,584**	**34.6**

* Figures do not include the unrecognized Trans-Dnistrian Republic.

Each of the three countries held elections within 8 months during 2000-2001, with very different results. In December 2000, Romanian voters elected former Communist Party official Ion Iliescu to the presidency in a runoff over nationalist Corneliu Vadim Tudor, elevating Iliescu to the post that he held from 1990 to 1996, as well as returning his Party of Social Democracy in Romania to power. In Moldova, the Communist Party, already the largest political party in the country, swept to a resounding victory in February 2001 elections, winning an absolute majority in the parliament and installing Vladimir Voronin as president. In Bulgaria, former King Simeon II, returning to his homeland for the first time since the monarchy was abolished by the Communists in 1946, led his National Movement for Simeon II to a victory in June 2001 parliamentary elections, swore allegiance to a republican constitution and accepted the post of prime minister. Georgi Parvanov then ousted incumbent President Petar Stoyanov in Bulgaria's presidential elections.

Since the elections, Romania, Moldova, and Bulgaria have achieved significant economic growth, with a regional average of 5.0 per cent growth in nominal gross domestic product (GDP) in 2001. Although the slowdown in the global economy has negatively affected Southeastern Europe as well, the region still is projected to post an average of 4.7 per cent growth in nominal GDP in 2002. At an historic summit in Prague in November 2002, Bulgaria and Romania, along with Slovakia, Slovenia, Lithuania, Latvia, and Estonia, were invited to join the North Atlantic Treaty Organization (NATO). Membership in NATO has been a goal of the Romanian and Bulgarian governments since the fall of the Berlin Wall and the collapse of communism in Eastern Europe in 1989. The two countries are expected to accede to the military alliance in 2004. Bulgaria and Romania are also in negotiations to join the European Union, although neither country is expected to become a full member of the organization before 2006.

Regional Energy Issues

Although none of the three countries is a major energy producer, Romania is an important regional oil and natural gas producer, and both Romania and Bulgaria hold substantial coal deposits. In addition, Bulgaria, with its coal-fired and nuclear power generators, is a major regional exporter of electricity, both to Turkey and to the Balkans. Nevertheless, it is the geographic location of Romania, Moldova, and Bulgaria–between major producers of energy in Russia and the Caspian Sea region and major consumers of energy in Turkey and Europe–that gives southeastern Europe its importance as a transit point for Russian natural gas supplies to Turkey in world energy markets. With Caspian Sea region oil production expected to increase, Romania and Bulgaria, in particular, may become important transit centers for oil destined for consumers in western Europe.

Table 78.2: Energy Consumption and Carbon Dioxide Emissions in Southeastern Europe, 2000

Country	*Total Energy Consumption (Quadrillion Btu)*	*Petroleum*	*Natural Gas*	*Coal*	*Nuclear*	*Hydro-Electric*	*Other Electricity*	*Net Electricity Imports*	*Carbon Dioxide Emissions (Million metric tons of carbon)*
Bulgaria	0.94	20.2%	20.2%	37.2%	16%	3.2%	0.4%	-4.4	15.0
Romania	1.59	25.8%	39.0%	19.5%	3.8%	11.9%	0%	-4.1	24.7
Moldova	0.11	18.2%	72.7%	0.4%	0%	2.7%	0%	0.4	1.6
Total/weighted aaverage	2.64	23.5%	33.7%	25.2%	8.0%	8.4%	0.01%	-8.1	41.3

Note: percentages may not add up to 100 per cent due to rounding.

Oil

Romania has proven oil reserves of 955 million barrels, but despite a steady decline in its crude oil production over the past 25 years, the country remains the largest oil producer in central and eastern Europe. From 294,000 barrels per day (bbl/d) in 1976, Romania's oil production has plummeted 58 per cent, sliding to 124,500 bbl/d in 2002. In a bid to increase the country's petroleum production, Romania is liberalizing its oil sector by privatizing the state-owned oil company SNP Petrom and luring foreign investment for exploration. Although Romania's oil consumption fell to just 180,000 bbl/d in 2001, the country remains a net oil importer.

Table 78.3. Energy Supply Indicators, Southeastern Europe

Country	*Proven Crude Oil Reserves, 1/1/02E (Million Barrels)*	*Natural Gas Reserves, 1/1/02E (Billion Cubic Feet)*	*Coal Reserves, 1/1/01E (Million Short Tons)*	*Petroleum Production, 2001E (Thousand Barrels Per Day)*	*Natural Gas Production, 2000E (Billion Cubic Feet)*	*Coal Production, 2000 (Million Short Tons)*	*Electric Generating Capacity, 2000E (Gigawatts)*	*Crude Oil Refining Capacity, 1/1/02E (Thousand Barrels Per Day)*
Bulgaria	15	210	2,988	1	2.1	29.8	12.1	115
Romania	955.6	3,560	1,606	124.5	501.5	32.2	22.2	504
Moldova	minimal	minimal	minimal	0	0	0	1.0	0
Total	**970.6**	**3,770**	**4,594**	**125.5**	**503.6**	**62.0**	**35.3**	**619**

With 10 refineries and an overall refining capacity of approximately 504,000 bbl/d, Romania has the largest refining industry in the region. Romania's refining capacity far exceeds domestic demand for refined petroleum products, allowing the country to export a wide range of oil products and petrochemicals,–such as lubricants, bitumen, and fertilizers–throughout the region. However, nearly all Romanian refineries are underutilized because of a lack of crude oil supplies, and the majority remain in the government's hands, running at 50 per cent of capacity or less and needing

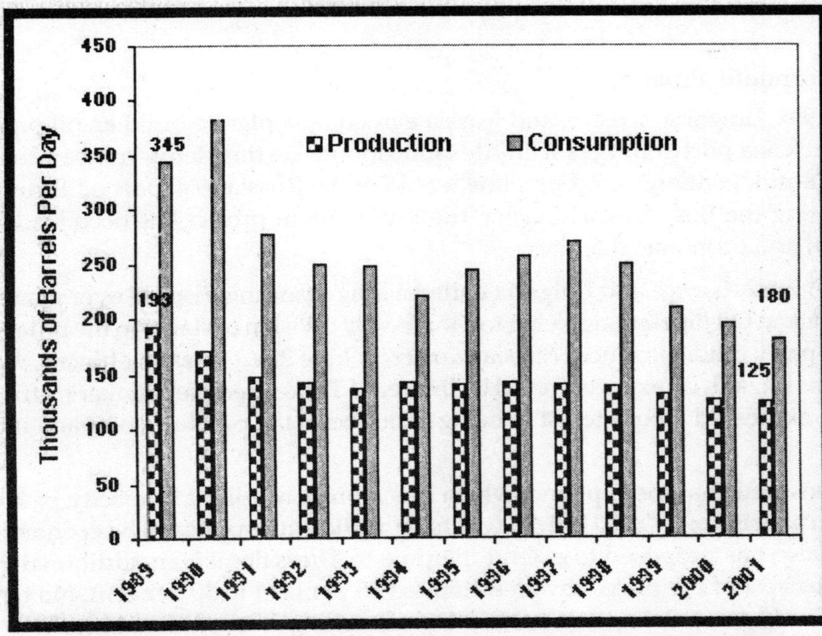

ROMANIAN OIL PRODUCTION AND CONSUMPTION, 1989-2001

repair. Years of low investment have left the country's refining industry in poor health, requiring massive amounts of capital to modernize and improve efficiency.

Bulgaria and Moldova both have minimal reserves of oil and thus each has only a very small oil industry. Bulgaria, which has one refinery with a 115,000-bbl/d capacity, consumed approximately 121,000 bbl/d in 2001, while Moldova consumed 21,000 bbl/d in the same year. Bulgaria currently is producing just 1,000 bbl/d, while Moldova does not produce any oil, making both countries nearly entirely dependent on imports to satisfy their domestic petroleum needs. Moldova, without any refineries of its own, must import all refined products.

Oil Transit

Increasing oil and natural gas production in and around the Caspian Sea, along with forecast increases of oil consumption in the European Union, means that Bulgaria, Romania, and Moldova may play a strategic role in the European transport corridor to bring Caspian oil exports to European markets. The March 2001 launch of the Tengiz-Novorossiisk Caspian Pipeline Consortium (CPC) pipeline from Kazakhstan to Russia means that additional oil will be transported via the Black Sea through the Bosporus Straits, which is already a major chokepoint for oil tankers. The difficulty in navigating the narrow straits, exemplified by a number of accidents, has led Turkey to raise environmental concerns over the increase in tanker traffic through the Bosporus.

The projected increase in oil exports from the Caspian Sea region in general, and Kazakhstan in particular, has led to the proposal of a number of Bosporus bypass options. Bulgaria, Romania, and Moldova all have made proposals to allow Caspian oil exports to bypass the Bosporus, although Moldova, lacking a Black Sea port, is highly unlikely to transport any Caspian oil exports coming via the Black Sea. Ukraine has an advantage over the countries of southeastern Europe in capturing Caspian oil export transit, since its Odesa-Brody pipeline already has been completed. Nevertheless, several Bosporus bypass pipeline options running through Bulgaria or Romania are being seriously considered.

Burgas-Alexandropoulis Pipeline

In January 1997, Bulgaria, Greece, and Russia agreed on a plan to build an oil pipeline linking the Bulgarian Black Sea port of Burgas with Alexandropoulis on the Mediterranean coast of Greece. The proposed 178-mile, underground pipeline would allow Russia to export oil through the Black Sea while bypassing the Bosporus. However, the $700 million project has been stalled by a wide range of technical and economic disputes.

In November 2002, Greece and Bulgaria ended a long-standing dispute over plans to build the so-called Trans-Balkan Oil Pipeline, agreeing to a three-way division of stakes in the pipeline company, along with third-party crude producer Russia. Earlier, in June 2002, Kazakh officials, worried about the possibility that Kazakh oil exports would be limited if Turkey reduced tanker traffic through the Bosporus Straits, expressed their interest in being included in the pipeline company if it were as an equal partner.

Russia has ensured that the pipeline, which has a proposed initial capacity of 300,000 bbl/d and a maximum capacity of 700,000 bbl/d, will have sufficient supplies to be economical. Russian oil major Lukoil also has proposed to extend the pipe to Thessaloniki, an additional 93 miles that would increase the cost of the project by an estimated 25 per cent to 30 per cent. No time frame for the start of pipeline's construction has been set, as financing is conditional on the conclusion of commercial agreements, including commitments of oil deliveries to the Burgas terminal for transit through the pipeline. Previous estimates on the timetable for construction suggested it would start in 2003 and would be completed in four years.

Burgas-Vlore Pipeline

A 750,000-bbl/d pipeline connecting Burgas with the Albanian Adriatic port of Vlore via Macedonia also has been proposed. This pipeline proposal has received letters of acceptance from the governments of Albania, Bulgaria, and Macedonia, and a $980,000 feasibility study, partially $588,000 of which was funded by the U.S. Trade and Development Agency, concluded that the 560-mile pipeline project was feasible. The Albanian-Macedonian-Bulgarian Oil (AMBO) Pipeline Corporation has been established with exclusive rights to construct the pipeline, which is estimated to cost approximately $1.1 billion. A joint venture to carry out the project was to be set up between AMBO and potential investors. Fundraising for the project already has begun, and once financing is secured, the pipeline could be constructed in four years.

Constanta-Trieste/SEEL Pipeline

Romanian government officials have advocated that a pipeline to transport crude oil from the Caspian Sea to European markets pass through its territory, claiming that Romania, which has sought to develop its infrastructure to increase its chances of sharing in the Caspian oil bonanza, offers the shortest route, best refining technology, and links via waterways to major ports in the West. The proposed 660,000-bbl/d Constanta-Trieste pipeline would allow crude oil from Kazakhstan to be shipped via the Novorossiisk port on the Black Sea to the Romanian port of Constanta, where it would then be piped to Italy across the Balkan Peninsula.

The pipeline, estimated to cost $900 million to construct, would be used mostly to provide oil to the countries along the route, and would incorporate existing pipelines connecting Constanta with 10 refineries. Several alternatives exist for the route, with a proposed northern route transiting southern Hungary and central Slovenia before terminating at Italy's oil terminal of Trieste. From there, the oil the Constanta-Trieste pipeline would be linked with the Trans Alpine Pipeline (TAP), which would carry the oil further to customers in Austria, Germany, and the Czech Republic. The southern route for the pipeline, sometimes known as the South-East European Line (SEEL), would transport Caspian oil from Constanta via a similar route as the northern route, but instead would pass through Yugoslavia and an intermediate transit point at Croatia's Adriatic port of Omisalj before crossing Slovenia and ending at Trieste. The SEEL pipeline also would link to the TAP to deliver oil to Central Europe.

Feasibility studies have shown that both proposed Constanta routes are viable, but neither pipeline has moved to the construction stage yet. In September 2002, officials from Romania, Serbia, and Croatia signed an agreement on the SEEL pipeline, but financing for the southern route remains a problem, as construction is expected to cost around $1 billion despite the fact that nearly two-thirds of the pipeline will be made up of existing pipelines. The Serbian section is expected to account for nearly 80 per cent of the budget, and Romanian officials have said that attracting financing for the 248-mile section in their country will be easier since the route will traverse easily accessible terrain.

In addition to serving as a transit point for Caspian oil, Romania is hoping to offload some Caspian crude at Constanta and deliver it to its own refineries in order to offset the country's declining domestic production. Romania and Kazakh officials already have an agreement to refine Kazakh crude oil at Romanian refineries, and Romania hopes to supply its own domestic market as well as transport refined products to Europe via barges on the Danube-Main-Rhine link. Romania also could use its own distribution network to transport refined products into other European lines.

Natural Gas

Since 1983, when Romania's production peaked at 1.4 trillion cubic feet (Tcf), the country's natural gas output has declined nearly 65 per cent, dropping to 501.5 billion cubic feet (Bcf) in 2000. In its difficult transition to a market economy, Romania's natural gas demand has decreased precipitously as well, with consumption decreasing 55 per cent from 1989 to 2000, from 1.4 Tcf to 621.5 Bcf. Romania has proven natural gas reserves of 3.56 Tcf, but additional exploration has been discouraged by the country's economic woes and the poor investment climate. Also, the slow pace of reform has prevented potential investors from entering the Romanian natural gas market to help boost current levels of production. As a result, Romania is reliant on imports to meet its natural gas consumption needs. In 2000, Romania began to reorganize Romgaz, the state-run natural gas utility, and in July 2001 Germany's Ruhrgas became the first foreign company to invest in Romania's s natural gas distribution network.

Bulgaria has minimal natural gas reserves of 210 Bcf, forcing it to rely on imports for almost all of its natural gas consumption. From 0.4 Bcf in 1989, Bulgaria's natural gas production inched up to a peak of just 2.5 Bcf in 1993. In 2000, the country produced 2.1 Bcf of natural gas, far short of Bulgaria's rate of natural gas consumption, which reached 192.8 Bcf in 2000. State-owned Bulgargas owns the country's 1,380-mile pipeline network and functions as the country's main importer of natural gas. Bulgaria is hoping that a concession to British Petreco will boost the country's natural gas production, and the Bulgarian government, in line with a European Union directive, is beginning to implement measures to liberalize the country's natural gas market.

Moldova is thought to contain small reserves of natural gas, but currently produces none, leaving it entirely dependent on Russia. Since Moldova became independent in 1992, the country's natural gas consumption has been wildly inconsistent, with consumption falling to just 49.4 Bcf in 1994 and jumping to 84.8 Bcf in 1997 before dropping to 75.2 Bcf in 2000. This pattern reflects the economic contraction and rise in fighting between Moldova and the breakaway Trans-Dnistrian Republic during

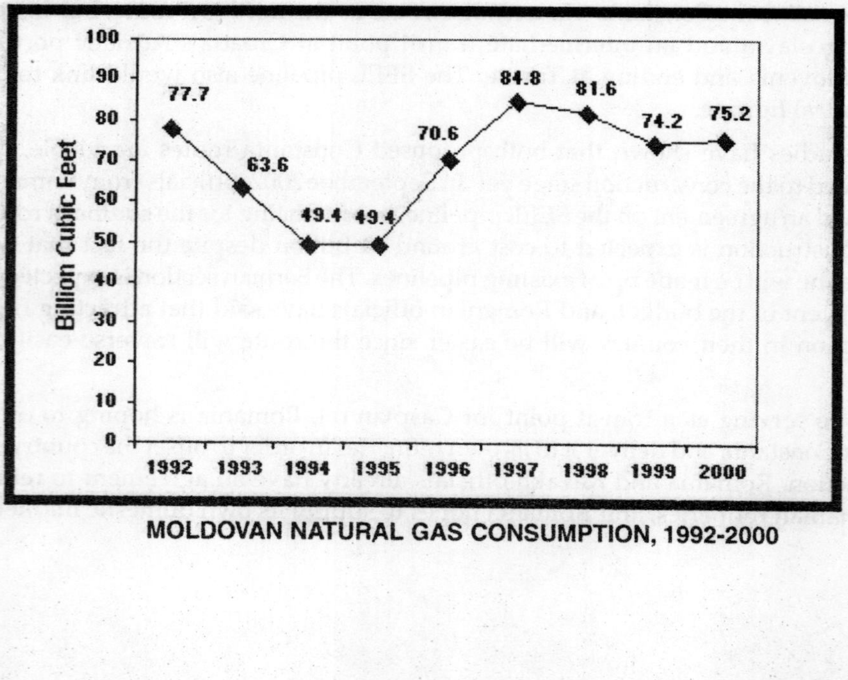

MOLDOVAN NATURAL GAS CONSUMPTION, 1992-2000

the mid-1990s, followed by the relative stability later in the decade as the fighting stalemated and gave way to negotiations.

Nevertheless, the Moldovan government's inability to reach a settlement with the Trans-Dnistrians, along with a below-market pricing policy that has failed to discourage natural gas consumption, has led to a mounting debt to Russia for natural gas already supplied. As of November 1, 2002, Moldova owed Russia approximately $913 million for natural gas supplies, with approximately $690 million of this debt accumulated by the Trans-Dnistria region.

Natural Gas Transit

In addition to oil, southeastern Europe also represents an important transit site for Russian natural gas exports, mainly to Turkey. Russia's Gazexport, the export arm of Gazprom, transports natural gas from Russia via Ukraine and Moldova to Romania and on to Bulgaria. From Bulgaria, Russian natural gas is delivered via to Turkey, Greece, and Macedonia. During the past few years, the countries of southeastern Europe have sought to upgrade their pipeline links and increase their natural gas transit capacity in order to ensure that Russian natural gas continues to flow their way. Although Russia is looking to deliver natural gas directly to Turkey via the Blue Stream pipeline below the Black Sea, the Blue Stream will be in addition to–rather than replacing–natural gas flowing to Turkey via southeastern Europe. With Russia seeking to increase its natural gas exports, the countries of southeastern Europe will remain important transit centers.

In March 2002, Romania opened a 124-mile pipeline linking the country's borders with Ukraine and Bulgaria, completing a project begun with Russia in 1996 to develop the natural gas transit corridor in southeastern Europe. A shortage of funds in Romania delayed the construction until 1999, when Russia's Gazprom offered credit (in the form of natural gas) to Romania to finance the pipeline's construction. With the Isaccea-Negru Voda pipeline now operational, Romania now has the ability to transit up to 988 Bcf of natural gas per year, up from 353 Bcf per year previously. In addition, by 2004, Romania will increase its underground capacity for storage of natural gas from 53 Bcf to approximately 159 Bcf.

Bulgaria also is increasing its natural gas transit capacity, but mainly by widening its existing network and building new compressor stations rather than by building new pipelines. In the last three years, Bulgargas, which owns and operates Bulgaria's pipeline network (which includes over 400 miles of transit pipelines), has enlarged the country's natural gas transiting network to pump more Russian natural gas to its Balkan neighbors. From a transit capacity of 283 Bcf of natural gas per year before the enlargement program began, in 2000 Bulgaria transported to Greece, Macedonia, and Turkey some 423 Bcf of Russian natural gas, up 14 per cent from 1999 and up 57 per cent from 1999, according to Bulgargas chief executive director Kiril Gegov. Nearly 388 Bcf of that natural gas went to Turkey, the region's biggest energy consumer. Under a 1998 agreement with Gazprom, Bulgaria's only natural gas supplier, transit volumes to Greece, Macedonia, and Turkey should increase to 494 Bcf after 2002 and to 670 Bcf by 2010.

Coal

Romania is rich in coal deposits, with an estimated 4 billion short tons, much of which is lignite and sub-bituminous coal. However, the country's ailing coal industry–which suffers from obsolete equipment, inefficient mines, labor strikes, safety issues, and mounting wage arrears to miners–is in dire need of major restructuring. Since the revolution of 1989, when Romanian coal production peaked at 66.4 million short tons (Mmst), the country's production has dropped over 50 per cent.

Romania's severe economic problems, combined with a parallel drop in coal demand and a lack of reform, have crippled the country's coal mining industry. After leveling off in the mid-1990s, the decline in Romania's coal production accelerated in the late 1990s as pits began to be shut down and miners periodically went on strike to protest poor working conditions and to demand payment of wage arrears owed to them by the government.

The coal-rich Jiu Valley region has been hit particularly hard by problems in the coal sector, with 18,000 miners losing their jobs in 1997 alone. Over 70,000 jobs in Romania's coal sector have been cut in the last five years, and the World Bank has stated that Romania must shut additional loss-making pits in the Jiu Valley. Starvation caused by the 1997-1998 job severance program led to bloody clashes, suicides and mass hunger strikes by Romanian miners, and in 1999, miners protesting the shutdowns and unhappy about wage arrears clashed with government forces as they marched to Bucharest to voice their concerns. Former Prime Minister Radu Vasile was forced to bargain with striking miners to negotiate a settlement to the confrontation before further violence erupted. In 2000, Romania's coal production rose to 32.2 Mmst from 25.2 Mmst the year before as the the country attempted to revive its coal mining industry by squeezing out as much coal as possible from existing mines.

Coal is Bulgaria's most significant natural resource, with reserves estimated at 3 billion short tons, almost all of which is lignite or sub-bituminous coal. The country's biggest coal deposit, with estimated lignite reserves of 2 billion short tons, is the Maritsa Iztok coal basin, located in the southeast of the country. The Maritsa coal fields produce low-quality lignite coal with high ash and high sulfur content, but the adjacent Maritsa Iztok power plants are designed to work with this coal. Of the 29.8 Mmst of coal mined in Bulgaria in 2000, 23.82 Mmst was mined at Maritsa Iztok, while Bobov Dol, the second largest coalfield, produced approximately 3 Mmst. Despite a continuing decline in coal consumption over the last decade, Bulgaria's coal consumption still exceeds the country's domestic production, which has also decreased since 1989.

Moldova has a small coal industry, with reserves estimated at approximately 10 Mmst. Since 1992, when Moldova's coal production stood at 290,000 short tons, the country's output of coal has gradually been reduced, with zero production since 1998, as markets for Moldova's low-grade bituminous coal have disappeared. The country's coal consumption, like its production, has dropped

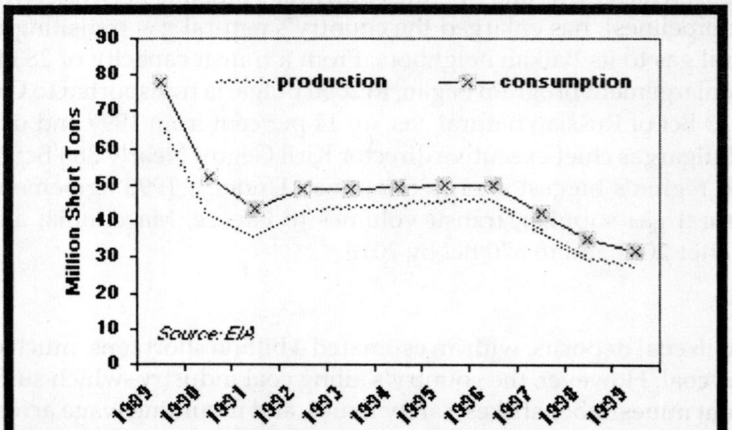

ROMANIAN COAL PRODUCTION AND CONSUMPTION, 1989-1999

significantly in the past decade, from 2.96 Mmst in 1992 to just 263,400 short tons in 2000–a 91 per cent reduction.

Electricity

With 22.2 gigawatts (GW) of installed electric-generating capacity, Romania has the largest power sector in southeastern Europe. However, approximately 60 per cent of Romania's existing power capacity is more than 20 years old, and about 8 GW will need to be rehabilitated or replaced by 2010. As a result, in 2000, Romania produced just 49.8 billion kilowatt hours (Bkwh), down from 71.9 Bkwh in 1989 but a slight increase over the low of 48.5 Bkwh that the country generated in 1999. In addition, technical losses in Romania's inefficient power transmission and distribution system means that an estimated 13 per cent of all electricity dispatched is lost before it reaches any customers. Of the 49.8 Bkwh produced in 2000, approximately 56 per cent came from thermal-fired (oil, natural gas, and coal) power plants and 34 per cent from the country's hydropower plants, with the remainder from Romania's sole nuclear power plant at Cernavoda.

Nevertheless, plummeting domestic electricity consumption, largely due to the collapse of industrial demand after the 1989 revolution and Romania's economic woes in the mid and late 1990s, has assured Romania's status as a net electricity exporter. In 2000, as Romania emerged from a prolonged recession, the country's electricity consumption increased for the first time in more than a decade, to 45.7 Bkwh–still 39 per cent lower than Romania's electricity consumption of 74.7 Bkwh in 1989.

Romania is reforming its power market, having already split up Conel, the state-owned power-generating company, and working to restructure its electricity distribution networks. According to the government's medium-term energy strategy, Romania is planning to rehabilitate 10 thermal power stations, with a combined capacity of 1.36 GW, between 2000 and 2005. Rehabilitation of these units will cost an estimated $460 million, while power-generating units with a total capacity of

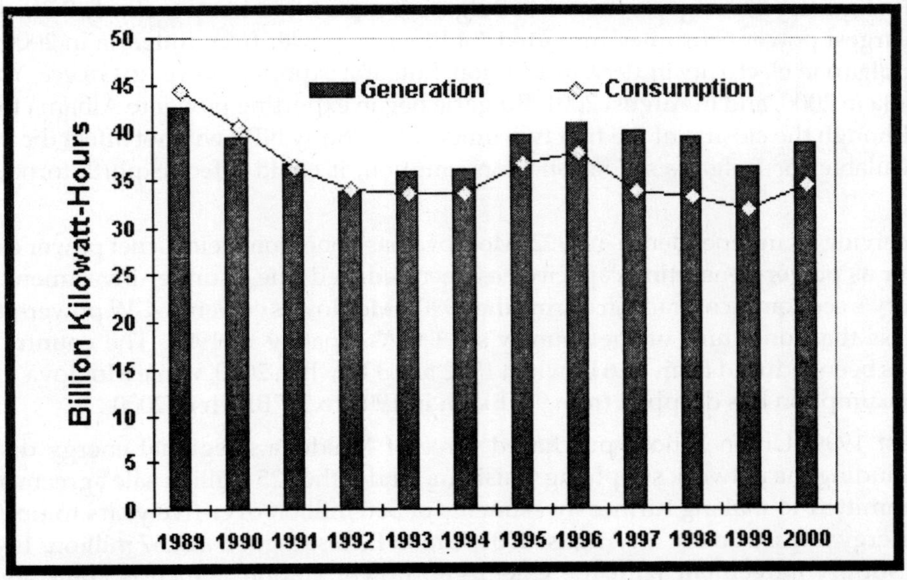

BULGARIAN ELECTRICTIY GENERATION AND CONSUMPTION, 1989-2000

5.9 GW are planned to be shut down. A second unit at the Cernavoda Nuclear Power Plant (NPP) also is under construction, with completion scheduled for 2005.

Bulgaria's installed electric capacity in 2000 was 12.1 GW, made up of 5.6 GW of thermal power (all coal), 3.8 GW of nuclear power, and 2.7 MW of hydropower. With domestic electricity consumption of 34.4 Bkwh in 2000, Bulgaria has significant spare capacity, even with nearly 2.6 MW of installed capacity currently inoperable. In 2000, Bulgaria produced 38.8 Bkwh of electricity, with coal-fired power plants generating 50 per cent, the Kozloduy NPP accounting for 40 per cent, and hydropower supplying the remaining power. The Maritsa Iztok complex, made up of three coal-fired power plants with combined capacity of 2,950 MW, accounted for nearly two-thirds of the power generated by coal-fired plants.

In 1998, the Bulgarian parliament began to liberalize the country's power sector by unbundling the generation, transmission, and distribution activities of the national electricity company, NEK. In the summer of 2000, the largest power plants and distribution networks, including the Kozloduy NPP, were separated from NEK, creating seven generation and seven distribution companies. Six of the seven independent power generators registered a profit in 2000, and some of them (but not Kozloduy) will be eligible for privatization.

Bulgaria has sealed investment deals with two U.S. firms to rehabilitate the Maritsa Iztok complex, and Bulgaria and Turkey are looking to launch a rehabilitation project at the Gorna Arda hydropower station. Bulgaria has agreed with the EU to shutdown two of six reactors at Kozloduy NPP by the end of 2002 out of safety concerns, and the country may close two additional units by 2006, although opponents claim that upgrades could allow the two reactors to remain safely operational until at least 2008.

The shutdown of reactors at Kozloduy NPP could affect Bulgaria's status as the regional power hub in the Balkan Peninsula. In 2000, Bulgaria more than doubled its electricity exports, sending 5.6 billion kilowatt-hours (Bkwh) of electricity to its neighbors and earning more than $105 million in the process. Preliminary 2001 figures show that Bulgaria's exports increased to nearly 7 Bkwh. Turkey, the region's largest power consumer, imported 3.4 Bkwh of power from Bulgaria in 2000, up from 2.2 Bkwh of Bulgarian electricity in 1999. In addition, Bulgaria exported power to Greece, Yugoslavia, and Macedonia in 2000, and in August 2001, Bulgaria began exporting power to Albania for the first time ever. Although the closure of the first two units at Kozloduy NPP will not affect the amount of electricity available for Bulgaria's domestic consumption, it could affect exports to neighboring countries.

Since receiving its independence in 1992, Moldova has gone from being a net power exporter to a net importer as power-generating capacity has been reduced due to under-investment, warfare, and the country's economic contraction during the 1990s. Moldova's current 1-GW power-generating capacity is less than one-third of the country's 3.1-GW capacity in 1992. The country's power generation has been reduced from 10.6 Bkwh in 1992 to 3.3 Bkwh in 2000, while Moldova's domestic electricity consumption has dropped from 9.8 Bkwh in 1992 to 3.7 Bkwh in 2000.

In August 1999, Union Fenosa purchased three of Moldova's regional energy distribution networks, including the network supplying Chisinau. Under the $25-million sale agreement, Union Fenosa is committed to making further investments of $60 million over five years to upgrade and modernize energy infrastructure. In August 2001, Union Fenosa signed a $267 million, 10-Bkwh, 5-year power supply agreement with the Cuciurgan power station, which is controlled by the secessionist Trans-Dnistrian Republic. The agreement is expected to cover 70 per cent to 80 per cent of the needs of the three power distribution grids. However, Moldova's power sector continues to suffer from consumer nonpayment of bills, leaving the countries' power-generating facilities short of cash for investment and leading to an energy crisis in northern Moldova in 2001. In April 2001, Moldova passed several urgent measures to resolve the crisis, including finally allowing electricity suppliers to cut off indebted customers.

79

Southern Africa and the Southern African Development Community

The following provides a brief economic and energy sector overview of Southern Africa, including the fourteen countries that make up the Southern African Development Community (SADC). SADC member-states are Angola, Botswana, Democratic Republic of Congo, Lesotho, Malawi, Mauritius, Mozambique, Namibia, Seychelles, South Africa, Swaziland, Tanzania, Zambia and Zimbabwe. Overviews of non-SADC countries Comoros & Madagascar are also included.

General Background

The Declaration and Treaty establishing the Southern African Development Community (SADC) was signed at the Summit of Heads of State or Government on July 17, 1992, in Windhoek, Namibia. SADC replaced the Southern African Development Coordination Conference (SADCC), in existence since 1980. South Africa joined SADC in 1994 followed by Mauritius (1995), the Democratic Republic of Congo–DRC (1997), and Seychelles (1997). Uganda's application for membership in SADC, submitted in the fall of 2000, is currently awaiting SADC approval. Comoros and Madagascar, with

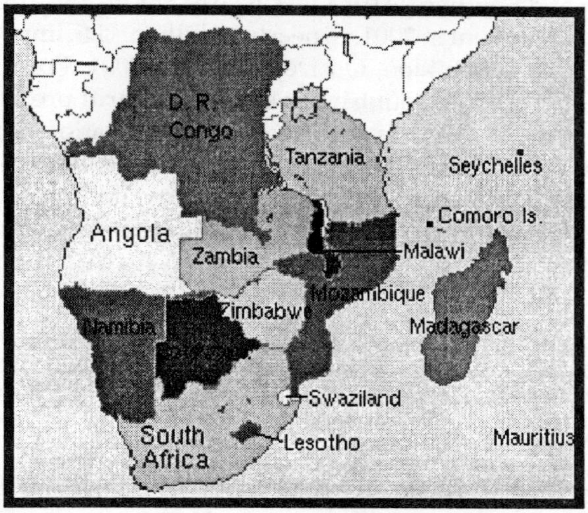

SADC-members Angola, DRC, Malawi, Mauritius, Namibia, Seychelles, Swaziland, Zambia and Zimbabwe, are members of the Common Market for Eastern and Southern Africa (COMESA).

Angola's President, Jose Eduardo dos Santos, is the current SADC Chairman. President dos Santos is continuing the process of restructuring the SADC mandate to focus on regional economic integration. Current SADC objectives include harmonization and rationalization of policies and strategies for sustainable development in all areas, as well as the successful implementation of the SADC Trade Protocol (Protocol). The Protocol, which calls for an 85 per cent reduction of internal trade barriers, went into effect on September 1, 2000. The full implementation of the Protocol is on track and the region hopes to attain a free trade area by 2008. In October 2000, COMESA launched its own free trade regime, creating overlapping schedules with SADC for internal tariff reductions. Within the SADC region, the national currencies of Namibia, Lesotho, and Swaziland are linked to the South African rand through the Common Monetary Area (CMA). SADC members are working to eliminate exchange controls in preparation for an eventual single currency in the region.

Economic Overview

In 2001, the combined Gross Domestic Product (GDP) for Southern Africa was estimated at \$173.8 billion (see Table 80.1). Regional economies are structurally diverse and at varying stages of development. South Africa, the region's most developed economy, with a GDP of \$113.3 billion nearly doubles the combined GDP of the other Southern African countries. While the region's economies grew at a combined rate of 2.2 per cent in 2001, the substantial external debt of individual states remains one of the region's greatest challenges. War (in Angola and DRC), and internal strife (Zimbabwe) have adversely affected economic performance in several states. The economies of Zimbabwe and DRC declined, and hyperinflation was rampant in Angola (153 per cent), DRC (337 per cent) and Zimbabwe (77 per cent).

Total regional exports, including intra-regional exports, were \$48.5 billion in 2001. Southern Africa had a \$5.3 billion trade surplus for 2001. The region's major export commodities were energy products (oil and coal) and various minerals including diamonds, gold, and copper. Trade growth between the United States and Southern Africa was nearly flat in 2001 (up only 0.7 per cent). U.S. exports to the region grew 5.7 per cent in 2001, to nearly $4 billion. U.S. imports from Southern Africa fell by 1.4 per cent in 2001, to $8.8 billion. On December 31, 2001, President Bush approved the designation of 35 sub-Saharan African countries as eligible for tariff preferences under the African Growth and Opportunity Act (AGOA). As required by the legislation, this annual determination signifies which countries are making continued progress toward a market-based economy, the rule of law, free trade, economic policies that will reduce poverty, and protection of worker's rights. Angola, Comoros, DRC and Zimbabwe were the only countries in the region not covered by AGOA. Although not included in AGOA, Angola was sub-Saharan Africa's third largest exporter to the United States in 2000 and 2001.

Table 79.1: Economic and Demographic Indicators

Country	*Gross Domestic Product (GDP), 2002E (Billions of U.S. $)*	*Real GDP Growth Rate, 2001 Estimate*	*Real GDP Growth Rate, 2002 Projection*	*Per Capita GDP, 2001E*	*Population 2001E (Millions)*
Angola	$8.3	3.4%	11.2%	$610	10.4
Botswana	$5.1	8.9%	4.2%	$3,057	1.6
Comoros	$0.2	1.5%	3.1%	$285	0.6
Democratic Republic of Congo	$7.0	-4.3%	2.7%	$134	53.6
Lesotho	$0.8	3.2%	2.9%	$386	2.2
Madagascar	$4.6	6.5%	3.0%	$283	16.0
Malawi	$1.5	3.0%	5.3%	$133	10.5
Mauritius	$4.5	6.3%	4.7%	$3,792	1.2
Mozambique	$2.4	10.5%	9.5%	$132	19.4
Namibia	$2.9	2.7%	4.0%	$1,538	1.8
Seychelles	$0.6	2.8%	3.3%	$8,050	0.1
South Africa	$113.3	2.2%	2.0%	$2,492	43.6
Swaziland	$1.2	2.0%	2.5%	$1,117	1.1
Tanzania	$8.9	5.1%	5.7%	$247	36.2
Zambia	$3.1	5.2%	4.3%	$301	9.8
Zimbabwe	$9.4	-7.9%	-15.0%	$761	11.4
Regional Total/Average	**$173.8**	**2.2%**	**2.1%**	**$792**	**219.5**

Energy

Southern Africa is a net energy exporter. In 2000, the countries of Southern Africa collectively consumed 5.47 quadrillion British thermal units (Btu) of commercial energy (1.4 per cent of total world consumption) and produced 9.32 quadrillion Btu (2.3 per cent of total world production). Also in 2000, the region generated 119.2 million metric tons of carbon emissions (1.9 per cent of the world total). The region's dominant economy, South Africa, accounted for 84.9 per cent (4.64 quadrillion Btu) of energy consumption, 77.4 per cent (7.21 quadrillion Btu) of energy production, and 88.8 per cent (105.85 million metric tons) of the region's carbon emissions.

Commercial energy resources in the region are diverse, with significant reserves of coal, petroleum, and natural gas. Electricity in Southern Africa is generated through thermal or hydroelectric resources (with one nuclear facility in South Africa). Natural gas is becoming more significant to the region's energy sector as fields off Mozambique, Namibia, South Africa and Tanzania are developed.

Table 79.2: Total Energy and Carbon Dioxide Emissions, 2000

Country	*Total Commercial Energy Consumption, (Quadrillion Btu)*	*Total Commercial Energy Production, (Quadrillion Btu)*	*Net Energy Exports, (Quadrillion Btu)*	*Carbon Dioxide Emissions (Million Metric Tons of Carbon)*
Angola	0.090	1.621	1.531	3.60
Botswana	0.053	0.025	-0.028	0.95
Comoros	0.001	0.000	-0.001	0.02
Democratic Republic of Congo	0.104	0.111	0.008	1.10
Lesotho	0.004	0.000	-0.004	0.05
Madagascar	0.025	0.005	-0.019	0.37
Malawi	0.020	0.010	-0.011	0.24
Mauritius	0.036	0.001	-0.035	0.67
Mozambique	0.031	0.073	0.042	0.36
Namibia	0.025	0.000	-0.025	0.32
Seychelles	0.007	0.000	-0.007	0.14
South Africa	4.643	7.213	2.570	105.85
Swaziland	0.022	0.009	-0.013	0.32
Tanzania	0.058	0.024	-0.034	0.68
Zambia	0.095	0.086	-0.010	0.59
Zimbabwe	0.252	0.144	-0.108	3.97
Regional Total	**5.465**	**9.321**	**3.856**	**119.22**

Due to the region's relatively small urban population (approximately 25.4 per cent), access to commercial energy sources is limited. The majority of Southern Africa's population still relies on the use of biofuel (wood and charcoal) as its primary source of energy. Biofuel accounts for approximately 75 per cent of total final energy demand in the region. The countries with the highest rates of biofuel consumption are Tanzania, Mozambique, Zimbabwe, Zambia and Malawi.

SADC representatives announced plans to elevate its energy technical unit (ETU) to a regional energy commission. The commission is being designed to facilitate regional integration in the energy sector. The ETU started its operations in April 2001 in Luanda, Angola.

Petroleum

Angola is Southern Africa's only significant oil producer, production averaged 742,000 barrels per day (bbl/d) in 2001. Angolan oil production has averaged 900,000 bbl/d for the first six months of 2002. Angola's estimated proven crude reserves of 5.4 billion barrels constitute 96 per cent of the region's estimated proven crude reserves. Smaller reserves are found offshore DRC and South Africa.

The region's refineries are concentrated in South Africa, with additional refining capacity located in Angola, DRC, Madagascar, Tanzania and Zambia. In 2001, South Africa produced about 163,000 bbl/d of synthetic oil from coal and natural gas. South Africa is the region's largest oil consumer (over 73 per cent of the region's total), and is the second largest oil consumer in Africa behind Egypt. South Africa also is the continent's largest net oil importer.

Table 79.3: Petroleum Overview

Country	*Petroleum Production, 2002 (Jan.–June) (Thousand Barrels Per Day)*	*Petroleum Consumption, 2001 (Thousand Barrels Per Day)*	*Petroleum Net Exports, 2001 (Thousand Barrels Per Day)*	*Crude Oil Reserves, 1/1/2002 (Million Barrels)*	*Crude Oil Refining Capacity, 1/1/2002 (Thousand Barrels Per Day)*
Angola	899.8	30.7	711.7	5,412.0	39.0
Botswana	0.0	9.2	-9.2	0.0	0.0
Comoros	0.0	0.6	-0.6	0.0	0.0
Democratic Republic of Congo	23.2	21.6	2.4	187.0	15.0
Lesotho	0.0	1.2	-1.2	0.0	0.0
Madagascar	0.0	9.1	-9.1	0.0	15.0
Malawi	0.0	5.0	-5.0	0.0	0.0
Mauritius	0.0	16.1	-16.1	0.0	0.0
Mozambique	0.0	8.2	-8.2	0.0	0.0
Namibia	0.0	7.9	-7.9	0.0	0.0
Seychelles	0.0	3.5	-3.5	0.0	0.0
South Africa	202.2	492.7	-292.4	15.7	468.5
Swaziland	0.0	3.5	-3.5	0.0	0.0
Tanzania	0.0	16.2	-16.2	0.0	14.9
Zambia	0.0	12.2	-12.2	0.0	23.8
Zimbabwe	0.0	33.7	-33.7	0.0	0.0
Regional Total/Average	**1,125.1**	**671.4**	**295.3**	**5,614.7**	**576.2**

Sources: Energy Information Administration, Oil & Gas Journal

Exploration and Production

Angola has the potential to more than double its crude oil production in the next few years as recent discoveries come on stream. Since 1996, nearly 30 major oil discoveries–most all located in deeper waters offshore Cabinda and Angola proper–have been announced. In December 2001, the Girassol field, located on Block 17 went into production. Full production of nearly 200,000 bbl/d from Girassol began in the first quarter of 2002. Dalia, the second big field on Block 17, is expected to be as big a producer as Girassol, with daily output close to 200,000 bbl/d and initial production anticipated in early 2006.

Production from other recent discoveries are planned to come online in the next few years. BP plans to utilize a floating production, storage and offloading (FPSO) vessel on its Greater Plutonio project on Block 18. First oil from the Greater Plutonio project is likely in late 2005, and it will have a production capacity of 250,000 bbl/d. ExxonMobil's $3 billion Kizomba A is targeted to have a peak production of 250,000 bbl/d. First oil is scheduled for late 2004. In August 2002, ExxonMobil commenced construction of the Xikomba deepwater development offshore Angola. An estimated 100 million barrels of recoverable resources exist in the field, with first oil production scheduled for late 2003. In January 2000, ChevronTexaco began production on Block 14's Kuito field. ChevronTexaco has made eight discoveries on the block dating back to the 1997. The Benguela, Belize, Tomboco and Lobito fields are expected to come onstream some time in 2005, with output in the first phase pegged at 140,000 bbl/d in 2005 and then rising to over 200,000 bbl/d once all the fields are in production.

South Africa's Oribi oil field began production from an FPSO in May 1997, the country's first conventional oil production. The field currently produces a light oil (API 42°) at the rate of 10,000 bbl/d with 15 million cubic feet per day (Mmcf/d) of associated gas, which is flared. The Oryx field, which lies 3.7 miles (6 kilometers) from Oribi, began production in May 2000, and currently produces at 12,000 bbl/d.

PetroSA (formerly Soekor) plans to concentrate its exploration efforts on South Africa's western and southern coasts. Several discoveries have been made within the Bredasdorp Basin on Block 9. Oribi and Oryx are located on Block 9. Another significant find, Sable, has been made on the block. PetroSA, and its Sable field partner Picneer Natural Resources (Pioneer) announced plans for the development of the Sable field in June 2001. The field will be developed with six subsea wells tied back to a FPSO with a capacity to process 60,000 bbl/d of oil, re-inject 80 Mmcf/d of natural gas, and recover natural gas liquids. Production of 40,000 bbl/d is expected to begin in the first quarter of 2003. PetroSA and Pioneer announced the results of their Boomslang discovery in February 2001. The find, located on the southern portion of Block 9, tested at a combined rate of 3,120 bbl/d of oil, 26 million cubic feet of natural gas per day (Mmcf/d), and 300 bbl/d of condensate.

In May 2000, Namibia renewed a stalled agreement granting Russian companies the right to explore for oil and natural gas along Namibia's northern coastline. In November 2000, Namibia formally granted an exploration license to the Russian parastatal, V/O Zarubezchgeologia. Namibia's government also awarded an oil and gas exploration license in June 2000 to a consortium led by U.S.-based Vanco International (Vanco), granting the consortium the rights to Block 1711 offshore of northern Namibia. Vanco has identified two prospects, Kunene and Hartmann, on Block 1711 and the company plans to obtain seismic data by the end of 2002.

In April of 2002, Canada's Antrim Energy Inc. and the Tanzanian Petroleum Development Corporation (TPDC) reached an agreement to lift force-majeure on their Pemba-Zanzibar production-sharing contract (PSC). Antrim, which holds a 100 per cent interest in the concession, said the PSC was granted in 1997 for a four-year term, but that a mutually agreed force-majeure was declared in 1998 when Antrim was denied access to the isiands of Zanzibar and Pemba. TPDC and the government of Tanzania concluded its second offshore licensing round in August 2002, putting 11 deepwater blocks and one shallow-water block up for bid. Tanzania expects to begin talks with Shell before the end of 2002 on the terms of the offshore exploration licence awarded in September 2002. Shell was awarded rights to offshore blocks 9-12, located on the northern edge of the east coast.

In July 2001, Vanco contracted for a seismic survey of its Majunga Offshore Profond Block offshore Madagascar. In March 2002, Vanco signed a PSC for this block, the first awarded in the Majunga Basin since 1997. The block covers 3.57 million acres. U.S.-based Xpronet announced that it has signed two PSCs with Madagascar's Office of National Mines and Strategic Industries in December

2000. The PSCs grant the company petroleum exploration and production rights for the Mavony and Rivomena blocks, offshore the west coast of Madagascar. The blocks were originally awarded as exploration blocks in December 1998. Xpronet holds 100 per cent of the operator's interest in the PSCs.

In 2001, ChevronTexaco's Misato exploration well made a marginal oil discovery on DRC's continental shelf. This was the first wildcat drilled in DRC since October 1998. It was completed and brought into production in March 2001. Onshore in DRC, Perenco successfully drilled a development well in the Kinkasi field during October 2001. It was completed, testing 1,800 bbl/d before stabilizing at 600 bbl/d. The well came online in November 2001. Perenco also bought Shell's interests in five onshore blocks in DRC. Perenco was already the operator of the blocks after it bought TotalFinaElf's stakes in 2000. Calgary-based Heritage Oil Corp. (Heritage) says it will be drilling exploratory oil wells in former rebel-held eastern DRC. Heritage began drilling Block 3 in Uganda in August 2001. The company is in talks with DRC over terms of a PSC in the area contiguous to Block 3.

Refining

Southenr Africa's petroleum refining is concentrated in South Africa. South Africa's four refineries have the capacity to process 468,547 bbl/d of crude. Other Southern African refineries are in Angola (Luanda, 39,000 bbl/d); DRC (Muanda, 15,000 bbl/d); Madagascar (Toamasina, 15,000 bbl/d); Tanzania (Dar es Salaam, 14,900 bbl/d) and Zambia (Ndola, 23,750 bbl/d). In September 2002, Sasol announced that capacity expansion of its Natref oil refinery was complete.

Angola's state oil company, Sonangol, has announced plans for the construction of Angola's second refinery. The facility, with a potential refining capacity of 200,000 bbl/d, is to be built near the coastal city of Lobito. The $3.6-billion refinery is expected to begin operations in 2006. In October 2002, Sonangol signed a deal with Dresdner Kleinwort Wasserstein to advise on the financing of the Lobito refinery. Diesel and gasoline produced in the refinery will meet technical and environmental specifications required in targeted markets such as the United States, Western Europe and South Africa. TotalFinaElf, operator of the Luanda refinery, plans to raise capacity to 60,000 bbl/d by 2004. Two processing units will be enlarged and a third scrapped to boost efficiency. The refinery needs to adapt by 2005 to product specifications set by SADC. These include phasing out lead and increasing the octane content in gasoline. The refinery produces almost all of Angola's domestic requirements of gasoline, kerosene and jet fuel.

TotalFinaElf acquired the refining and marketing assets of Agip in Zambia in April 2002 as part of consolidating its position in Africa. Through the transaction, TotalFinaElf has acquired a 50 per cent interest in Zambia's Indeni refinery and will provide technical support for the refinery. The Zambian government will retain control of the remaining half. The marketing assets covered by the agreement consist of 20 service stations, a motor fuel depot, and an interest in a lubricant plant. Citing the relative low cost of imported refined petroleum products in the region, the World Bank has recommended that the Zambia National Oil Company (ZNOC) be dissolved and that the Indeni refinery be permanently shut down.

Consumption

Petroleum consumption in Southern Africa averaged 655,570 bbl/d in 2000. The vast majority of petroleum consumed in the region is imported; Angola and DRC are the only net exporters of petroleum. The transportation sector consumes approximately 50 per cent, followed by the industrial sector, which accounts for approximately 15 per cent. Kerosene is extensively used in rural areas for lighting and, in the urban areas, for cooking and lighting as well. Liquefied petroleum gas (LPG) is also used for cooking, particularly in Angola.

Several countries in the region have recently experienced periodic, sometimes severe, petroleum shortages. In September 2002, Zimbabwe and Libya signed an agreement for Libya to supply $360-million worth of oil to the fuel-starved country. In August 2001, Zimbabwe secured a similar deal with Libya. Fuel shortages first began in December 1999 when the Zimbabwean government was unable to meet payments for supplies from foreign oil companies. Companies in Zimbabwe's oil industry have urged the government to dissolve the National Oil Company of Zimbabwe (Noczim). Internal strife has blocked petroleum shipments and created massive shortages in inland Madagascar, especially in the capital, Antananarivo. Several fires at Zambia's Indeni refinery helped to create petroleum shortages in the country.

Oil Integration

The 1,069-mile (1,710-kilometer) Tanzania-Zambia Tanzama Pipeline transports crude from the oil depot at Dar es Salaam, Tanzania to Zambia's Indeni refinery in Ndola. The pipeline, jointly owned by the governments of Zambia (67 per cent) and Tanzania (33 per cent), has a capacity of 22,000 bbl/d (1.1 million metric tons annually). In April 2002, Zambian President Levy Mwanawasa stated that Tanzania and Zambia are considering privatizing the Tanzama Pipeline. He said a technical committee was carrying out a feasibility study on the privatization.

The Mozambique-Zimbabwe Petrozim Petroleum Products Pipeline runs from the Mozambican port city of Beira to Feruka, Zimbabwe and from there to Msasa, which is located near the capital city of Harare. Noczim imports 80 per cent of Zimbabwe's petroleum through the pipeline. Petrozim is a joint-venture between Noczim and the South African-based Lonhro. Zimbabwe's Noczim is planning to construct an additional oil-product pipeline from Beira to Msasa. The 500-mile (800-kilometer) pipeline would help to meet Zimbabwe's growing oil demand.

Natural Gas

There are significant reserves of natural gas in Southern Africa. Field discoveries have been confirmed and reserves have been proven in Angola (1.6 trillion cubic feet–Tcf); DRC (35 billion cubic feet–Bcf); Madagascar (70 Bcf); Mozambique (4.5 Tcf); Namibia (2.2 Tcf); South Africa (780 Bcf) and Tanzania (800 Bcf). Southern Africa contains approximately 2.5 per cent of Africa's natural gas reserves. Although natural gas is still in early stages of use in the region, several projects for the expansion of its use are under way.

Angola is developing projects to utilize associated natural gas, which is currently flared or re-injected (approximately 83 per cent). ChevronTexaco and a consortium of oil companies, including ExxonMobil, BP, TotalFinaElf and Norsk Hydro, are planning to gather associated natural gas from deepwater fields and develop shallow water fields in the Congo Basin as part of the Angola liquefied natural gas (LNG) project. ChevronTexaco, the operator, will hold a 32 per cent stake, while TotalFinaElf, Norsk Hydro, ExxonMobil, and BP will each hold a 12 per cent interest. Angola's state oil firm, Sonangol will retain the remaining 20 per cent. In March 2002, participation agreements between the oil companies and Sonangol were signed. The project would utilize natural gas reserves located offshore Angola, south of the Congo River. Associated gas from the huge deepwater projects in Blocks 15, 17 and 18 including Girassol, Kizomba and Greater Plutonio will be piped to a central gas compression complex located over the Atum field in Block 2. The gas will then be pumped from the complex to Luanda where the consortium will construct an LNG plant. The Angola LNG project is expected to come onstream in 2007. The LNG plant is expected to consist of a single train with capacity of producing four million tons per year of LNG. The plant design will include the potential to add a second four million-ton train. ChevronTexaco is researching methods of laying a pipeline across the Congo River to utilize gas offshore Cabinda. Large associated and non-associated gas reserves are found in Cabinda, and if the pipeline is constructed the gas can be utilized by the LNG project. Angola is exploring further uses for the gas, including power generation and a gas-to-liquid fuels project.

Table 79.4: Natural Gas Overview (Billion Cubic Feet)

Country	*Production, 2000*	*Consumption, 2000*	*Reserves, 1/1/2002*
Angola	20.48	19.78	1,620
Botswana	0.00	0.00	0
Comoros	0.00	0.00	0
Democratic Republic of Congo	0.00	0.00	35
Lesotho	0.00	0.00	0
Madagascar	0.00	0.00	70
Malawi	0.00	0.00	0
Mauritius	0.00	0.00	0
Mozambique	2.12	2.12	4,500
Namibia	0.00	0.00	2,200
Seychelles	0.00	0.00	0
South Africa	49.44	49.44	780
Swaziland	0.00	0.00	0
Tanzania	0.00	0.00	800
Zambia	0.00	0.00	0
Zimbabwe	0.00	0.00	0
Regional Total	**72.04**	**71.34**	**10,005**

Tanzania plans to develop two offshore natural gas fields to provide fuel for power generation. One project plans to exploit natural gas in the country's largest known field on Songo Songo Island, located in the Indian Ocean southeast of the capital Dar-es-Salaam. Natural gas from the Songo-Songo field will be transported to Dar es Salaam by a 130-mile (217-kilometer) pipeline. Upon completion of the project, five liquid fuel turbines at the 112-megawatt (MW) Ubungo power plant will be converted to natural gas, and the power generated by the plant will be fed directly into the national grid. U.S.-based AES Corporation, the project's developer, received $250-million credit financing package from the International Development Association and the European Investment Bank. The World Bank also has approved $183 million in financing for the Songo Songo project. Plans call for other industrial users to utilize Songo-Songo gas, and the pipeline could be extended to the Kenyan port city of Mambasa to supply gas for industrial usage and power generation.

A second Tanzanian project would utilize natural gas reserves from the Mnazi Bay field. The gas will be piped to the southern Tanzanian town of Mtwara for use in power generation. The proposed 15-MW generating plant could later be expanded to 50 MW. In July 2001, the World Bank announced it has shelved its plan to fund the Mnazi Bay gas project until the government completes energy sector reforms.

South Africa's FA field currently produces at a rate of 194 Mmcf/d of natural gas and 9,500 bbl/d of condensate. An offshore natural gas discovery was made in March 2000 close to South Africa's border with Namibia. Located in Block 2, off South Africa's western coast, the find is reportedly part of the same reservoir, which extends to Namibia's Kudu prospect. Denver-based Forest Oil Corporation (Forest) reported that the find, AK-1, flowed at a rate of 52.8 Mmcf/d of natural gas and 342 bbl/d of condensate. Forest's initial estimates placed recoverable reserves at 200 Bcf of natural gas. Appraisal drilling conducted in 2000-2001 caused Forest to revise its estimate of total reserves for the discovery, renamed the Ibhubezi field, to 2.5 Tcf.

South Africa's synthetic fuels and chemicals producer Sasol will soon begin utilizing natural gas reserves in neighboring Mozambique. Sasol estimates that the switch to natural gas will reduce investment expenditures in its coal mining operations and the high costs of compliance with environmental regulations associated with coal. The project is expected to start delivering natural gas to South Africa during the first half of 2004. The project consists of several major components including the development of the Pande and Temane gas fields in Mozambique. The 536-mile (865-kilometer) transport pipeline will run from the Mozambican fields to Secunda. Sasol will convert its existing pipeline-gas network to natural gas, and supply natural gas to industries in South Africa, including its own facilities. Sasol will switch its Sasolburg plants from coal to gas feedstock and utilize natural gas at Secunda to supplement coal-based growth there.

The pipeline will be owned by a joint venture between Sasol, and the governments of Mozambique and South Africa. The parties have made provision for the future inclusion of Black empowerment shareholders as well as privatization initiatives. Sasol has signed memoranda of understanding with two empowerment groups for local gas distribution companies in Mpumalanga and KwaZulu-Natal. On May 3, 2002, President Chissano of Mozambique and President Mbeki of South Africa officially launched the project at a ceremony held at the Temane gas field in the Inhambane Province of Mozambique.

In June 2002, Malaysia's Petronas, announced that it had signed an exploration and production deal for the offshore Zambezi Delta Block, together with Empresa Nacional de Hidrocarbonetos de Mozambique (ENH), Mozambique's national oil company. Under terms of the deal, Petronas will hold an 85 per cent stake in the block and ENH will own the remaining 15 per cent. Petronas officials stated that the block has proven hydrocarbon reserves and several onshore gas field discoveries were made recently in that area. In the event of commercial discovery, ENH will have the option to increase its stake to 20 per cent during the development and production phase.

In September 2002, Shell announced that it was pulling out of Namibia's offshore Kudu gas project. The decision was made after the failure to find sufficient reserves to build an offshore LNG export facility. Kudu's reserves are estimated at 1.3 Tcf, well short of the 5 Tcf needed for the proposed $2.5-billion floating LNG facility. ChevronTexaco and Energy Africa, the remaining partners, say there are sufficient reserves to develop gas-fired power schemes utilizing Kudu gas. The Cape Power Project would include a 1,200-2,000 MW combined-cycle, gas-turbine power station near Cape Town. The power plant would serve the metropolitan area and industrial plants at Saldanha. The plan also would include a 400-MW power plant at Oranjemund, Namibia that will supply Namibia and western South Africa with electricity.

Coal

Coal resources are abundant in Southern Africa, especially in South Africa, where recoverable reserves are estimated at 54.6 billion short tons, representing 6 per cent per cent of world recoverable coal reserves. The total recoverable coal reserves in the region amount to 60.7 billion short tons. In 2000, regional coal production was 253.6 million short tons (mmst), of which South Africa produced 247.2 mmst. The primary use of coal in the region is in the generation of electricity. Coal generates a significant portion of the electricity in Botswana, South Africa and Zimbabwe. South Africa also uses coal to produce synthetic fuels and chemicals, while Zambia uses coal in the mining transformation process. South Africa consumed the vast majority (96 per cent) of the region's coal in 2000. Total coal consumption in Southern Africa was 182.8 mmst in 2000.

Table 79.5: Coal Overview (Million Short Tons)

Country	*Production, 2000*	*Consumption, 2000*	*Reserves, 1/1/2002*
Angola	0.00	0.00	0.00
Botswana	1.06	1.07	4,739.94
Comoros	0.00	0.00	0.00
Democratic Republic of Congo	0.11	0.28	97.00
Lesotho	0.00	0.00	0.00
Madagascar	0.00	0.02	0.00
Malawi	0.06	0.07	2.20
Mauritius	0.00	0.07	0.00
Mozambique	0.02	0.00	233.69
Namibia	0.00	0.01	0.00
Seychelles	0.00	0.00	0.00
South Africa	326.08	176.30	54,586.46
Swaziland	0.32	0.32	229.28
Tanzania	0.01	0.01	220.46
Zambia	0.21	0.20	11.02
Zimbabwe	4.63	4.46	553.36
Regional Total	**332.49**	**182.80**	**60,717.51**

South Africa's Industrial Development Corporation will study the possibility of re-developing a coal project in Mozambique. The Moatize mine in northwestern Mozambique was initially operated by the South African-steel producer, Iscor, in the 1970's. High production costs and lack of infrastructure led to Iscor's withdrawal in 1977. The re-development project would include the development and improvement of mine facilities and the establishment of a rail line to supply coal to Zimbabwe, Malawi, Zambia and DRC. Port facilities at Beira would also be upgraded. The Moatize mine would have the capacity to produce 3.3 mmst of coking coal annually, as well as quantities of steam coal. A feasibility study for the project, including analysis of establishing a 1,000-MW power plant to utilize steam coal reserves, is expected to commence before the end of 2002.

Coal production from Botswana's Morupole Collieries (coal mines) could increase in the near term. The production expansion will hinge on the Botswana Power Corporation's (BPC), decision to build two new coal-fired plants with a combined capacity of 240 MW. The Morupole Collieries are owned by Debswana, a joint venture between De Beers and the government of Botswana.

Malawi's Mwabvi coal mine plans to triple its current output by 2003. The mine, which currently produces 0.06 mmst annually, hopes to raise output to at least 0.20 mmst by 2003. Malawi currently consumes approximately 0.07 mmst of coal annually.

Zimbabwe's petroleum shortages also have affected the coal industry. It was reported, in April 2001, that the National Railways of Zimbabwe (NRZ) which transports the majority of the coal produced at Zimbabwe's Wankie Colliery, had been forced to ground some of its trains because of diesel fuel shortages. Industrial facilities, including Zimbabwe's Hwange Power Station, were forced to scale back production.

Electricity

Southern Africa's total installed electric generating capacity was 54,183 MW at the beginning of 2000, the majority of which is thermal. Total electricity generation for the region in 2000 was 229 billion kilowatthours (bkwh). Net hydroelectric generation was 28.76 bkwh, with Zambia (7.78 bkwh), Mozambique (6.77 bkwh) and the DRC (5.30 bkwh) being the largest generators. In 2000, total regional electricity consumption was 213.29 bkwh, led by South Africa's 181.52 bkwh (85.1 per cent). Zimbabwe (10.48 bkwh, 4.9 per cent), Zambia (5.8 bkwh, 2.7 per cent) and the DRC (4.58 bkwh, 2.2 per cent) were the next largest electricity consumers.

Table 79.6: Electricity Overview, Billion Kilowatthours except where noted

Country	*Consumption, 2000*	*Generation, 2000*	*Installed Capacity, 1/1/2000 (gigawatts)*	*Exports, 2000*	*Imports, 2000*
Angola	1.11	1.12	0.586	0.00	0.00
Botswana	1.45	0.50	0.217	0.00	0.99
Comoros	0.02	0.02	0.005	0.00	0.00
Democratic Republic of Congo	4.58	5.40	2.473	0.50	0.06
Lesotho	0.10	0.00	0.00	0.00	0.10
Madagascar	0.76	0.82	0.285	0.00	0.00
Malawi	0.77	0.83	0.308	0.00	0.00
Mauritius	1.20	1.29	0.365	0.00	0.00
Mozambique	0.93	7.02	2.388	5.70	0.10
Namibia	0.89	0.03	0.00	0.00	0.86
Seychelles	0.15	0.16	0.028	0.00	0.00
South Africa	181.52	194.38	43.110	4.55	5.29
Swaziland	0.90	0.36	0.131	0.00	0.56
Tanzania	2.62	2.77	0.620	0.00	0.05
Zambia	5.84	7.82	1.786	1.54	0.10
Zimbabwe	10.48	6.43	1.881	0.00	4.50
Regional Total	**213.29**	**229.00**	**54.183**	**12.28**	**12.61**

Created in 1995, the South African Power Pool (SAPP) aims to link SADC member states into a single electricity grid. The national utilities currently participating in the SAPP are Angola's Empresa Nacional de Electricidade (ENE), the Botswana Power Corporation (BPC), the DRC's SNEL, the

Lesotho Electricity Corporation (LEC), Malawi's Electricity Supply Commission (Escom), Mozambique's Electricidade de Mocambique (EDM), Namibia's NamPower, South Africa's Eskom, the Swaziland Electricity Board (SEB), Tanzania Electric Supply Company (Tanesco), Zambia's ZESCO, and Zimbabwe's ZESA. SAPP's coordination center is located in Harare, Zimbabwe.

The Inga hydroelectric facility in the DRC supplies power to country and the neighboring Republic of Congo along a 220 kilovolt (KV) connection. The interconnection supplies nearly one-third of the electricity consumed in Congo. Power from Inga is also transmitted to the Zambian grid along a 500-KV DC line from Inga to Kolwezi in southern DRC, and a 220-KV line from Kolwezi to Kitwe in northern Zambia.

Mozambique's Cahora Bassa hydroelectric facility, with a nominal capacity of 2,000 MW, supplies electricity domestically, to Zimbabwe and South Africa. Cahora Bassa is operated by Hidroelectrica de Cahora Bassa (HCB), a joint-venture between Portugal and EDM. In October 2002, HCB cut-off power supplies to South Africa's Eskom over a dispute in pricing. HCB stated that it will resume supplies of power to Eskom when the company agrees to pay at least 3.6 South African cents (about 0.36 U.S. cents) per kilowatthour. This is the temporary price that was in effect for 2001. At the end of 2001, when no definitive arrangement on pricing was forthcoming, Eskom resumed payments at the old price, two South African cents per kilowatthour. HCB spokesman stated that this price is completely unacceptable, particularly since the dams other major client, Zimbabwe's ZESA, is paying the equivalent of 15 South African cents per kilowatthour.

The Mozambican government is also seeking investors for a second hydroelectric facility on the Zambezi River. The $1.3- billion Mepanda N'cua dam is be to built south of the existing Cahora Bassa dam. The new facility will have a capacity of 2,400 MW. The government expects construction to begin in 2005, and generation to begin in 2010. The Mepanda dam will also help to reduce the impact of floods in the Zambezi valley.

Paris-based Alstom has been selected by ZESCO to rehabilitate the Victoria Falls power station. The facility consists of three stations, Station A (8 MW), Station B (60 MW) and Station C (40 MW). The plant is situated on the Zambezi river, near Livingstone and within the UNESCO World Heritage site. Alstom will provide both equipment and services for the turnkey rehabilitation of all three stations and the associated high-voltage switchyard.

Botswana plans to provide electricity to all of its citizens by 2016. Currently, only 22 per cent of Botswana's population has access to electricity. Botswana is continuing talks with Eskom and Nampower concerning the importation of additional electricity into the country. BPC plans to spend $700 million to extend the company's transmission and distribution systems.

Renewable Energy/Environment

The Compagnie Thermique de Belle Vue (CTBV), a joint-venture composed of Harel Freres (51 per cent) of Mauritius, France's Cidec (27 per cent), the Sugar Investment Trust of Mauritius (14 per cent) and the State Investment Fund (8 per cent), built a 70-MW IPP facility north of the Mauritian capital of Port Louis. The CTBV plant utilizes bagasse (biomass refuse from the processing of sugar cane) as its primary fuel. Representatives from Uganda and Swaziland took fact-finding tours of the CTBV plant; Swaziland currently produces about 500,000 tons of sugar per year and is heavily dependent upon South Africa for its energy needs.

In November 2001, a 20-MW steam-turbine/generator for a sugar mill near Chirezdi, Zimbabwe was commissioned. The facility, owned by Hippo Valley Estates, will use steam generated from burning bagasse to generate the mill's electrical power. The exhaust steam will be used in the sugar refining process.

Solar energy is being viewed by several countries in the region as a prime tool to speed up rural electrification programs, which have been slowed by the high costs of grid extension services. Zambia, Namibia and South Africa are developing programs to utilize solar energy for off-grid (rural) electrification. Zambia's government has taken steps to encourage investment in solar energy by eliminating all import duties on solar panels and waiving the otherwise obligatory annual license fees for solar energy projects. In April 2002, Namibia's President Nujoma officially opened a solar module factory in northern Namibia. President Nujoma stated the project is important because it provides Namibians with opportunities to promote the usage of solar energy, and the domestically produced photovoltaic dishes will be more affordable to the public.

Mauritius also launched an $800,000 solar energy project in February 2000; 125 solar lamps and 960 photo-voltaic panels will light up Mauritian streets and government office buildings as part of the government's strategy to promote renewable energy.

Southern Africa faces various environmental problems, including pollution of water supplies, deforestation, desertification, pollution associated with oil and gas development, and dramatic decline in biodiversity throughout the region.

In April 2002, BP, the third-largest oil company operating in South Africa, unveiled a lead-free gasoline for older motor vehicles. The new fuel has been specially created for vehicles that were originally designed to use only leaded fuel, specifically pre-1986 models and those without catalytic converters. BP's decision to launch a lead-free gasoline for older model cars comes four years ahead of a government imposed deadline that will see the use of lead in fuel outlawed in South Africa.

80

Spain

Spain is one of the fastest growing European economies but has very limited domestic energy resources. As a result, Spain is expected to become an increasingly important energy importer.

General Background

Spain has been a member of the European Union (EU) since 1986. Over the last decade, Spain has enjoyed rapid economic expansion. Economic growth has slowed over the past few years, but at a forecast real growth rate of 1.9 per cent in 2002 (the lowest since 1993), Spain will remain one of the fastest growing of the large EU economies. Spain's unemployment rate has decreased significantly, from 19.8 per cent in 1994 to 11 per cent as of the third quarter, 2002. The government budget returned to a deficit in 2002 and the deficit is expected to increase in 2003. Inflation is expected at 2.8 per cent in 2002, roughly the same as in 2001.

In March 2000, Prime Minister Jose Maria Aznar's center-right Popular Party (PP) was reelected with an absolute majority, obtaining control in both houses of parliament. Aznar has continued his liberalization of Spanish industry. In June 2000, the government passed legislation aimed at eliminating monopolies (state-held or private) in the energy, telecommunications, and services industries. Oil, natural gas, and electricity markets are key targets in Aznar's liberalization program. The Popular Party, which has been in control since 1996, will face elections in the spring of 2004. PP will have to find a new party chief, as Prime Minister Aznar is schedule to step down in 2003.

Continued economic instability in Argentina has adversely affected Spanish companies, which invested 45 billion euros there over the last decade. Five large Spanish companies, including oil company Repsol-YPF and power company Endesa, that alone account for about three-quarters of the trading volume on the Madrid stock exchange, have lost billions of euros because of the default on Argentine government debt and the devaluation of the Argentine peso. Recently, Argentina's economic situation has slightly improved, allowing Repsol-YPF to make up for some of its losses in 2002.

Economic growth and accelerated industrialization associated with EU membership have led to increased Spanish energy demand, up 75 per cent since the mid-1970s. Electricity demand is growing at a particularly rapid rate of 6 per cent per year, necessitating greater investment in this section. Spain is highly dependent on imported oil and natural gas, leaving the country economically vulnerable to world oil price fluctuations. Increased natural gas imports are expected to meet a large share of future energy demand. Spain currently imports 60 per cent of its natural gas from Algeria. With an extensive gas network now in place, Spain's demand for natural gas is expected to increase by 45 per cent by 2010, according to International Energy Agency's forecasts. In order to meet these demands, the Spanish Cabinet approved a 10-year energy plan in September 2002, which focuses on meeting the countries future energy requirements. The plan reflects an increased reliance on natural gas but also revamps the country's basis energy infrastructure, as well as encourages investment in electricity and natural gas distribution networks. According to the plan, natural gas and renewables, which are to account for 33.1 per cent and 28.4 per cent respectively, will be favored, while nuclear, coal and oil would represent smaller percentages. The plan still requires approval from the parliament.

Spain continues to privatize its energy sector, a process which it began in 1994 with the LOSEN Electricity Act. The Electricity Act 54 in 1997 and the Hydrocarbons Act 34 in 1998 accelerated and bolstered the process. The Hydrocarbons Act created the Comisión Nacional de Energía (National Energy Regulatory Commission). The commission is a regulatory body, which is tasked with ensuring the existence of competition in Spain's electricity market and liquid and gaseous hydrocarbons market.

These programs place Spain in compliance with EU Directives 96/92/EC and 98/30/EC, which require EU members to privatize their electricity and natural gas markets.

Oil

Oil plays a major role in the Spanish energy sector, although its share of primary energy is decreasing. In 2001, oil accounted for roughly 52 per cent of Spain's primary energy consumption, down from 73 per cent in 1973. Nonetheless, oil consumption has increased. In 2002, Spain consumed about 1.5 million barrels per day (bbl/d) of oil, an increase of 50 per cent since 1990. Spain imports nearly 99 per cent of its crude oil, mainly from Nigeria, Mexico, Libya and Saudi Arabia.

Furthermore, Spain has limited proven domestic oil reserves, 157 million barrels as of January 2003. Its production is therefore marginal. In 1990, Spain was producing 27,000 bbl/d and in 2002, averaged an estimated 6,600 bbl/d, a decrease of over 76 per cent. Spain has five major oil fields: Lora; Casablanca-Montanazo; Rodabello; Chipirón; and Angula-Casablanca. Chipirón produced the most oil in 2002 with 2,978 bbl/d. With limited domestic reserves, the Spanish government has considered other regions and in late 2002, gave Repsol-YPF permission to explore for oil deposits near the Canary Islands.

Until 1993, the Spanish oil industry was state controlled. Today, formerly state held (now private) Repsol still dominates the Spanish oil sector. Repsol acquired the top Argentine oil company, YPF, in 1999, changing the company name to Repsol-YPF. The merged company is responsible for over 50 per cent of Spain's oil production. Worldwide, the company has reserves of 5.6 billion barrels of oil equivalent and and a daily production of about 1 million barrels of oil equivalent per day. The company owns the majority of Spain's refineries, its distribution network (through Compania Logistica de Hidrocarburos, CLH, in which it holds a majority stake), and its gasoline stations (through its trademarks Repsol, Campsa, and Petronor). Divestments in the wake of the merger are working to

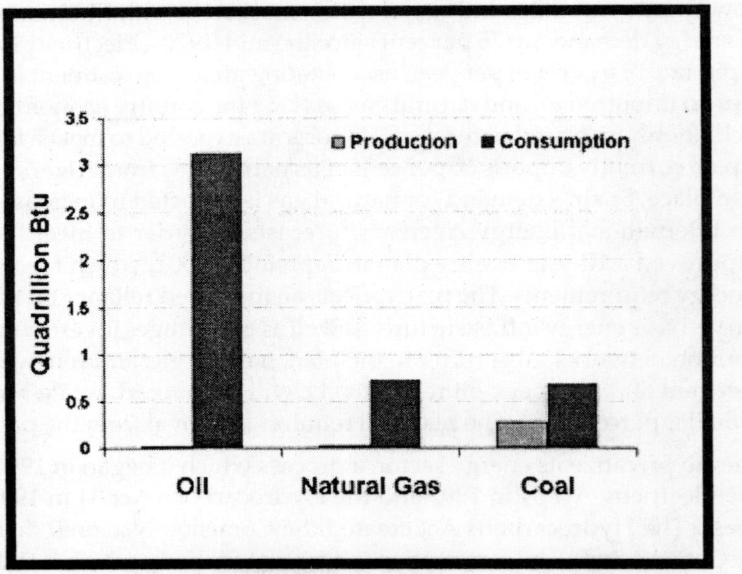

SPANISH FOSSIL FUEL PRODUCTION AND CONSUMPTION, 2001

lessen Repsol-YPF's control in the industry. Economic liberalization plans, which begun in June 2000, also work toward this end; the company reduced its shares in CLH from 62 per cent to 26.45 per cent. Repsol-YPF's profits have been lower for both 2001 and 2002 than the record $2.10-billion profit achieved in 2000 because of exposure to the Argentine economic situation. Repsol-YPF derives 45 per cent of its operating income from Argentina's oil and natural gas fields.

Compania Espanola de Petroleos (Cepsa), established in 1929, is Spain's oldest private oil and gas company. The company has exploration and production activities in Colombia and Algeria and is aiming to expand its operations in Yemen and Iran. It is the second largest oil group in Spain, with a 25 per cent retail market share. BP Oil is also active in Spain. Repsol, Cepsa, and BP Oil account for almost all of the activity in the Spanish oil sector.

Refining

Spain has nine major refineries. Four are owned by Repsol-YPF, and another is owned by a Repsol-YPF subsidiary, Petronor, in which Repsol-YPF has an 86 per cent stake. Petronor is also the largest oil refinery in Spain. Cepsa owns three, and one is owned by BP. Because of state regulation of the industry, Spain has avoided developing the excess refining capacity that characterizes some other countries in southern Europe. Spain's total crude oil refining capacity stands at 1.3 million bbl/d.

Natural Gas

Natural gas is expected to account for a much larger share of Spain's total energy consumption in coming years, especially as new pipelines and natural gas-fired power plants come on line. Within the European Union, Spain is regarded as having the fastest growing natural gas market. Natural gas consumption has grown from 2 per cent of total energy consumption in the 1970s to 12.8 per cent in 2001. Preliminary estimate of consumption for 2001 is about 627 billion cubic feet (Bcf). Some estimates predict natural gas consumption growing at a 10 per cent annual rate in this decade. Spain's National Energy Regulatory Commission predicts demand for natural gas rising to 1.2 Tcf by 2005. Almost all of this consumption will be satisfied with imports, as Spain has extremely limited natural gas reserves (94 billion cubic feet). The country's largest natural gas field went out of production in 1995, and only two smaller fields remain in production, Valle de Guadalquivir and Poseidón, with the latter producing 17.8 Bcf in 2002.

The Gas Natural Group (GN) is the leading natural gas conglomerate in Spain. However, in the market for industrial customers, which was partially opened in 2000, GN's market share has been decreasing as other companies vie for customers. GN comprises: Gas Natural SDG, the main natural gas distributor in Spain; Gas Natural Aprovisionamientos (supplies); Gas Natural Comercializadora (commercialization); Gas Natural Servicios, the services company of the Group; Gas Natural Overseas Trading Company; 14 natural gas distribution companies in Spain; and Gas Natural Internacional, which brings together in a single business unit the interests of GN in Gas Natural BAN (Argentina), Gas Natural ESP (Colombia), Companhia Distribuidora de Gas do Rio de Janeiro-CEG, CEG RIO and Gas Natural SPS (Brazil), in addition to Gas Natural México. Also, GN has minority holdings in three natural gas distribution companies in the region of Aragon and in the Basque Country. Last year, GN spun off 65 per cent of Enagás, a gas distribution company, in order to comply with privatization legislation. Previously, GN controlled 100 per cent of the company until the Spanish government requested that it reduce its share to 35 per cent.

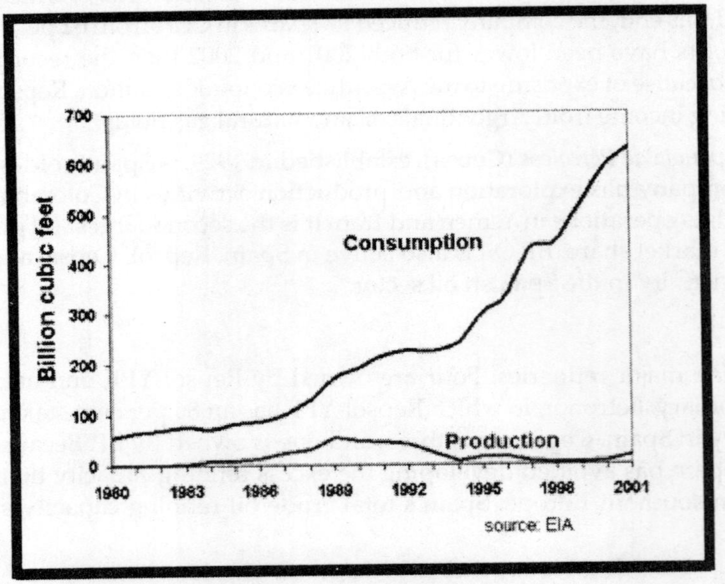

SPANISH NATURAL GAS PRODUCTION AND CONSUMPTION, 1980-2001

According to liberalization legislation passed in June 2000, no single operator may command over 70 per cent of the Spanish natural gas market by 2004. Since June 2000, large industrial consumers have been able to choose suppliers, and as of January 1, 2003, all consumers are able to choose their suppliers. Several additional regulatory measures were taken in 2001. In July, Spain's Economy Ministry published the terms under which GN must auction off one-third of its 580 Bcf per year Algerian pipeline natural gas imports. In September, Spain's Economy Ministry detailed new natural gas sector regulations that include a revised system for calculating pipeline tariffs and procedures for accessing the national grid. Finally, in October, the government ended GN's monopoly of natural gas imports when a contract for Algerian gas imports equivalent to about 25 per cent of Spain's total annual consumption was awarded to Spain's four largest electricity companies (Endesa, Iberdrola, Union Fenosa, and Hidrocantabrico), BP, and Royal Dutch/Shell. This is part of a strategy being pursued by Spanish electricity companies to enter into the natural gas market. As Algeria supplies about 75 per cent of Spain's imports, these companies now control about 19 per cent of the market. These companies have until 2004 to sell the natural gas to their industrial clients.

The expected growth of natural gas consumption and the continued liberalization of the market have attracted outside and domestic companies to invest in the market. Eni, an Italian oil and gas company, purchased last year 50 per cent of Union Fenosa Gas, the gas unit of Spain's electricity company Union Fenosa. Domestically, Endesa plans to increase its share of the natural gas market by integrating its gas and electricity distribution subsidiaries into a new company called Endesa Red. In November 2002, Cepsa signed an agreement with Sonatrach, the Algerian state oil company, to import 35.3 billion cubic feet of natural gas per year, which will be transported via the Medgaz pipeline.

Enagás transports natural gas imports to the Iberian Peninsula via gas pipelines connected to international networks (or via methane carriers for liquefied natural gas. discussed below). There are two international gas pipelines in Spain: Lacq-Calahorra in the north and the Pedro Duran Farell pipeline (formerly the Mahgreb-Europe line) in the south. The Lacq-Calahorra gas pipeline is the main Spanish connection to the European network, linking to Norway's North Sea gas sources. The Pedro Duran Farell pipeline, which crosses through Algeria and Morocco and travels under the Strait of Gibraltar, is about 870 miles (1,400 kilometers) long and connects the Algerian deposits with the Spanish gas pipeline network in Córdoba. This pipeline made its first Spanish delivery in 1996. Work is underway to expand the Pedro Duran Farell pipeline's annual capacity from 282.5 Bcf to 388.5 Bcf by adding a compressor station. Completion is expected in late 2003.

There are two new projects underway as well. Spain will have an additional connection with France via Irun in the Basque Country as a new trans-frontier connector is being built with completion expected by the end of 2003. A further extension of the pipeline network coming to and from Irun, France is planned to be ready between 2005 and 2008. It will have 111 miles in Spain and 93 miles in France, and possess an annual capacity of 144.8 Bcf. In July 2001, Cepsa and Sonatrach signed an agreement for the construction of the new Medgaz undersea natural gas pipeline between Algeria and Almeria, Spain. A feasibility study was completed in late 2002, giving the green light for construction. The pipeline would have a length of 137 miles and have a capacity 564.8 Bcf per year. Natural gas would be allotted in proportion to each shareholder's equity ownership. At present, Sonatrach and Cepsa each hold 20 per cent, while BP, Endesa, Eni, Gaz de France, and TotalFinaElf each hold 12 per cent. TotalFinaElf and GdF are expected to take their gas into France, and ENI its share on to Italy.

Liquefied Natural Gas

Spain is Europe's second-largest liquefied natural gas (LNG) importer, behind France. Enagás operates Spain's three regasification terminals which are located in Barcelona, Cartagena, and Huelva. Algeria is Spain's largest LNG supplier. Spain also imports LNG from the United Arab Emirates and Qatar, as well as, more recently, from Trinidad and Tobago, Nigeria and Oman. In November 2002, Union Fenosa finalized a contract with Oman LNG to buy 63.5 Bcf, which Oman will deliver during 2004-2005. Union Fenosa also signed a long-term agreement with Oman to purchase 50 per cent of the output from its third LNG train for 20 years.

Over the last years, a number of companies have been increasing their capacity to export LNG to Spain. Enagás plans to invest in new infrastructure in order to handle increased LNG imports for rising domestic consumption. It aims to expand the capacity of its three regasification terminals from 734 Bcf per year to 1.2 Tcf per year by 2005. Union Fenosa signed a contract with the Egyptian General Petroleum Corporation in July 2000 to purchase LNG from a new liquefaction terminal under construction at Damietta, Egypt. The terminal is expected to come on stream in 2005 with a capacity of 243.5 Bcf per year. Endesa, Union Fenosa and Iberdrola are constructing a new receiving gasification terminal at the Sagunto port in Valencia. The plant, to be completed in winter 2004, will be able to process 194 Bcf per year and will be linked to new combined cycle gas turbine plants being constructed by Fenosa nearby and Iberdrola in Castellon. Some of the natural gas will also transit to other locations.

The Bahia de Bizkaia Gas group, a consortium led by BP and including Repsol-YPF, Iberdrola, and EVE (the Basque Energy Authority) is building a new import facility in conjunction with a new power station in Bilbao, Spain. The regasification facility is expected to begin operations in 2003.

In 2002, Endesa, Union Fenosa, the Tojeiro Group and Algeria's Sonatrach, forming the Reganosa group, began building the El Ferrol LNG import facility. Algerian LNG will supply the new terminal, located in Galicia in northwest Spain, for ten years following the terminal's projected 2004 commissioning. In conjunction with the terminal, which will have an initial capacity ranging from 88 to 123.5 Bcf per year, a new pipeline will be constructed to connect the terminal to power plants located about 60 miles away.

Coal

Coal is Spain's most plentiful indigenous energy source. Production has fallen in recent years, and the decline is expected to continue as Spain works to meet environmental standards. Currently, 95 per cent of coal is used to generate electricity, and 42.3 per cent of Spain's electricity was generated from coal, according to Spain's National Energy Regulatory Commission. All of the major coal companies are state-owned.

Spanish coal is too expensive to be competitive in a free energy market, with about 80 per cent of the coal costing at least twice international prices to produce, so the Spanish government subsidizes coal production. According to new EU regulations that took effect in July 2002, Spain must lower its coal production by 65 per cent over the next ten years. Also, coal mines that do not improve their economic viability will only be able to receive production subsidies until 2008. Spain is one of three EU countries that will be permitted to continue coal production for reasons of economic security, and hence will continue to receive subsidies for more competitive mines. There is increased pressure on coal, however, as the electricity market privatizes, and as electricity generation will no longer be a captive market for domestic coal. Imports of foreign coal already are on the rise, and electricity generators are looking more to natural gas.

The coal sector now employs only half the number of people it did a decade ago. Most of those employed are in the Asturias region, where the jobs are badly needed. It would be difficult to completely phase out coal mining because of this region's dependence on the industry for employment.

Electricity

Spain has the fifth largest electricity market in Europe (behind Germany, France, the United Kingdom, and Italy), and it is growing quickly. Electricity demand for 2001 was estimated to be 210.4 billion kilowatthours (bkwh), a 5 per cent increase over 2000. It is estimated that Spain's electricity demand will increase 30 per cent by 2010.

To meet Spain's increased demand for electricity, domestic utility companies have invested in generation capacity and distribution. Red Electrica de España (REE), for instance, invested heavily in the network in 2001, dedicating 78.4 million euros to expanding the electricity network. REE also announced plans in October 2001 to invest between 60.2 and 72.2 million euros to improve the electricity connection with France. Spain's three largest electricity groups–Endesa, Iberdrola, and Union Fenosa–have dedicated 34 billion euros of investments from August 2001 to 2005, with much of that in Latin America and other European countries, but nevertheless including 8 billion euros for new generating plants in Spain.

Endesa is in the process of building a natural-gas-fired, 400 MW, combined-cycle generating turbine (CCGT) plant in Huelva in addition to two other gas-fired 400-MW CCGTs the company already has under construction in Barcelona and Tarragona. Endesa recently completed the construction of a plant with 8,000 MW in Cadiz. Union Fenosa plans to add 5,000 MW of new capacity by 2005, mostly in Spain, of which 2,800 MW would be natural-gas-fired. Piemsa, an affiliate of Petronor, is planning to construct an 800-MW integrated gasification combined-cycle (IGCC)

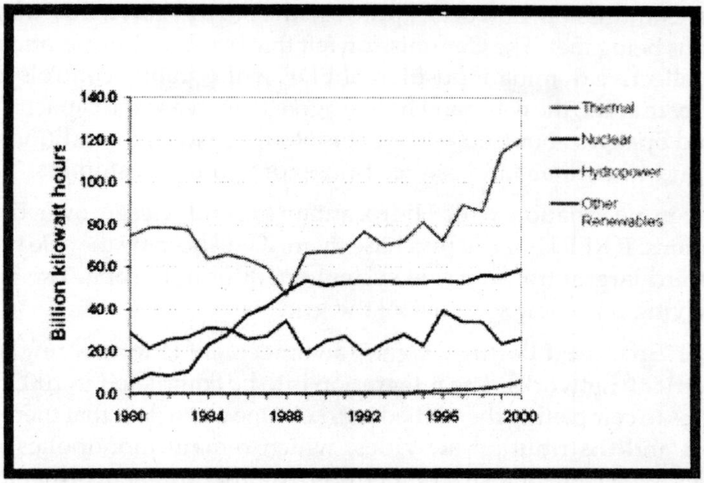

SPANISH ELECTRICITY GENERATION, 1980-2000

complex at a refinery near Bilbao that will make use of heavy refinery stocks. The plant will be one of the largest and most advanced of its kind in the world.

Spain's electricity market is privatizing ahead of the schedule mandated by the EU. A 1996 EU directive required that at least 26.48 per cent of electricity sales in member countries be open to competition, beginning in February 1999. This requirement increased to about 28 per cent in February 2000 and will grow to 33 per cent in 2003. Spain already has surpassed the 2003 requirement. As of January 1, 2003, domestic customers have been able to choose their electricity supplier. However, customers will most likely experience limited benefit from privatization for the time being, as the Spanish government has mandated that all utilities must charge the same amount per unit, rendering true competition almost nonexistent. Furthermore, the government has mandated that between 2003-2010 prices can rise by a maximum of 2 per cent a year. The actual percentage will be adjusted according to demand, interest rates, the treatment of renewable energy and gas prices.

The Spanish electricity sector is in the midst of restructuring. There are five major utility companies in Spain, in descending order of size: the formerly state held Endesa; Iberdrola; Union Fenosa; Hidrocantabrico; and the newly independent Viesgo. Viesgo's acquisition by Enel of Italy from Endesa was completed in January 2002, and Viesgo has a 5 per cent market share. This is part of Enel's strategy of regaining market share abroad after selling its Elettrogen utility at home to Endesa in 2001.

In 2001, Energie Baden-Wurttemburg (EnBW), Eletricidade de Portugal (EdP), and Caja de Ahorros de Asturias (Cajastur), a Spanish bank, took joint control of Hidrocantabrico. However, this initial sale came into controversy because the Spanish government threatened to block France's state-owned utility, Electricité de France (EdF), which has a 34.5 per cent share in EnBW, from entering the Spanish electricity market. The Spanish government enacted the Law 55/1999, which includes provisions that make the exercise of voting rights by public entities directly or indirectly taking control of or acquiring 3 per cent or more of the equity in Spanish energy companies subject to the prior authorization of Spain's Council on Ministers.

The European Commission, however, approved the bid by EnBW, EDP and Cajastur but insisted on certain conditions being met. The Commission felt that because the operation could have led to the strengthening of collective dominant position of EDF, which jointly controls EnBW, on the Spanish wholesale electricity market, the company had to agree to increase the capacity on the interconnector between France and Spain to about 4,000 MW. This would create the conditions for greater electricity trade volumes to and from Spain and reduce prices for Spanish customers.

REE has been in negotiation with Hidrocantabrico and Viesgo over buying their electricity distribution operations. If REE is able to purchase them, it will become the sole transporter of electricity in Spain and the third largest transmission system operator in Europe. The Spanish government is also supporting having one company control the grid.

In August 2001, Spain and Portugal signed an agreement to form a single electricity market by unifying their electricity networks. The integration is to be completed in 2003. There remain several unresolved obstacles to completing the unified market. One obstacle is that there is minimal separation between transport and distribution activities, which remain monopolies, and production and marketing activities, which are open to competition. Another problem is that in Portugal production is sold to the state-held REN, which transports the electricity, whereas in Spain producers compete to sell electricity, but receive compensation payments for market liberalization called CTCs. The Spanish government in March 2001 reiterated its support for CTCs, but these payments are under investigation by the EU. The opposition PSOE party has called for their end. However, electricity companies have called for an end to tariff privileges enjoyed by several large industrial companies that they believe have made these companies non-competitive internationally. In addition, electricity companies would like to raise their rates, arguing that prices have fallen 17 per cent in the past five years, while inflation for the period has been 14 per cent. In December 2003, a 2 per cent rate increase was authorized for industrial customers.

As electricity demand has increased rapidly in Spain in the past years, domestic supply has not kept up and Spain began importing electricity from Morocco for the first time in December 2001 when cold temperatures created a surge in demand. Union Fenosa and Endesa have signed agreements with Moroccan power company ONE. Spain granted ONE the status of an "external operator" in 1998, giving the company the right to deal directly with Spanish electricity companies or on the Spanish spot market. The power exchange between ONE and Spanish companies is through the Spain-Morocco grid interconnection, which became operational in 1998. Two power connections between Algeria and Spain are also planned, one of which will run along side the Medgaz pipeline.

Nuclear Power

Spain has nine nuclear reactors in operation with one, Vandellos-1, being shut down in 1990. According to Spain's National Energy Regulatory Commission (CNE), the country derived 33.9 per cent of its gross production of electricity from nuclear power in 2002. For the year, Spain's nuclear power plants produced 63 bkwh, a decrease of 1.1 per cent compared to 2001, according to CNE.

The Popular Party supports nuclear power, but the PSOE has indicated that it supports a gradual shutdown of Spain's nuclear plants. The government announced in October 2002 that it will allow nuclear power plants across the region to increase their output in order to compensate for the planned shutdowns of the Guadalajara (2006) and the Ante Maria de Gabon (2009) plants.

Country Overview

Independence: 1492 (expulsion of the Moors and unification)

Capital City: Madrid

Population (2002E): 40.1 million

Location/Size: Southwestern Europe, bordering the Bay of Biscay, Mediterranean Sea, North Atlantic Ocean, and Pyrenees Mountains, southwest of France/504,750 sq. km (slightly more than twice the size of Oregon)

Language: Castilian Spanish 74 per cent, Catalan 17 per cent, Galician 7 per cent, Basque 2 per cent

Religion: Roman Catholic 99 per cent, other 1 per cent

Economic Overview

Currency: Euro

Exchange Rate (3/05/03): US$1= 0.9187 Euro

Gross Domestic Product (GDP, nominal, 2002E): $657 billion

Real GDP Growth Rate (2002E): 1.9 per cent *(2003F):* 2.2 per cent

Inflation Rate (consumer prices, 2002E): 3.5 per cent *(2003F):* 3.3 per cent

Unemployment Rate (2002E): 11.4 per cent *(2003F):* 10.6 per cent

Merchandise Exports (2002E): $120.9 billion

Merchandise Imports (2002E): $158.7 billion

Major Trade Partners: France, Germany, Italy, United Kingdom, United States, Portugal

Major Export Products: Automobiles, tourism, power generation equipment, electrical machinery, petroleum and chemical products, foodstuffs

Major Import Products: Crude petroleum, vehicle and automobile parts, capital goods, and food

Foreign Exchange Reserves (excluding gold) (2002E): $31.0 billion

Energy Overview

Proven Oil Reserves (1/1/03E): 157 million barrels

Oil Production (2002E): 20,583 barrels per day (bbl/d), of which 6,583 bbl/d was crude oil

Oil Consumption (2002E): 1.5 million bbl/d

Net Oil Imports (2002E): 1.5 million bbl/d

Crude Oil Refining Capacity (1/1/03E): 1.3 million bbl/d

Natural Gas Reserves (1/1/03E): 94 billion cubic feet (Bcf)

Natural Gas Production (2001E): 18 Bcf

Natural Gas Consumption (2001E): 627 Bcf

Net Natural Gas Imports (2001E): 609 Bcf

Coal Reserves (2000E): 728 million short tons (Mmst) (of which 507 Mmst is lignite and subbituminous, 220 Mmst anthracite and bituminous)

Coal Production (2001E): 25 Mmst

Coal Consumption (2001E): 45.19 Mmst

Electric Generation Capacity (2000E): 46.2 million kilowatts (55.4 per cent Thermal), (25.5 per cent Hydro), (15.9 per cent Nuclear)

Net Electricity Generation (2001E): 222.5 billion kilowatthours (bkwh)

Electricity Consumption (2001E): 210.4 bkwh

Environmental Overview

Total Energy Consumption (2001E): 5.7 quadrillion Btu* (1.4 per cent of world total energy consumption)

Energy-Related Carbon Emissions (2001E): 82.7 million metric tons of carbon (1.3 per cent of world carbon emissions)

Per Capita Energy Consumption (2000E): 136.7 million Btu (vs U.S. value of 351.0 million Btu)

Per Capita Carbon Emissions (2000E): 2.1 metric tons of carbon (vs U.S. value of 5.6 metric tons of carbon)

Energy Intensity (2000E): 7,684 Btu/$1995 (vs U.S. value of 10,918 Btu/$1995)**

Carbon Intensity (2000E): 0.12 metric tons of carbon/thousand $1995 (vs U.S. value of 0.17 metric tons/thousand $1995)

Fuel Share of Energy Consumption (2001E): Oil (52.3 per cent), Coal (15 per cent), Natural Gas (13 per cent), Nuclear (13 per cent)

Fuel Share of Carbon Emissions (2000E): Oil (66.2 per cent), Coal (22.6 per cent), Natural Gas (11.2 per cent)

Status in Climate Change Negotiations: Annex I country under the United Nations Framework Convention on Climate Change (ratified December 21st, 1993). Signatory to the Kyoto Protocol (signed April 29th, 1998 and ratified on May 31, 2002–not yet entered into force).

Major Environmental Issues: Pollution of the Mediterranean Sea from raw sewage and effluents from the offshore production of oil and gas; water quality and quantity nationwide; air pollution; deforestation and desertification.

Major International Environmental Agreements: A party to Conventions on Air Pollution, Air Pollution-Nitrogen Oxides, Air Pollution-Sulphur 94, Air Pollution-Volatile Organic Compounds, Antarctic-Environmental Protocol, Antarctic Treaty, Biodiversity, Climate Change, Endangered Species, Environmental Modification, Hazardous Wastes, Law of the Sea, Marine Dumping, Marine Life Conservation, Nuclear Test Ban, Ozone Layer Protection, Ship Pollution, Tropical Timber 83, Tropical Timber 94, Wetlands and Whaling. Has signed, but not ratified: Air Pollution-Persistent Organic Pollutants, Desertification.

The total energy consumption statistic includes petroleum, dry natural gas, coal, net hydro, nuclear, geothermal, solar, wind, wood and waste electric power. The renewable energy consumption statistic is based on International Energy Agency (IEA) data and includes hydropower, solar, wind, tide, geothermal, solid biomass and animal products, biomass gas and liquids, industrial and municipal wastes.

81

Sudan

With the completion of a major oil export pipeline in July 1999, Sudanese crude oil production and exports have risen rapidly over the past three years. Sudan's estimated reserves of crude oil have doubled over the past year, and production is expected to reach 300,000 bbl/d by 2004.

General Background

Sudan gained its independence from Egypt and the United Kingdom in 1956. The current government, led by General Omar Hassan Ahmad al-Bashir, came to power in 1989 after overthrowing a transitional coalition government. A new constitution was adopted on January 1, 1999. Multi-party presidential and parliamentary elections were held in December 2000, and President Bashir and his party won an overwhelming victory, tainted by a boycott by all main opposition parties.

Despite considerable natural resources, Sudan is among the world's (and region's) poorest countries. Traditionally, its economy has been mainly agricultural–a mix of subsistence farming and production of cash crops such as cotton and gum arabic. With the start of significant oil production (and exports) beginning in late 1999, however, Sudan's economy is changing dramatically, with oil export revenues now accounting for around 70 per cent of Sudan's total export earnings. Sudan no longer relies on expensive imported oil products, which has helped the country's trade balance, while foreign investment has started to flow into the country. Sudan's currency, the dinar, has remained relatively stable since 1999, at around 259 dinars per U.S. dollar.

Sudan's economic performance has been strong over the past few years. In 2001, the country's real GDP grew by 5.4 per cent, though growth is estimated to have slowed to 3.6 per cent in 2002. Meanwhile, inflation has slowed dramatically over the past few years, from an average 110 per cent between 1991 and 1996 to 4.9 per cent in 2001 and 6.7 per cent in 2002. Exports have grown sharply since 1999, when the oil export pipeline was completed, turning the country's trade balance from negative to positive. Sudan's government has been negotiating the payment of its substantial debts

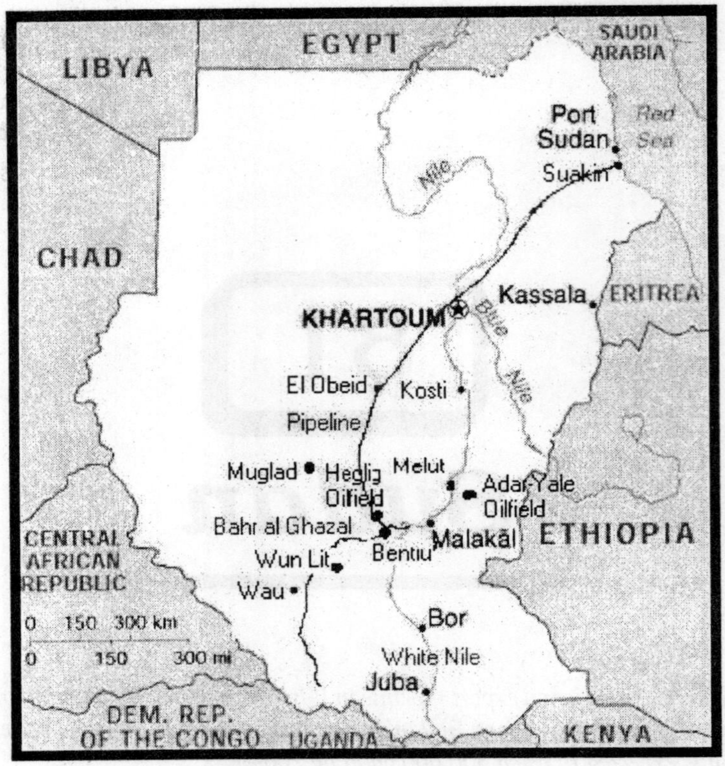

to the International Monetary Fund so that the nation can improve its relations with the institution. Sudan's debt to the IMF was rescheduled in 2002, and the IMF deferred Sudan's arrears payments so that the nation can give priority to repaying loans from the Arab Fund for Socioeconomic Development. Some of those funds will be used to finance a new hydroelectric dam in Merowe. As part of the rescheduling agreement, Sudan agreed to reduce its military spending as well as make the management of its oil revenues more transparent

Despite its economic progress, Sudan still faces developmental obstacles, including a limited infrastructure and an external debt estimated to be around $15 billion in 2001, representing a debt-to-GDP ratio of just over 122 per cent. Furthermore, the government remains embroiled in the long-running conflict with rebel movements in the primarily non-Muslim south of the country. This costly and bloody conflict has, over the past two decades, claimed (directly or indirectly through famine) as many as 2 million Sudanese lives. Sudan's new reliance on oil export revenues makes the country vulnerable to world oil price fluctuations, thereby making it necessary for the government to spend its revenues wisely.

The United States has been imposing economic sanctions against Sudan since November 1997, prohibiting trade between the two countries, as well as investment by U.S. businesses in Sudan. In February 2000, the sanctions were broadened to include a prohibition against U.S. citizens and companies conducting business with the Greater Nile Petroleum Operating Company (GNPOC), an international consortium of petroleum companies currently extracting oil from Sudan. The sanctions, however, do not apply to the foreign individual parent companies of GNPOC, which include Calgary-based Talisman Energy, Malaysia's Petronas, and the Chinese National Petroleum Corporation (CNPC). Since the September 11, 2001 terrorist attacks on New York and Washington, however, U.S.-Sudanese relations appear to have improved somewhat, as the United States is attempting to encourage Sudanese cooperation in the war against terrorism. However, Sudan reacted unfavorably to the passage of the Sudan Peace Act in October 2002, which outlines stiff sanctions, ranging from a downgrading of diplomatic relations to a UN arms embargo, that could be imposed on the Sudanese government if it negotiates in bad faith with the main southern rebel force, Sudan People's Liberation Army (SPLA).

Oil

As of January 2003, Sudan's estimated proven reserves of crude oil stood at 563 million barrels, more than twice the 262.1 million barrels estimated in 2001. Crude oil production averaged 227,500 barrels per day (bbl/d) during 2002, a figure that has been rising steadily since the completion of the export pipeline in July 1999. By 2003, oil output could surpass 300,000 bbl/d, with plans to reach 450,000 bbl/d by 2005. In August 2001, in recognition of Sudan's growing significance as an oil exporter, OPEC granted the country observer status at OPEC meetings.

Petroleum exploration in Sudan began in the early 1960s. The activity was originally concentrated offshore in the Red Sea. The only significant offshore discovery was Chevron's Suakin gas discovery in 1976. Chevron's exploration in the 1960s and 1970s led to several oil finds in southern Sudan near the towns of Bentiu, Malakal and Muglad. Chevron abandoned its concessions in Sudan in 1985, due to their location in an area where fighting was taking place between government and rebel forces. The French firm, Total (now TotalFinaElf), also suspended its onshore exploration activities, but retained the rights to its concessions. The Sudanese government sub-divided Chevron's concessions into smaller exploration blocks, and Canadian independent Arakis Energy (Arakis) acquired the portion of Chevron's concession north of the town of Bentiu in 1993.

Greater Nile Oil Project

Arakis began development of the Heglig and Unity fields within its concessions, and started production on a small scale (around 2,000 bbl/d) in 1996; this oil was processed and consumed within Sudan. The remote location of the field, aproximately 930 miles from the Red Sea coast, meant that it would require a very substantial capital investment to transport the oil to a seaport. To attract the necessary capital and spread the risks, Arakis entered into a consortium in December 1996 with the Greater Nile Petroleum Operating Company (GNPOC), consisting of the China National Petroleum Corporation (CNPC) (40 per cent), Petronas of Malaysia (30 per cent), Sudanese national firm Sudapet (5 per cent), and Arakis (25 per cent, and the field operator). Construction on the pipeline from the fields to an export terminal near Port Sudan began in May 1998 on an accelerated schedule. Originally built to move 150,000 bbl/d, the pipeline has a current capacity of 250,000 bbl/d and can be expanded to 450,000 bbl/d.

Arakis' involvement in Sudan, even after the formation of the GNPOC consortium raised $700 million, remained hindered by a lack of capital. U.S. sanctions against Sudan prevented investment in the project by U.S. corporations and persons, and the high-risk nature of investment in Sudan also had an effect. In October 1998, Arakis agreed to be purchased by another Canadian independent, Talisman Energy, for $277 million in Talisman stock. The Talisman acquisition provided an infusion of capital which allowed the project to be completed on schedule in 1999. In July 1999, the pipeline began filling with crude, and the first cargo of "Nile Blend" departed the export terminal in early September 1999.

Production

The fields in the Muglad area produce crude oil with a 33 ° to 42° API range, with only 0.5 per cent sulfur content. The crude is highly parafinnic, which requires heating to maintain flow in the pipeline. Recoverable reserves from the Heglig and Unity fields have been estimated at 660 million to 1.2 billion barrels. The area around these two fields also is suspected to contain oil, but estimates of reserves vary. The Swedish firm, Lundin Oil (partly bought out by Talisman in June 2001), reported a discovery at the "Thar Jath 2" exploration well in the adjacent Block 5A in March 2001, with oil flows of 2,000 bbl/d. Petronas, the Austrian firm OMV, and Sudapet have minority stakes in the block. In July 2000, Petronas was awarded a 40 per cent share in Block 5B, and in October 2000, Petronas agreed to raise Sudan's oil output by 50,000 bbl/d by mid-2002. The increased production will result from Petronas' development of two untapped oil fields in Monga and Bambo in the Mujlad Basin of western Sudan, plus the construction of three additional oil pumps along the pipeline. In March 2001, OMV announced that it had found oil at its exploratory well, Jath 1." Since January 2002, production has been hampered somewhat by the decision by Petronas, Lundin Oil, and other members of the Greater Nile Oil Project, to temporarily suspend drilling operations on Block 5A due to safety concerns regarding personnel and drilling sites. The suspensions seem unlikely to be lifted until a cease-fire between the government and the SPLA can be negotiated and maintained. Once an agreement has been reached, restarting production is likely to take three months.

In October 2002, Talisman Energy agreed to sell its oil assets in Sudan to ONGC Videsh, which is a subsidiary of Oil and Natural Gas Corporation, the Indian state oil company. The sale was supposed to have been completed by the end of December 2002, but has been delayed by one month. ONGC Videsh will pay $758 million for Talisman's holdings, including a 25 per cent interest in the

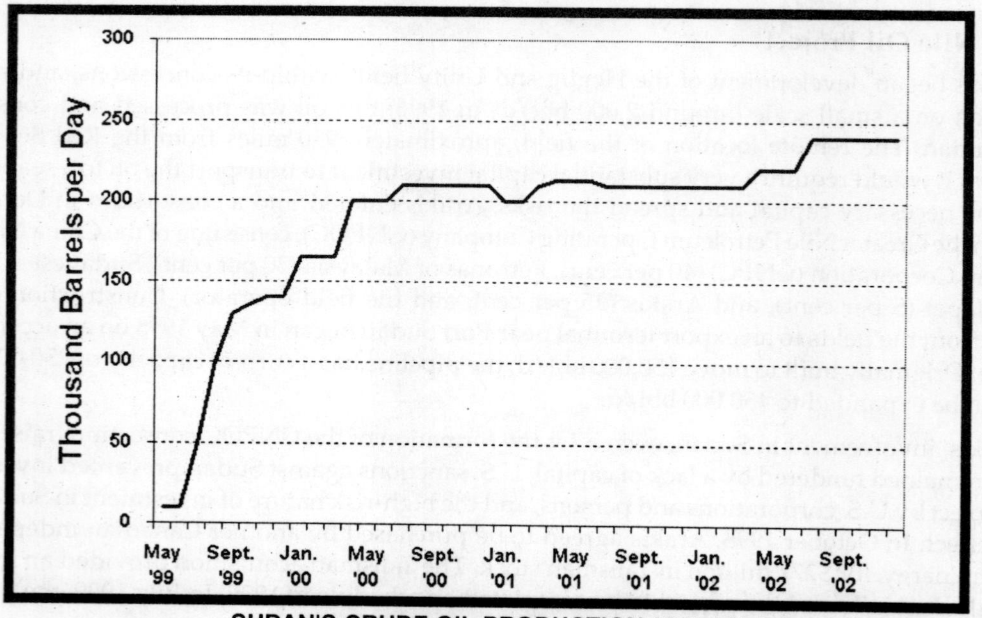

SUDAN'S CRUDE OIL PRODUCTION 1999-2002

GNPOC.

Broadened U.S. sanctions imposed against GNPOC in February 2002 reportedly had little impact on production or distribution of Sudanese oil. In September 2002, Sudan's Minister of Energy and Mining announced that a number of firms, including the British Vitol Group of Companies, had expressed a desire to invest in oil projects in Sudan. Petrel Resources (PLC) is also considering some projects in Sudan, and reported that it had a joint venture proposal with Sudapet, Sudan's state oil company, to explore a block on the Red Sea. In August 2002, a memorandum of understanding (MOU) was signed between Canadian oil firm Roland and Sudapet, to cooperate in the drilling of Sudanese oil. No specific block was mentioned in the agreement.

Petrodar, a consortium that was granted a production agreement by the Sudanese government in the Melut Basin in November 2000, is said to be ready to begin development work on blocks 3 and 7 in 2003. The consortium is currently carrying out exploration and drilling work on those blocks, which are a total of 44,700 square miles in size, and contain the Adar Yeil field, which produces 5,000 bbl/d. The consortium is comprised of Qatar's Gulf Petroleum Company (46 per cent), China National Petroleum Company International Ltd.(41 per cent), the Sudanese company Al Thani (5 per cent), and Sudapet (8 per cent).

In June 2000, Sudanese officials announced plans to begin oil exploration in northwest Sudan, the Blue Nile Basin in southeastern Sudan, and the Red Sea area in eastern Sudan. Oil exploration in Sudan previously was limited largely to the central and south-central regions, which, according to Khartoum officials, represent only 15 per cent of the national oil reserves. Sudanese Energy Ministry representatives place estimated total reserves in the country at 3 billion barrels and estimated proven reserves at 700 million barrels. Government spokesmen said that unnamed Japanese, European and Middle East companies had expressed interest in the new oil concessions.

Development of Sudan's oil resources has been highly controversial. Numerous international human rights organizations have accused the Sudanese government of financing wide-scale human rights abuses with oil revenues, including the mass displacement of civilians living near the oil fields. The SPLA has declared that it considers oil installations a "legitimate military target," as oil development has provided the Sudanese government the financial resources to expand its war effort. The SPLA says it destroyed the main oil well on the Heglig oil field in September 2002. In November 2001, southern rebels claimed to have ambushed an army convoy traveling near GNPOC facilities, and stated that such attacks would continue until "oil exploration, exploitation and development come to a halt." In August 2001, an attempt by rebels to blow up Sudan's oil export pipeline was thwarted, but rebels claimed to have killed 42 government soldiers in an attack earlier in the month, and also to have inflicted "extensive damage" to oil facilities at Heglig. The government and a representative of Talisman Energy both denied the latter claim. Rebels also claimed to have launched a successful attack on oil facilities in Bentiu in mid-October 2001, but this claim also was refuted by the government.

Refining and Downstream

Sudan has been self-sufficient in producing petroleum products (except jet fuel) since the June 2000 opening of the 50,000-bbl/d Khartoum Oil Refinery in the Jayli area, 30 miles north of Khartoum. The Khartoum refinery, built and jointly operated by CNPC, produces benzene and butane gas for domestic consumption and export, as well as gasoline for local consumption. A portion of the surplus gas eventually will be used in the production of electricity, according to government officials. The Khartoum refinery is expected to save the Sudanese government over $100 million per year in refined petroleum product imports. Following the opening of the Khartoum refinery, the price of gasoline was reduced considerably throughout the country and the price of gas cylinders, which Sudanese use for cooking, dropped from $5.30 to $2.60.

In January 2003, the Sudanese government announced plans to expand the capacity of its oil refineries, at Khartoum and El Obeid. Output at the Khartoum refinery will be increased by 80 per cent, to 90,000 bbl/d, while output at the El Obeid refinery will increase by 50 per cent, to 15,000 bbl/d. According to Sudan's Energy Minister, the country is currently conducting a study on the feasibility of expanding its third refinery at Port Sudan, which is located near the Red Sea. The Port Sudan facility is Sudan's second largest refinery and has a current capacity of 21,700 bbl/d. A petroleum products pipeline runs from the Port Sudan refinery to Khartoum. A fourth facility located in central Sudan near El Obeid has a capacity of 10,000 bbl/d.

Also in January 2003, Sudan agreed to begin providing Ethiopia with approximately 3,100 bbl/d of oil, with exports expected to begin before February 2003. Plans call for Ethiopia to eventually import as much as 60 per cent of its oil needs from Sudan, which will save Ethiopia an estimated $7 million per year. Sudan has also agreed to provide Ethiopia with approximately 15.5 square miles of land on which to build a storage depot for its fuel imports. In August, 2000, the Sudan's National Petroleum Company (NPC) announced plans to lay pipelines to supply Eritrea and Ethiopia with petroleum derivatives from the Khartoum refinery. If approved, the pipelines would pass through Sudan's Gezira, Sennar and Gedaref states. NPC is also studying the feasibility of running another pipeline to export crude oil from the Adar Yeil oil fields in southern Sudan to Ethiopia.

It was reported in March 2002 that Malaysian state company Petronas and its affiliate Engen have plans to buy Mobil Oil Sudan's oil distribution network in Sudan. Mobil Oil Sudan currently has a 20 per cent share in the country's oil product market. Petronas is already involved in Sudan's oil market through its 30 per cent share in the GNPOC consortium. Although the deal has yet to be finalized, Africa Intelligence reported that the sides were close to a successful conclusion in September 2002.

Sudan also has plans to export oil to fellow members of COMESA (the Common Market for Eastern and Southern Africa), including neighboring Kenya. Exports may be delayed, however, by concerns over human rights issues in Sudan, and some Kenyan officials have called for a boycott of Sudanese oil. In April 2002, Sudanese and Kenyan government officials announced that they are working on logistics for the construction of a new pipeline that would link oil fields in Sudan to the Kenyan port in Mombasa. Under COMESA, trade within the zone is not subject to tariffs, which means that Sudanese oil likely will be cheaper for COMESA members than other alternatives.

Electricity

Sudan currently has installed electric generation capacity of 580 megawatts (MW), managed by the state owned National Electricity Corporation (NEC).It is composed almost equally of thermal (mainly oil) and hydropower. Hydroelectric power generation varies greatly over time, due to rainfall patterns. The main generating facility is the Roseires dam located on the Blue Nile river basin approximately 500 kilometers (315 miles) southeast of Khartoum. Roseires has an installed capacity of 280 MW, but output varies greatly as water levels on the river rise and fall throughout the year.

Sudan's total electricity generation was 1.97 billion kilowatthours (kwh) in 2000. Two interconnected electric grids exist–the Blue Nile grid and the Western grid. Much of Sudan, however, is not served by the grids, and blackouts and brownouts are common. Some towns outside the grids are served by their own small-scale diesel-fired plants.

Faced with a power shortage, Sudan has plans to add additional generating capacity. The largest projects are the proposed Merowe and Kajbar hydroelectric facilities in northern Sudan. The Merowe facility will have a capacity of 1,000 MW, and will be built at the Nile's fourth cataract. Egypt has not voiced major objections about the issue of Nile water diversion, which Sudan's hydroelectric projects would entail. Construction is scheduled to begin in 2003, and to be completed in 2008 or 2009. In August 2001, the Merowe project received a boost when several Arab development funds when the Arab Fund for Economic and Social Development pledged a total of $780 million to the project. In 2002, the dam project received major funding commitments from Saudi Arabia, Qatar, and other Arab investors, each of whom pledged loans of $15 million or more. Two consortia met the October 2002 deadline to bid for the contract to build three packages of the civil works portion of the dam, which is estimated to cost $550 million. The consortia are a joint venture of China International Water & Electric and China National Water Resources & Hydropower Engineering Corporation, and a joint venture of Consolidated Contractors International Company (CCC) of Greece and Salini Costruttori of Italy.

The Kajbar Dam, located at the Nile's second cataract, is currently under construction, and will have a 300-MW capacity. An agreement to finance the Kajbar project was signed between Sudan and China in September 1997. Under terms of the agreement, China is financing 75 per cent of the project (approximately $200 million) and Sudan is to provide the remaining 25 per cent. Environmental groups have expressed concern about Kajbar Dam, citing potential damage both to the Nile ecosystem and to the culture of the displaced Nubian residents of the area.

The director of the NEC said in April 2001 that Sudan and Ethiopia had agreed to link their power grids. A related report in April 2001 indicated that Ethiopia had agreed to export power to Sudan (and Djibouti). In September 2002, Ethiopian television reported that the two nations had agreed to do what was necessary to implement previously proposed development projects through joint use of Nile waters, though the report did not give details about what was expected from each nation.

Country Overview

Independence: January 1, 1956 (from Egypt and the United Kingdom)

Population (2002E): 32.5 million

Location/Size: Northern Africa, bordering the Red Sea between Egypt (on the north); Eritrea and Ethiopia (on the east); Kenya, Uganda and the Democratic Republic of Congo (on the south); and Libya, Chad and the Central African Republic (on the west) / 2,505,810 square kilometers (967,000 square miles), slightly larger than the combined size of Texas, New Mexico, Arizona, Nevada, California, Oregon, and Washington

Major Cities: Khartoum (capital), Juba, Kassala, Medani, Nyala, El-Obeid, Omdurman, Port Sudan

Languages: Arabic (official), Dinka, Bedawi, Nuer, Fur, Hausa, Zande, English, and various other Nilo-Saharan, and Afro-Asiatic languages

Major Ethnic Groups: Arab, Beja, Kinka, Fur, Nuba, Nubian, Nuer, Zande (Azande)

Religion: Muslim (Sunni, in north) 70 per cent, traditional beliefs 25 per cent, Christian 5 per cent (mostly in south)

Economic Overview

Currency: Sudanese dinar (SD)

Market Exchange Rate (01/10/03): US$1 = 260 SD

Gross Domestic Product (GDP) (2001E): $12.5 billion

Real GDP Growth Rate (2001E): 5.4 per cent *(2002E):* 3.6 per cent

Inflation Rate (2001E): 4.9 per cent *(2002E):* 6.7 per cent

Current Account Balance (2001E): -$10.2 billion

Major Trading Partners: Saudi Arabia, China, France, Italy, Germany, Egypt, Japan

Merchandise Exports (2002E): $1.8 billion

Merchandise Imports (2002E): $1.6 billion

Merchandise Trade Balance (2002E): $0.2 billion

Major Export Products (2002): Crude oil, sesame, livestock, cotton, gum arabic

Major Import Products (2002): Machinery and equipment, manufactured goods, oil products, transport equipment, chemicals, wheat

Total External Debt (2001E): $15.3 billion

Energy Overview

Proven Oil Reserves (1/1/03E): 563 million barrels

Oil Production (2001E): 209,000 barrels per day (bbl/d) *(2002E):* 227,500 bbl/d

Oil Consumption (2002E): 33,000 bbl/d

Net Oil Exports (2002E): 194,500 bbl/d

Natural Gas Reserves (1/1/02E): 3 trillion cubic feet (tcf)

Natural Gas Production (2000E): None

Natural Gas Consumption (2000E): None

Electric Generation Capacity (2000E): 580 megawatts (49.2 per cent thermal, 50.8 per cent hydroelectric)

Electricity Generation (2000E): 1.97 billion kilowatthours (kwh)

Electricity Consumption (2000E): 1.83 billion kwh

Environmental Overview

Total Energy Consumption (2000E): 0.07 quadrillion Btu* (<0.1 per cent of world total energy consumption)

Energy-Related Carbon Emissions (2000E): 1.25 million metric tons of carbon (<0.1 per cent of world total carbon emissions)

Per Capita Energy Consumption (2000E): 2.3 million Btu (vs. U.S. value of 351.0 million Btu)

Per Capita Carbon Emissions (2000E): 0.0 metric tons of carbon (vs. U.S. value of 5.6 metric tons of carbon)

Energy Intensity (2000E): N/A

Carbon Intensity (2000E): N/A

Fuel Share of Energy Consumption (2000E): Oil (86.22 per cent), Natural Gas (0.0 per cent), Coal (0.0 per cent), Hydroelectricity (13.78 per cent)

Fuel Share of Carbon Emissions (2000E): Oil (100.0 per cent), Natural Gas (0.0 per cent), Coal (0.0 per cent)

Renewable Energy Consumption (1998E): 278 trillion Btu* (67 per cent increase from 1997)

Status in Climate Change Negotiations: Non-Annex I country under the United Nations Framework Convention on Climate Change (ratified November 19th, 1993). Not a signatory to the Kyoto Protocol.

Major Environmental Issues: Inadequate supplies of potable water; wildlife populations threatened by excessive hunting; soil erosion; desertification.

Major International Environmental Agreements: A party to Conventions on Biodiversity, Climate Change, Desertification, Endangered Species, Law of the Sea, Nuclear Test Ban, Ozone Layer Protection and Whaling.

82

Syria

With proven oil reserves expected to last only about 10 more years and a population growing at 2.5 per cent per year, Syria may become a net importer of oil within the next decade. Thus, the exploration for oil and natural gas is a top priority in Syria.

Recent Developments

On June 10, 2000, President Hafiz al-Asad died, and was replaced by his son, Bashar al-Asad, who promised to undertake a variety of economic, administrative, and political reforms. Progress in implementing economic reforms has been slow, however, due in part to the influence of holdovers from the regime of the elder Asad. Growth in real gross domestic product (GDP) was 3.3 per cent per cent in 2002, only slightly higher than the rate of population growth. Real GDP growth for 2003 is projected at 3.0 per cent. According to the Syrian government, the country's unemployment rate is under 10 per cent, although according to foreign diplomats, the rate is more likely over 20 per cent.

The ultimate goals of economic reform in Syria are to move the country away from a state-run economy and towards a more market-based, modernized one, to reduce the country's high rate of unemployment (particularly given the country's rapid population growth), and to diversify the country's economic base for the day when oil reserves run out. In recent years, Syria has moved extremely slowly and cautiously in this direction. For the most part, however, as of early 2003, large state corporations continue to control all strategic sectors, including oil, electricity, banking, and chemicals. The Syrian government announced in early 2002 that it had put on hold any immediate moves to privatize large state-owned enterprises, concentrating instead on increasing their efficiency. The main economic reform of the past year was the decision in October 2002 to allow banks to trade foreign currencies at rates closer to the "market-based" rates against the Syrian pound prevailing in neighboring Lebanon.

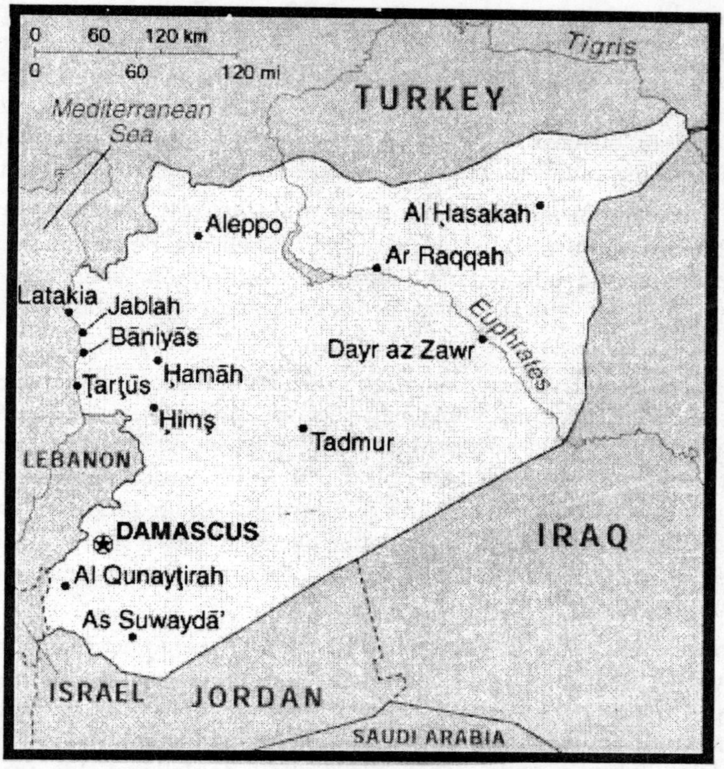

Syria's relations with Iraq have improved significantly over the past few years, particularly in the economic sphere. In June 1997, Syria reopened its border with Iraq for businessmen, and since then a series of trade delegations has traveled between the two countries. On July 14, 1998, Syria and Iraq signed a memorandum of understanding on reopening an important oil pipeline–closed in 1982–from northern Iraqi oil fields to Syria's Mediterranean port of Banias. The pipeline is now in operation. On February 27, 2000, Iraq opened an interests section in Syria–the first direct bilateral diplomatic link in over 19 years.

Oil

Syria's oil industry faces many challenges in the years to come. Oil output and production continues to decline due to technological problems and depletion of oil reserves. Since peaking at 590,000 barrels per day (bbl/d) in 1996, Syria's oil output has fallen steadily, to an estimated 525,682 bbl/d in 2002, as older fields, especially the large Jebisseh field discovered in 1968, have reached maturity. Syrian oil production is expected to continue its decline over the next several years, while consumption rises, leading to a reduction in Syrian net oil exports. If this trend continues, it is feared that Syria could become a net oil importer within a decade.

Syria hopes to reverse this trend through intensified oil and natural gas exploration and production efforts, plus a switch from oil-fired to natural-gas fired electric power plants. Syria also has opened up new blocks for oil and natural gas exploration, with the Oil and Mineral Resources Ministry receiving bids from several international companies in December 2001 on five exploration areas. Awards for these blocks were made in January 2003, with Shell receiving exploration rights in the Damascus-Palmyra area and India's ONGC Videsh receiving another onshore block. Independents Ocean Energy and Stratic Energy also received awards. Since the first round of bids closed, bids for a second round of 11 blocks closed in October 2002. Bids for a third round of 11 blocks were solicited in December 2002. The current third bid round includes five blocks from the second round for which bids were not received, as well as six new blocks. No awards have yet been made from the second round.

Syria's main oil producer (by far) is al-Furat Petroleum Co. (AFPC) a joint venture established in 1985 and owned by the Syrian Petroleum Company (SPC), Shell, and PetroCanada. AFPC's fields are located in the northeastern Syria–particularly the Deir ez-Zour region, where commercial quantities of oil were discovered in the late 1980s–and are producing about 400,000 bbl/d of high quality light crude.

AFPC's main oil field is al-Thayyem, although production there has been declining since 1991. Another important field–Omar/Omar North–began production in February 1989 at 55,000 bbl/d. Shortly thereafter, operator Shell was pressed by the cash-strapped Syrian government to step up production (against Shell's advice) to 100,000 bbl/d. The result was serious reservoir damage, and in April 1989, output plummeted to 30,000 bbl/d. Currently, Omar produces about 15,000 bbl/d from natural pressure and 30,000 bbl/d from water injection. Other AFPC fields include al-Izba (light oil), Maleh (34° API gravity oil), Sijan, and Tanak. Production from fields run by SPC peaked in the late 1970s at more than 165,000 bbl/d.

SPC's fields include: 1) Karatchuk–Syria's first discovery, located near the border with Iraq and Turkey; 2) Suwaidiyah–a giant heavy oil field located south of Karatchuk in the Hassakeh region (and extending into northwestern Iraq) which currently produces around 85,000 bbl/d; 3) Jibsah–a major field producing both oil and gas; 4) Rumailan–a small field near Suwaidiyah which produces heavy oil; and 5) Alian, Tishreen, and Gbebeh–three small, depleting fields producing heavy oil. China's CNPC signed a contract with SPC in March 2003 to undertake an enhanced oil recovery project for Gbebeh, which is to increase production from the current 4,500 bbl/d to 10,000 bbl/d.

Other Syrian oil fields include Maleh, Qahar, Sijan, Azraq, and Tanak. Jafra, discovered in late 1991 and located near Deir ez-Zour, is operated by TotalFinaElf and has current production of around 60,000 bbl/d. Besides conventional oil reserves, Syria also has major shale oil deposits in several locations, mainly the Yarmouk Valley stretching into Jordan.

Oil exploration activity in Syria has been slow in recent years due to unattractive contract terms by SPC, and poor exploration results. For these reasons, only a few companies out of more than a dozen operating in the country in 1991 remain in Syria at present. The recent bid rounds are an attempt to reverse this trend, but it is unclear how successful this will be. Officials of TotalFinaElf have publicly expressed their intention to scale down their Syrian operations in May 2002.

Refining/Downstream

Syria's two refineries are located at Banias and Homs. Total current production from these refineries is 239,865 bbl/d (132,725 bbl/d and 107,140 bbl/d, respectively). Syria is planning to construct a third refinery, with an initial capacity of 60,000 bbl/d (possibly increasing to 120,000 bbl/d), at Deir ez-Zour to supply products to the eastern part of the country. A feasibility study on this project reportedly was completed in January 1998, but it has not been implemented. In addition, Syria plans to upgrade its two current refineries, both of which are in urgent need of overhauling, to replace output of fuel oil with lighter products.

Syria markets all of its crude oil, including that produced by foreign companies, solely through state marketing company Sytrol. Prices for Syrian Light, a blend of light and sweet crudes produced primarily from the Deir ez-Zour and Ash Sham fields, and heavy Suwaidiyah crude, produced from the Suwaidiyah and Jebisseh fields, are tied to the price of dated Brent and are adjusted monthly. Since January 1994, Sytrol has had a clause in its term contracts prohibiting customers from re-selling Syrian crudes without written permission from Sytrol. This is intended to curb spot trading in Syrian crudes and especially sales to Israel. Besides crude oil, Syria also exports fuel oil and other products. Syria is a member of OAPEC (the Organization of Arab Petroleum Exporting Countries), although not of OPEC.

Syria's major oil export terminals are at Banias and Tartous on the Mediterranean, with a small tanker terminal at Latakia. Banias can accommodate tankers up to 120,000 dead weight tons (dwt), and has a storage capacity of 437,000 tons of oil in 19 tanks. Tartous can take tankers up to 210,000 dwt, and is connected via a pipeline to the Banias terminal. Latakia can handle oil tankers up to 50,000 dwt. All three terminals are operated by the Syrian Company for Oil Transport (SCOT), a sister company of SPC.

SCOT also is in charge of Syria's pipelines, including: 1) a 250,000-bbl/d export line from SPC's northeastern fields to the Tartous terminal, with a connection to the Homs refinery; 2) a 500,000-tons/year refined products pipeline system linking Homs refinery to Damascus, Aleppo, and Latakia; 3) a 100,000-bbl/d spur line from al-Thayyem and other fields to the T-2 pumping station on the old Iraqi Petroleum Company (IPC) pipeline; 4) a spur line from the al-Ashara and al-Ward fields to the T-2 pumping station.

In early March 2000, both the Iraqi and Syrian sections of the IPC pipeline, which links the Kirkuk oil fields in northern Iraq with Syria's port of Banias on the Mediterranean, were reported as ready for operation. The 552-mile, 1.1-1.4 million-bbl/d (nameplate capacity) pipeline was closed in 1982 after a break in diplomatic ties, then severely damaged during the 1991 Gulf War. Beginning in late 2000, numerous press reports indicated that Syria and Iraq had reopened the IPC pipeline, and were using it to export Iraqi crude oil (most likely Basrah Light) to Syria, where it reportedly is refined and consumed domestically, freeing up Syrian crude oil for export. These exports earn Syria extra hard currency revenues, plus provide Saddam Hussein with potentially significant revenues outside the UN "food-for-oil" program. Since Iraq remains under UN sanctions, with oil exports allowed only under the "food-for-oil" arrangement via approved export routes, the use of the Kirkuk-Banias line represents a potentially serious breach of UN sanctions against Iraq. According to the International Energy Agency's *Monthly Oil Market Report*, indications are that Iraqi oil flows through Syria ceased in January 2001, but apparently increased in early February back to 150,000 bbl/d. On January 23, 2001, the Bush administration offered to allow Iraqi oil exports through Syria as long as they were regulated by the UN food-for-oil program. In late January and early February 2002 at the UN Security Council, the United Kingdom accused Syria of importing 100,000 bbl/d or more. Syria denies the charges, but its export levels since 2001 would have been impossible if much of its own petroleum consumption needs were not supplied from Iraq. Independent analysts have put the probable volume of imports from Iraq in the range of 150,000-200,000 bbl/d.

As a replacement for the aging IPC, Syria and Iraq reportedly signed an agreement in November 2001 on building a new, $200 million pipeline from Kirkuk to Banias. The two countries also have discussed plans for a joint, 140,000-bbl/d refinery at Banias to handle the Iraqi (and possibly Syrian) crude being pumped through the pipeline

Natural Gas

Syria's proven natural gas reserves are estimated at 8.5 trillion cubic feet (Tcf). Most (around three-quarters) of these reserves are owned by SPC, including about 3.6 Tcf in the Palmyra area, 1.6 Tcf at the al-Furat fields, 1.2 Tcf at Suwaidiyah, 0.8 Tcf at Jibsah, 0.7 Tcf at Deir ez-Zour, and the remainder at al-Hol, al-Ghona, and Marqada. About half of Syria's gas is non-associated, with the rest either associated (with oil) or "cap" gas. In June 1999, a new natural gas field, called North al-Faydh, reportedly was discovered by SPC. The field reportedly has production potential of 35 million cubic feet per day (mmcf/d).

In 2001, Syria produced about 206 billion cubic feet (Bcf) of natural gas, a slight decline from the 213 Bcf produced in 2000 after a steady increase over the last decade. Syria plans to increase this production in coming years as part of a strategy to substitute natural gas for oil in power generation in order to free up as much oil as possible for export. A number of new gas-fired power projects are currently under construction or being planned. Another possible source of natural gas is imports. Syria signed agreements with Egypt, Jordan, and Lebanon in early 2001 for an onshore pipeline network (the "Arab Gas Pipeline") which would link the four countries and make Syrian imports of natural gas from Egypt a possibility. The section of the pipeline running from Egypt to Jordan currently is in the final stages of construction, but it remains to be seen whether the extension to Syria will be built. The primary original rationale for including Syria–possible exports to Turkey–looks less likely due to Turkey's economic problems. Another possible link into this network would be Cyprus, which has held talks about a possible subsea extension of the pipeline network from Lebanon or Syria.

Meanwhile, Syria itself is set to begin exporting a small quantity of natural gas to Lebanon. In May 2001, Syria signed a deal with Lebanon to build a 26-mile pipeline that would supply power stations in Lebanon with natural gas from Syria.

A key challenge for the Syrian natural gas industry is logistical, with reserves located mainly in northeastern Syria, while population is centered in western and southern Syria. SPC currently is working to increase Syria's natural gas production through several projects. The Palmyra area in central Syria is the site of much of this activity, including development of the Al Arak gas field, which came onstream at the end of 1995. Other gas fields in the Palmyra area include Al Hail and Al Dubayat–both of which are "sweet gas" and two "sour gas" fields–and Najib and Sokhne, which came onstream in 2000. Syria is attempting to expand output at Najib through its central area gas project. Foreign energy companies have been invited to submit proposals on gas projects in the Palmyra region. Several have reportedly held discussions with the Syrian government on the project, but no contracts have been awarded.

In October 1997, Syria announced discovery of a large new natural gas field in the Abi Rabah area of the Palmyra region. In addition to supplying a new (completed in 1997), 375-megawatt, power plant at Zaisoun in central Syria, the Palmyra fields have been linked with a new pipeline to Aleppo, as well as to the Tishreen power plant in Damascus and the Mhardeh power plant in Homs. Najib, the fourth and final field to be developed in the Palmyra region, started production in 2000 at a capacity of 100 mmcf/d. A modest-sized new discovery was reported in the Palmyra area in August 2002 by the Croatian company INA Naftaplin, which tested at about 9 mmcf/d.

Syria's Jibsah natural gas treatment plant, which came online in 1988, accounts for more than one-quarter of the country's total natural gas processing capacity. Jibsah's capacity was increased 88 per cent in a project completed during the first half of 1997, and now is being increased again (to 105 mmcf/d from 60 mmcf/d currently). Other Syrian gas processing plants include: the Deir ez-Zour

Gas Treatment Plant (since 1991); the Jafra Gas Separation Plant (late 1996); and the Palmyra Gas Processing Plant (late 1996).

In September 2001, several months ahead of schedule, an important new, integrated natural gas project (called "Desgas") was completed in the Deir ez-Zour region, three years since a $430-million service agreement was signed between SPC on the one hand, and Conoco and TotalFinaElf on the other. The new complex utilizes approximately 175 mmcf/d of previously-flared, "associated" (found together with oil) natural gas, in the Deir ez-Zour oil fields. TotalFinaElf and Conoco each hold 50 per cent interest in the project, with Conoco as lead operator. The Deir ez-Zour complex now includes a natural gas gathering system and processing plant, plus a 155-mile pipeline to carry 150 mmcf/d of natural gas to the grid serving western Syria. Natural gas also is being transported to the Tabiyeh oil field for processing and reinjection/enhanced oil recovery.

As increased volumes of natural gas feedstock become available, and given abundant phosphate reserves, Syria is adding capacity to produce fertilizer. At present, Syria has two nitrogenous fertilizer plants and one phosphate-based unit, both located at Homs.

Electricity

As of January 2001, total installed Syrian electric generating capacity was reportedly around 4.7 gigawatts (GW), about half of which is oil-fired, two-fifths natural-gas-fired, and 10 per cent hydroelectric. With Syrian electric power demand growing rapidly, adding electricity supply capacity is an important national priority. Also, Syria is aiming to replace its oil-fired power plants with natural-gas-fired plants, in order to free up oil for export and to avoid becoming a net oil importer in a few years. In September 1993, with Syria suffering a severe electricity shortage, and with the country's power demand growing at a rapid, 9 per cent annual rate, President Hafiz al-Asad declared that a secure supply of electricity was the right of every Syrian. Since then, existing power stations have undergone maintenance and four new generating plants have been built (including the 600-MW al-Zara gas/oil plant near Hama, completed by Mitsubishi in November 2000). Also planned are the 300-MW Zeizoun plant and the 630-MW Tishreen hydro station. Overall, Syria hopes to add 3,000 MW of power generating capacity by 2010, at a probable cost of around $2 billion, but progress toward implementing these projects has been slowed by a lack of investment capital. Foreign-owned power projects are still not under consideration.

While power generation capacity in Syria now appears adequate, the country's power distribution network remains a problem. Transmission losses are estimated as high as 25 per cent of total generated capacity due to a variety of factors including poor quality wires and transformer stations. In December 2000, the European Investment Bank (EIB) agreed to lend Syria 75 million Euro for expansion and upgrading of the country's power transmission network (in February 2001, the EIB agreed to lend another 115 million Euro). The project, scheduled for completion by 2005, will involve construction, upgrading, and expansion of sub-stations, overhead power lines, and underground cables. In addition to the EIB, funds for this project is coming from Arab Gulf states and the Syrian government. Also, in March 2001, Kuwait agreed to loan Syria $65 million to help double the size of a natural-gas-fired power plant at Nasriyeh, northeast of Damascus. Nasriyeh currently has capacity of 300 MW.

In March 2001, a project to link the electric power grids of Syria, Egypt, and Jordan was completed, with an inauguration ceremony attended by Egyptian President Husni Mubarak and Jordan's King Abdullah. Linking the three countries' grids together creates a network of approximately 45 GW. In the future, the intention is to link up with Lebanon, Turkey, and Iraq as well. A 217-mile connection to Turkey is tentatively scheduled for completion sometime in 2002. The link between Jordan and Syria is via 400-kilovolt cable, while the Jordanian-Egyptian link is via underwater cable in the Red Sea. In August, 1998 Turkey reported that its 115-mile transmission line from the Ataturk Dam to the border region with Syria was completed, but that the electricity could not be provided due to Syria's delayed construction on the connecting grid. In June 2001, Syria and Iran signed an agreement on joint cooperation in power production, transmission, and distribution.

As part of its strategy to save oil for hard currency exports, Syria has plans to build several natural gas, combined-cycle power plants, and to convert the country's major oil-fired plants to natural gas. Two of Syria's largest power stations–the Mahrada and Banias plants–have been converted from fuel oil to natural gas in recent years. Natural gas for these two plants comes from the Palmyra fields. Syria also plans to increase natural gas usage at the dual-capacity (fuel oil or natural gas) Tishreen power plant. Gas for Tishreen is to come from the Omar treatment plant. In addition to these plants, Suwaidiyah Station II had five new natural gas turbines installed in 1989, while Suwaidiyah I operates mainly on associated gas from nearby fields.

On May 19, 1999, the director-general of Syria's Atomic Energy Commission signed an agreement with Russia on cooperation in peaceful uses of nuclear power, including construction of two nuclear reactors in Syria. On February 23, 1998, Syria and Russia had signed an agreement on the peaceful use of nuclear energy, and in July, 1998 the two countries had agreed on a timetable for a 25-MW light-water nuclear research center project in Syria with the participation of Russia's Atomstroyeksport and Nikiet. The project appears to have been abandoned, however, due to lack of financing. Russia's Minister of Atomic Energy stated in January 2003 that the project is no longer under discussion.

Country Overview

Independence: 17 April 1946 (from League of Nations mandate under French administration)

Population (07/2002E): 17.2 million

Location/Size: Middle East, at eastern end of the Mediterranean Sea, between Turkey and Lebanon/71,498 sq. miles (slightly larger than North Dakota)

Major Cities: Damascus (capital), Aleppo, Latakia, Homs

Languages: Arabic (official), Kurdish, Armenian, Aramaic, Circassian, French and English widely understood

Ethnic Groups: Arab 90 per cent; Kurd, Armenian, other 10 per cent

Religion: Sunni Muslim 74 per cent, Alawite, Druze, and other Muslim sects 16 per cent, Christian (various sects) 10 per cent, Jewish (tiny communities in Damascus, Al Qamishli, and Aleppo)

Economic Overview

Economy and Foreign Trade Minister: Ghassan al-Rifai (since 12/01)

Currency: Syrian Pound

Market Exchange Rate (3/11/03): $1 = 47.8 Syrian pounds

Gross Domestic Product (GDP) (2002E): $21.0 billion

Real GDP Growth Rate (2002E): 3.3 per cent *(2003F):* 3.0 per cent

Inflation Rate (Consumer Prices, 2002E): 2.4 per cent *(2003F):* 6.4 per cent

Unemployment Rate (2002E): under 10 per cent (official statistic); 20 per cent or higher (unofficial estimate)

Merchandise Exports (2002E): $9.1 billion

Merchandise Imports (2002E): $7.8 billion

Merchandise Trade Balance (2002E): $1.3 billion

Major Trading Partners: Germany, Italy, Lebanon, France, Saudi Arabia, Japan, Spain

Major Export Products: Petroleum, fruits and vegetables, textiles, cotton

Major Import Products: Manufactured goods, machinery, food and livestock, chemicals and chemical products

Oil Export Revenues/Total Export Revenues (2002E): around 60 per cent

Energy Overview

Proven Oil Reserves (1/1/03E): 2.5 billion barrels

Oil Production (2002E): 525,682 barrels per day (bbl/d), of which 517,682 bbl/d was crude oil

Oil Consumption (2002E): 268,000 bbl/d

Net Oil Exports (2002E): 257,682 bbl/d (not counting smuggled Iraqi oil, which reportedly is in the 150,000 bbl/d-200,000 bbl/d range)

Crude Oil Refining Capacity (1/1/03E): 239,865 bbl/d

Major Crude Oil Customer: European Union

Major Ports: Latakia, Banias, Tartus

Natural Gas Reserves (1/1/03E): 8.5 trillion cubic feet (Tcf)

Natural Gas Production/Consumption (2001E): 206 billion cubic feet (Bcf)

Electric Generation Capacity (2001E): 4.7 million kilowatts (80 per cent thermal, 20 per cent hydroelectric)

Electric Generation (2001E): 23.3 billion kilowatthours (58 per cent thermal, 42 per cent hydroelectric)

Environmental Overview

Total Energy Consumption (2001E): 0.9 quadrillion Btu* (0.2 per cent of world total energy consumption)

Energy-Related Carbon Emissions (2001E): 14.0 million metric tons of carbon (0.2 per cent of world carbon emissions)

Per Capita Energy Consumption (2001E): 51.6 million Btu (vs. U.S. value of 341.8 million Btu)

Per Capita Carbon Emissions (2001E): 0.8 metric tons of carbon (vs. U.S. value of 5.5 metric tons of carbon)

Energy Intensity (2001E): 14,139 Btu/$1995 (vs U.S. value of 10,736 Btu/$1995)

Carbon Intensity (2001E): 0.23 metric tons of carbon/thousand $1995 (vs U.S. value of 0.17 metric tons/thousand $1995)

Fuel Share of Energy Consumption (2001E): Oil (65.1 per cent), Natural Gas (23.0 per cent), Coal (0.0 per cent)

Fuel Share of Carbon Emissions (2001E): Oil (78.6 per cent), Natural Gas (21.4 per cent), Coal (0.0 per cent)

Status in Climate Change Negotiations: Non-Annex I country under the United Nations Framework Convention on Climate Change (ratified January 4th, 1996). Not a signatory to the Kyoto Protocol.

Major Environmental Issues: Deforestation; overgrazing; soil erosion; desertification; water pollution from dumping of raw sewage and wastes from petroleum refining; inadequate supplies of potable water.

Major International Environmental Agreements: A party to Conventions on Biodiversity, Climate Change, Desertification, Hazardous Wastes, Nuclear Test Ban, Ozone Layer Protection, Ship Pollution. Has signed, but not ratified, Environmental Modification.

The total energy consumption statistic includes petroleum, dry natural gas, coal, net hydro, nuclear, geothermal, solar and wind electric power. The renewable energy consumption statistic is based on International Energy Agency (IEA) data and includes hydropower, solar, wind, tide, geothermal, solid biomass and animal products, biomass gas and liquids, industrial and municipal wastes.

Oil and Gas Industries

Organization: The state-owned SPC (SPC) controls all oil resources, and directly produces about one-fourth of Syrian output. Al-Furat Petroleum Company (AFPC), which is owned by SPC, Shell, and PetroCanada, is responsible for about 65 per cent of Syrian output. Sytrol is the state oil marketing company.

Major Foreign Energy Company Involvement: Conoco, PetroCanada, Shell International, TotalFinaElf

Major Oil Fields: Deir ez-Zour and Jafra in eastern Syria; Karatchuk in the far northeast

Major Refineries: Syria's two refineries are located at Homs and Banias, with crude refining capacities of 107,140 bbl/d and 132,725 bbl/d, respectively.

Major Oil Export Terminals: Banias, Tartous, Latakia

83

Taiwan

Located across the Taiwan Strait from mainland China (80 miles at the closest point), Taiwan is a leading economic and trading center, with one of the busiest ports in the world (Kaohsiung). As Taiwan lacks sufficient domestic energy sources, it is almost totally dependent on energy imports.

General Background

Taiwan's economy currently is recovering from its recession of 2001, but still has not returned to the strong economic growth of the late 1990s. After contracting by 2.2 per cent in 2001, Taiwan's real gross domestic product (GDP) increased by 3.5 per cent in 2002. Real GDP growth for 2003 is projected at 3.8 per cent. Taiwan's economy is heavily oriented toward the manufacturing of consumer electronics products, and weak demand for these products as a result of slowing growth in key export markets contributed to Taiwan's recession. Conversely, recovery in export demand in late 2002 and early 2003 has helped the country's eeconomic recovery.

Oil is by far the dominant fuel in Taiwan's energy mix, accounting for 51 per cent of total primary energy consumption. Coal also plays an important role (32 per cent of total energy consumption), followed by nuclear power (8 per cent), natural gas (6 per cent), and hydroelectric power (2 per cent). Taiwan has very limited domestic energy resources and relies on imports for most of its energy requirements. The country's industrial sector accounts for more than half of total energy demand, but this share is expected to decline slightly, since Taiwan's economy is moving toward newer, less energy-intensive industries. The transportation sector accounts for one-quarter of total energy demand.

Taiwan was admitted to membership in the World Trade Organization (WTO) in November 2001, concurrently with China's admission. Unlike China, Taiwan has been admitted to the WTO as a "developed country," which imposes more stringent requirements for reducing barriers to foreign competition. Taiwan recently has lifted some restrictions on direct trade with and investment in mainland China, which is expected to increase cross-strait commercial ties.

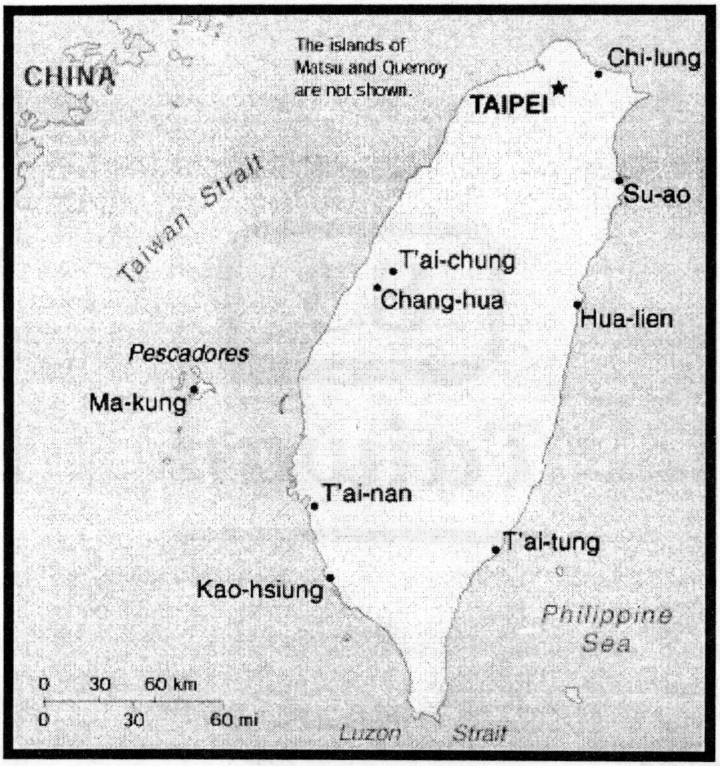

Oil

Chinese Petroleum Corporation (CPC), Taiwan's national oil company, is the dominant player in all sectors of the country's petroleum industry, including exploration, refining, storage, transportation, and marketing. However, significant competition began in July 2000 with the opening of a refinery at Mailiao owned by Formosa Petrochemical Company (FPC), a subsidiary of the private Taiwanese petrochemical firm Formosa Plastics. The first phase of production from FPC's Mailiao refinery began in mid-2000 at 150,000 barrels per day (bbl/d), and it reached its full capacity of 450,000 bbl/d in 2002.

Prior to the construction of the FPC Mailiao refinery, Taiwan imported a significant quantity of refined petroleum products. Now the country's refining capacity exceeds its domestic consumption of petroleum products, making it an exporter. The global and Asian economic slowdowns, however, have reduced demand, which has had a negative impact on Taiwanese refiners, which have been operating at somewhat below full capacity. FPC signed an export agreement with China's state-owned oil trader Sinochem in December 2002, which has opened up mainland China to petroleum products from Taiwan.

Taiwan's current crude oil production is under 1,000 bbl/d. Most of Taiwan's crude oil imports come from the Persian Gulf, though West African countries also are important suppliers for Taiwan.

To ensure against a supply disruption, Taiwan's refiners are under a regulatory requirement to maintain stocks of no less than 60 days of consumption. Refiner-held strategic petroleum stocks are the norm in Asia, and Taiwan's policy is similar to those of Japan and South Korea.

Taiwan's government has announced plans to further liberalize the petroleum sector. Taiwan's legislature passed the Petroleum Administration Act in September 2001, which removed CPC's quasi-governmental policy implementation functions, and which will permit the eventual sale of a majority stake in the firm. In January 2003, the Taiwanese government announced that it would accelerate the timetable for the sale, offering equity in the firm to private investors in 2003, with the process to be completed in 2004. Foreign firms will be allowed to acquire stakes in CPC on an equal basis with domestic investors.

Despite the lack of formal ties between Taipei and Beijing, Taiwan and mainland China have developed a cooperative relationship in the field of energy. CPC and Beijing's state-owned China National Offshore Oil Corporation (CNOOC) signed a deal in 1996 to jointly explore a 5,939-square-mile area in the Tainan Basin of the Taiwan Strait. Taipei officially ratified the deal in March 1998, and the first round of seismic surveys by the two companies was completed in October 1999. A joint venture agreement was signed between the two companies in May 2002, but drilling has yet to commence.

Natural Gas

Besides oil, CPC also is responsible for Taiwan's natural gas exploration, production and imports. Taiwan had net imports of 238 billion cubic feet (Bcf) of liquefied natural gas (LNG) in 2001–up over 9 per cent over the figure for 2000. Indonesia and Malaysia are Taiwan's two LNG suppliers. Total natural gas consumption in 2001 was 234 Bcf.

CPC operates Taiwan's only LNG receiving terminal–at Yungan, Kaohsiung. CPC anticipates an increase in natural gas demand due to the construction of additional power plants. The government plans to triple LNG consumption by 2010 for environmental reasons, as natural gas is the cleanest burning fossil fuel.

Meanwhile, Taiwan's energy sector market liberalization program has made possible the construction of competing LNG import terminals. Several private Taiwanese and Japanese firms have formed a consortium, Tung Ting Gas, to pursue an LNG regasification project in Tao-Yuan county. Construction of the terminal was started in mid-2001 by Japan's Chiyoda and Taiwan's CTCI, and it is scheduled for completion in 2006. A preliminary agreement has been signed with Qatar's RasGas for LNG supplies. Most of the natural gas from the terminal is slated to be used for electric power generation. CPC also is reportedly discussing the construction of its second LNG import terminal, which could be needed around the end of the current decade if natural gas demand in Taiwan rises as projected in the current forecast.

Coal

In Taiwan, coal is used for electric power generation as well as for the steel, cement and petrochemical industries. Taiwan ceased producing coal in 2000, and its 52.9 million short ton (Mmst) demand is met with imports, with Australia and China as major suppliers.

Electricity

Taiwan Power Company (Taipower), the state-owned electric power utility, currently dominates Taiwan's electric power sector. However, Taipower's monopoly status has been undermined by a

1994 measure which allowed independent power producers (IPP's) to provide up to 20 per cent of Taiwan's electricity.

Independent power producers are required to sign power purchase agreements with Taipower, which distributes the power to consumers. However, regulations issued by the government in July 1998 allow foreign investors to play a greater role in Taiwan's electric transmission and distribution sector. The rules raised the level permitted of foreign investment in these sectors to 50 per cent, from 30 per cent previously.

Taiwan's first major IPP, the coal-fired Mailiao plant owned by Formosa Plastics, began operation in 1999. It currently has a capacity of 2,400 megawatts (MW) in four 600-MW generating units, and sells about three-quarters of its output to Taipower. A second, coal-fired IPP plant, Ho-Ping Power, is to begin commercial operation of its first 660-MW unit in March 2002, with a second unit beginning operation in July 2002. Ho-Ping is a joint venture including Taiwan Cement Corporation and Hong Kong's China Power and Light Corporation. Several other IPP projects have been approved.

At the end of 2001, Taipower operated 70 power plants with a total installed capacity of 30,136 MW, of which 68 per cent was thermal and 17 per cent nuclear, according to Taipower's published figures. Taipower retains exclusive control over nuclear and hydropower plants. The government will continue to own these assets after Taipower is privatized. Taipower currently has 5,144 MW of nuclear generating capacity at 3 plants (Kuosheng and Chinshan stations in the north and Maanshan station in the south).

Taipower also has plans for a 4,000-MW LNG-fired complex, the Tatan Power Plant. The facility was scheduled to begin operation in late 2003, but this date will slip to 2006 due to lack of available LNG supplies pending the completion of a second import terminal.

One major controversy which has affected Taiwan's electric power over the past two years concerns the future role of nuclear power. Taiwan's Democratic Progressive Party government came into office in early 2000 promising to approve only LNG-fired power projects in the future, and to increase LNG's share of Taiwan's power generation to roughly one-third by 2010. This raised the question of what to do about the 2,700-MW Kungliao nuclear power plant, which is currently under construction. The DPP government attempted to terminate the project in October 2000, but action by the legislature and courts forced a resumption of construction in February 2001. The plant currently is scheduled to begin operation in July 2006. After that, the current government does not plan to support additional nuclear generating capacity. Some officials have reportedly started to question the designation of LNG as the "fuel of choice" for electricity generation, due to costs, which might mean the approval of additional coal-fired capacity for incremental demand after the completion of the Kungliao and Tatan projects in 2006.

A new electricity law has been under consideration in Taiwan's parliament, the major feature of which is expected to be the privatization of Taipower. Progress, though, has been slow, and the legislation is being reworked by the new Democratic Progressive Party government which came into office in May 2000. Under the basic framework envisioned, Taipower would retain a monopoly on transmission and distribution networks, but Taipower's generation assets would be split into several firms. The planned privatization is to an extent the result of rapidly rising power demand in Taiwan, and Taipower's inability to build sufficient capacity to keep pace with demand, which led to a power crisis during the summer peak-demand months in 1999. The expected date for completion of Taipower's privatization was originally 2001, but several issues, most notably the controversy over nuclear power, have pushed the expected date back to 2006.

Environment

Taiwan currently is grappling with the environmental ramifications of building one of Asia's richest economies through a decades-long commitment to economic growth. Environmental issues include: the pollution of air and water in urban areas, stores of nuclear and toxic wastes, loss of fisheries and coastal ecosystems, and an overall degradation of the country's natural landscape.

Per capita energy use in Taiwan is on par with several of its neighboring countries in Asia. However, energy intensity levels in Taiwan compared to other developed countries tend to be relatively high. This is due primarily to the country's heavy concentration of energy-intensive manufacturing industries.

Taiwan's per capita carbon emissions have more than doubled since 1980, and in 2001 represented more than five times the amount of per capita carbon emissions in China (0.65 metric tons). Compared to Taiwan's neighbors, carbon intensity (the amount of carbon consumed per dollar of GDP) in Taiwan has experienced very little growth since 1980.

Air pollution in Taiwan is most obvious in Taipei–the country's capital and largest city. The primary cause of urban air pollution in Taipei is the large number of motorbikes and scooters which are the main source of transportation for millions of the city's residents. Nuclear power plays an important role in Taiwan's energy sector, making up 17 per cent of Taipower's electric generation.

Country Overview

Area: 14,000 square miles (slightly larger than Maryland and Delaware combined)

Major Cities: Taipei (Capital), Kaohsiung, Taichung, Tainan

Major Religions: Mixture of Buddhist, Confucian, and Taoist 93 per cent, Christian 4.5 per cent, other 2.5 per cent

Languages: Mandarin (official), Taiwanese (Min), Hakka Chinese dialects

Ethnic Groups: Taiwanese, 84 per cent; mainland Chinese, 14 per cent; aborigine, 2 per cent

Armed Forces (8/98): Army, 240,000; Air Force, 68,000; Navy, 68,000

Economic Overview

Currency: New Taiwan Dollar (NT$)

Exchange Rate (4/29/03): US$1 = NT$35.1

Gross Domestic Product (2002E): $282 billion

Real GDP Growth Rate (2002E): 3.5 per cent *(2003F):* 3.8 per cent

Inflation Rate (consumer prices, 2002E): -0.2 per cent *(2003F):* 0.4 per cent

Unemployment Rate (2002E): 5.2 per cent

Total Reserves, Non-Gold (2002E): $161.7 billion

Current Account Balance (2002E): $25.7 billion

Merchandise Exports (2002E): $129.9 billion

Merchandise Imports (2002E): $105.2 billion

Merchandise Trade Surplus (2002E): $24.7 billion

Major Exports: Consumer electrionics, machinery, chemicals & allied products, iron & steel, plastics

Major Imports: Crude oil, capital goods, consumer goods, agricultural and industrial raw materials

Major Trading Partners: United States, Japan, Europe, Hong Kong (China)

Energy Overview

Proven Oil Reserves (1/1/02E): 4 million barrels

Oil Production (2002E): 3,300 barrels per day (bbl/d), of which 800 bbl/d is crude oil

Oil Consumption (2002E): 985,000 bbl/d

Net Oil Imports (2002E): 982,700 bbl/d

Crude Oil Refining Capacity (1/1/02E): 1,220,000 bbl/d

Natural Gas Reserves (1/1/02E): 2.7 trillion cubic feet (Tcf)

Natural Gas Production (2001E): 26 Billion cubic feet (Bcf)

Natural Gas Consumption (2001E): 234 Bcf

Natural Gas Imports (2001E): 208 Bcf

Coal Reserves (12/31/96): 1.1 million short tons (Mmst) (active coal production ceased in 2000)

Coal Consumption (2001E): 52.9 Mmst

Net Coal Imports (2001E): 52.9 Mmst

Electric Generating Capacity (1/1/01E): 30 gigawatts

Electricity Generation (2001E): 151 billion kilowatthours

Environmental Overview

Total Energy Consumption (2001E): 4.1 quadrillion Btu* (1.0 per cent of world total energy consumption)

Energy-Related Carbon Emissions (2001E): 68.9 million metric tons of carbon (1.1 per cent of world total carbon emissions)

Per Capita Energy Consumption (2001E): 181.5 million Btu (vs. U.S. value of 341.8 million Btu)

Per Capita Carbon Emissions (2001E): 3.2 metric tons of carbon (vs. U.S. value of 5.5 metric tons of carbon)

Energy Intensity (2001E): 12,229 Btu/$1995 (vs U.S. value of 10,736 Btu/$1995)

Carbon Intensity (2001E): 0.21 metric tons of carbon/thousand $1995 (vs U.S. value of 0.17 metric tons/thousand $1995)

Fuel Share of Energy Consumption (2001E): Oil (51.4 per cent), Coal (32.2 per cent), Natural Gas (5.8 per cent)

Fuel Share of Carbon Emissions (2001E): Oil (48.7 per cent), Coal (46.5 per cent), Natural Gas (4.7 per cent)

Renewable Energy Consumption (1998E): 110 trillion Btu* (10 per cent increase from 1997)

Major Environmental Issues: Air pollution; water pollution from industrial emissions, raw sewage; contamination of drinking water supplies; trade in endangered species; low-level radioactive waste disposal.

Major International Environmental Agreements: Taiwan is not a party to any of the selected agreements.

The total energy consumption statistic includes petroleum, dry natural gas, coal, net hydro, nuclear, geothermal, solar and wind electric power. The renewable energy consumption statistic is based on International Energy Agency (IEA) data and includes hydropower, solar, wind, tide, geothermal, solid biomass and animal products, biomass gas and liquids, industrial and municipal wastes.

Energy Industries

State Energy Companies: Chinese Petroleum Company (CPC), Taiwan Power (Taipower)

Oil Refineries (1/1/03 capacity): Kaohsiung (270,000 bbl/d), Ta-Lin (300,000 bbl/d), Taoyuan (200,000 bbl/d); Mailiao (450,000 bbl/d)

Major Ports: Kaohsiung, Keelong, Hwalien, Taichung, Suao

LNG Terminal: Yungan

84

Thailand

Thailand is a significant net oil importer. Energy consumption is growing rapidly as a result of strong economic growth.

General Background

Thailand's economy has experienced strong growth over the past year, fuelled by improved demand for Thai exports, as well as by growth in domestic demand. The country's real gross domestic product (GDP) grew 5.3 per cent in 2002, up from only 1.9 per cent in 2001. Real GDP growth for 2003 is projected at 6.4 per cent. Longer-term annual growth rates in 2004 and beyond are projected in the range of 5.8 per cent -6.6 per cent.

Still, the risks confronting Thailand's economic recovery are serious. The Thai economy is burdened by a relatively weak banking sector with a high proportion of non-performing loans. Delays in the restructuring of corporate debt also have been worrisome enough to prompt warnings from the International Monetary Fund (IMF) and international credit rating analysts. Any worldwide economic downturn could affect Thailand rapidly due to these structural weaknesses.

Thailand's energy sector is undergoing a period of restructuring and privatization. The Thai electric utility and petroleum industries, which historically have been state-controlled monopolies, are currently being restructured.

Oil

Thailand contains 583 million barrels of proven oil reserves. In 2002, Thailand produced about 193,162 barrels per day (bbl/d) of oil, an increase of about 18,000 bbl/d from the previous year. Of that production, about 127,162 bbl/d was crude oil. Most of the remainder was natural gas liquids (NGLs). Oil consumption in 2002 was 821,000 bbl/d, up from 785,000 bbl/d in 2001. Preliminary figures indicate that consumption has continued to grow rapidly in 2003, despite the Thai

government's moves to increase taxes on petroleum products.

The oil industry in Thailand is dominated by PTT, formerly the Petroleum Authority of Thailand. PTT Exploration and Production (PTTEP) is the main upstream subsidiary of PTT. Thai Oil, the country's largest refiner, is also controlled by PTT. The company underwent a partial privatization in November 2001, in which 32 per cent of its equity was sold through the Bangkok Stock Exchange. The Thai government still owns a 68 per cent stake in PTT, and does not plan to sell its controlling interest in the near future.

Despite the industry's financial problems, there have been a number of significant recent Thai oil discoveries, most notably offshore in the Gulf of Thailand. The country's modest proven oil reserves rose in 2002, from 516 million barrels in January 2002 to 583 million barrels by the end of the year, as a result of small new finds. Chevron is investing heavily in developing Block B8/32 in the Gulf of Thailand. Unocal also is investing in offshore oilfield development, and reported a significant new find in October 2000. Unocal announced in September 2003 that new investments in developing its offshore fields in the Gulf of Thailand should double current production, reaching 40,000 bbl/d by mid-2005. Seven new exploration blocks, including both onshore and offshore acreage, were awarded in May 2003 as a result of the most recent licensing round. Companies receiving exploration rights included ChevronTexaco and China's CNPC.

Refining/Downstream

Thailand has four oil refineries, with a combined capacity of 703,100 bbl/d. The three main refineries are Shell Co. Of Thailand Ltd. (275,000 bbl/d) located in in Rayong, Thai Oil Co. Ltd., in Sriracha (192,850 bbl/d), and Esso Standard Thailand Ltd. (173,500 bbl/d), also located in Sriracha.

The Thai government has been discussing a proposal to attempt to turn the country into a regional processing and transportation hub for the oil industry. One possibility would be the amendment of regulations to create a bonded-processing zone for export-oriented refineries. The targeted export market would likely be cities in south-central China, which are closer to Thai ports than to the Pacific coast of China. Another proposal would see the construction of a pipeline across the isthmus of Kra, which would allow oil shipments from the Persian Gulf to East Asia to bypass the congested Strait of Malacca.

Thailand also plans to reduce its consumption of petroleum and imports of gasoline additive methyl tertiary butyl ether (MTBE) in the future by promoting ethanol. The Thai government approved a package of tax incentives in December 2000 to encourage more production of ethanol for fuel use, but as of 2003, it is not yet being blended into gasoline in commercial quantities. Marubeni of Japan reportedly plans to build Thailand's first ethanol plant by 2005, but construction has not yet begun.

Natural Gas

Thailand contains about 13.3 trillion cubic feet (Tcf) of proven natural gas reserves, of which it produced (and consumed) 845 billion cubic feet (Bcf) in 2001, up sharply from the 2000 figure of 705 Bcf. Much of the country's natural gas is used for generating electricity. In 2001, Thailand completed its program for the conversion of almost all oil-fired electric power plants to natural gas. Demand for natural gas is expected to rise at a 6 per cent annual rate over the next five years, which represents a substantial revision downward from previous official estimates. Bongkot is Thailand's largest gas field, located 400 miles south of Bangkok in the Gulf of Thailand. Thailand began imports of gas from Burma in late 2000, used mainly at the Ratchaburi power plant. PTT also is in the process of building an extensive gas distribution network around Bangkok, which will provide fuel for power plants as well as large industrial consumers.

Thailand's economic difficulties in 1997-1998, which reduced natural gas demand, along with rising domestic production, forced the country to re-examine two natural gas deals signed with Oman and Indonesia. Planned imports of liquefied natural gas (LNG) from Oman and piped natural gas from Indonesia's Natuna gas fields, for which preliminary agreements had been signed in the mid-1990's, were delayed. Development of Thailand's domestic natural gas resources and the imports from Burma are expected to cover anticipated Thai demand for the next several years, though LNG remains a long-term option for Thailand.

Unocal Thailand is the country's largest natural gas producer, and has continued to increase its production with the development of new reserves. The Pailin gas field, which came onstream in August 1999, added 165 million cubic feet per day (MMcf/d) to Thailand's gas production. Unocal also started production at the Trat field in 1999. Unocal is undertaking a second phase of development at its Pailin field, which will eventually incease its production to around 330 MMcf/d.

Chevron is currently producing about 145 MMcf/d from its offshore Block B8/32. The company has put its estimated gas reserves in the block at 2.5 Tcf, and has plans to expand production in the future to about 250 MMcf/d. Amerada Hess reported a new onshore natural gas find in northeastern Thailand in early 2003, which currently is under evaluation. The company has projected a startup date for production from the Phu Horm field sometime in mid-to-late 2005.

The $1 billion, 416-mile Thai-Burmese natural gas pipeline, running from Burma's Yadana gas field in the Andaman Sea to an Electricity Generating Authority of Thailand (EGAT) power plant in Ratchaburi province, was completed in mid-1999. A new connecting line also has been built linking Ratchaburi to the Bangkok area, which provides for other uses for imported Burmese gas in addition to the Ratchaburi power plant.

Joint Development Area

One of Thailand's most active areas for gas exploration is the Malaysian-Thailand Joint Development Area (JDA) located in the lower part of the Gulf of Thailand, and governed by the Malaysia-Thailand Joint Authority (MTJA). The JDA covers blocks A-18 and B-17 to C-19. A 50:50 partnership between Petronas Carigali and Triton Energy Ltd. (now a subsidiary of Amerada Hess) is developing the Cakerwala field in block A-18, while PTTEP and Petronas Carigali also share equal interests in the remaining blocks. An agreement was signed in October 1999 for sales of gas from the block to PTT and Petronas, for use in both Thailand and Burma. PTT has agreed to purchase 390 MMcf/d of gas over 10 years from the Cakerawala field, the first JDA field to come on stream, beginning in 2006. Nautural gas deliveries prior to then will be solely to Malaysia. Cakerawala contains estimated reserves of 2 Tcf.

As the project has moved forward, however, it has become controversial in Thailand. The pipeline is to come ashore in Songkla province in Thailand with a connection overland to Malaysia. Strong opposition to the project developed in 2000 among residents of Songkla, who have voiced concerns about the environmental impact of the project. The Thai government announced a decision in May 2002 to proceed with construction of the pipeline, but on a slightly different route which will avoid local population centers. Construction began in mid-2003, and the project is scheduled to begin commercial operation in mid-2005.

Electricity

Thailand had 21 gigawatts (GW) of power generation capacity as of January 2001, from which it produced approximately 98 billion kilowatt-hours (Bkwh) of electricity. The decline of the Thai economy as a result of the Asian financial crisis resulted in a decline in domestic demand for electricity of about 3 Bkwh in 1998, before rebounding in 1999. This situation compelled EGAT, the state-owned electricity company, to revise its electricity demand projections. EGAT postponed or delayed a number of projects including: delaying the commissioning of the third and fourth 300-MW thermal units of the Ratchaburi power complex by three years to 2004 and 2005, respectively; postponing the start-up of the second 300-MW thermal unit at the Krabi power plant from 2001 to 2005; and delaying power purchases from three Laotian projects–the lignite-fired Hongsa project and the Nam Ngum 1-2 hydro projects to 2004 and 2005, respectively. While demand growth has recovered in step with Thailand's economic recovery, EGAT decided to lower its planned generating capacity reserve from 25 per cent to 15 per cent, which has diminished the immediate need for additional generating capacity. In recent months, reserve capacity has still been over 25 per cent, though demand growth forecast at 4 per cent is expected to produce requirments for additional power plants.

The Ratchaburi power plant, Thailand's largest power project, has moved forward despite the slowdown in power demand growth. The complex eventually will have a capacity of 3,200 MW, including 1,800 MW in six combined cycle gas-fired generators and 1,400 MW in two conventional thermal units which can burn either natural gas or fuel oil. The first combined-cycle unit began operation in January 2000, and the current capacity of the plant is 2,125 MW. Ownership of the plant was transferred from EGAT to Ratchaburi Electric Generation in October 2000, and a successful initial public offering of stock was carried out, only the second IPO on the Thai market since the crisis of 1997-98.

One other independent power producer (IPP) also began operation in August 2000, Tri Energy, which has a 700-MW plant at Ratchaburi. The company is owned by a consortium including Edison Mission Energy, Texaco, and local Thai firms. Additional IPP capacity may be added later in the decade, once the power generation capacity reserve ratio declines. The Thai government has stated that it plans to eventually privatize EGAT, but it is still studying the options for structuring the privatization process. EGAT selected six firms in July 2003 to provide advice on structuring the privatization process, and current plans call for the initial public offering (IPO) of stock in EGAT to take place in March 2003.

The Thai government announced a decision in May 2002 to postpone two new coal-fired power projects in southern Thailand at Bo Nok and Hin Krut by at least two years. The two projects, originally scheduled for completion in 2002, had been met by opposition from local communities and environmentalists. The Thai government also announced its intention to assess the possibility of changing the location of the new power plants and switching to natural gas as a fuel. The question of possible overdependence on natural gas, however, is a growing concern for the Thai government.

Environment

The 1997-1998 Asian financial crisis brought an end to Thailand's economic boom and turned the spotlight on to the effects of rapid industrialization on the country's environment. Perhaps the most visible environmental side effect of Thailand's industrial development is the growing problem of air pollution, where thick smoke often chokes city streets in Bangkok. Although traffic congestion has proven to be a difficult problem to tackle, the Thai government has instituted a number of measures to address urban air pollution problems, including phasing out leaded gasoline, and Bangkok's air quality is improving. Marine pollution is another threat, with increased risks if an offshore gas pipeline linking Thailand and Malaysia proceeds as planned.

Although slowed by the 1997-1998 financial crisis, Thailand's rate of energy consumption has picked up speed once again, continuing to grow at a fast pace. Thailand's carbon emissions have mirrored the growth in energy consumption, doubling between 1990 and 2001. Thailand is a non-Annex I country under the Kyoto Protocol, meaning it is not required to reduce its carbon emissions below 1990 levels by 2008-2012, but, owing to concerns for Bangkok and the country's other low-lying coastal areas in the event of rising oceans due to climate change, the Thai government ratified the agreement in August 2002.

As with other East Asian countries, both Thailand's energy intensity and carbon intensity levels have increased over the past 15 years. Thailand hopes to reduce its carbon intensity by diversifying its fuel share of energy consumption to emphasize more renewable energy sources, chiefly solar power. A number of solar-powered projects are in development, and the Thai government has given incentives for other non-conventional, alternative energy production. Thailand's environmental outlook is improving, especially as the government conducts required environmental impact assessments (EIAs) and allows for more public participation in the development of infrastructure projects. Still, better enforcement of existing environmental laws and regulations will be necessary to boost Thailand's environmental record.

Country Overview

Independence: 1238 (traditional founding date)

Population (2003E): 64.3 million

Location/Size: Southeastern Asia/514,000 square kilometers (198,455 square miles), about twice the size of Wyoming

Major Cities: Bangkok (capital)

Language: Thai, English, ethnic and regional dialects

Ethnic Groups: Thai (75 per cent), Chinese (14 per cent), other (11 per cent)

Religions: Buddhism, 95 per cent; Muslim, 4 per cent; Other, 1 per cent

Economic Overview

Currency: Baht (Bt)

Exchange Rate (10/3/03): US$1 = Bt39.7

Gross Domestic Product (2002E): $126.5 billion *(2003E):* $140.1 billion

Real GDP Growth Rate (2002E): 5.3 per cent *(2003F):* 6.4 per cent

Inflation Rate (consumer prices) (2002E): 0.6 per cent *(2003F):* 1.8 per cent

Current Account Balance (2002E): $7.6 billion

Merchandise Exports (2002E): $66.8 billion

Merchandise Imports (2002E): $57.0 billion

Trade Balance (2003E): $9.8 billion

Major Export Products (2003): Textiles, canned food, integrated circuits, rice, tapioca, rubber, maize, precious stones

Major Import Products (2003): Food and beverages, household appliances, chemicals, base metals, machinery, fuel and lubricants

Major Trading Partners (2003): Japan, United States, Malaysia, Singapore, EU

External Debt (2002E): $52.3 billion

Energy Overview

Proven Oil Reserves (1/1/03E): 583 million barrels

Oil Production (2002E): 193,162 barrels per day (bbl/d), of which 127,162 bbl/d is crude oil

Oil Consumption (2002E): 821,000 bbl/d

Net Oil Imports (2002E): 627,838 bbl/d

Crude Oil Refining Capacity (1/1/03E): 703,100 bbl/d

Natural Gas Reserves (1/1/03): 13.3 trillion cubic feet

Natural Gas Consumption/Production (2001E): 845 billion cubic feet (bcf)

Recoverable Coal Reserves (12/31/96): 2.2 billion short tons

Coal Production (2001E): 21.6 million short tons

Coal Consumption (2001E): 27.1 million short tons

Electric Generation Capacity (1/1/01E): 21 gigawatts

Electricity Generation (2001E): 98 billion kilowatthours

Environmental Overview

Total Energy Consumption (2001E): 2.9 quadrillion Btu* (0.7 per cent of world total energy consumption)

Energy-Related Carbon Emissions (2001E): 48.5 million metric tons of carbon (0.7 per cent of world total carbon emissions)

Per Capita Energy Consumption (2001E): 46.2 million Btu (vs. U.S. value of 341.8 million Btu)

Per Capita Carbon Emissions (2001E): 0.8 metric tons of carbon (vs. U.S. value of 5.5 metric tons of carbon)

Energy Intensity (2001E): 8,126 Btu/ $1995 (vs U.S. value of 11,014 Btu/ $1995)**

Carbon Intensity (2001E): 0.14 metric tons of carbon/thousand $1995 (vs U.S. value of 0.17 metric tons/thousand $1995)

Fuel Share of Energy Consumption (2001E): Oil (55.8 per cent), Natural Gas (28.4 per cent), Coal (12.7 per cent)

Fuel Share of Carbon Emissions (2001E): Oil (56.0 per cent), Natural Gas (24.5 per cent), Coal (19.5 per cent)

Status in Climate Change Negotiations: Non-Annex I country under the United Nations Framework Convention on Climate Change (ratified December 28th, 1994). Signatory to the Kyoto Protocol (February 2nd, 1999- not yet ratified)

Major Environmental Issues: Air pollution from vehicle emissions; water pollution from organic and factory wastes; deforestation; soil erosion; wildlife populations threatened by illegal hunting

Major International Environmental Agreements: A party to Conventions on Climate Change, Endangered Species, Hazardous Wastes, Marine Life Conservation, Nuclear Test Ban, Ozone Layer Protection, Tropical Timber 83 and Tropical Timber 94. Has signed, but not ratified, Biodiversity and Law of the Sea

The total energy consumption statistic includes petroleum, dry natural gas, coal, net hydro, nuclear, geothermal, solar, wind, wood and waste electric power. The renewable energy consumption statistic is based on International Energy Agency (IEA) data and includes hydropower, solar, wind, tide, geothermal, solid biomass and animal products, biomass gas and liquids, industrial and municipal wastes.

Energy Overview

Major Foreign Oil Company Involvement: Chevron; Shell; Texaco; Total; Unocal

State Energy Companies: PTT is the state oil company. Electricity Generating Authority of Thailand (EGAT), Thailand's state electric power authority, has spun off Electricity Generating PCL (EGCOMP), but still owns a 41 per cent share. EGAT also has sold off the Electricity Generating Public Co. Ltd. of Thailand (EGCO). Ratchaburi Electric Generating Company is an IPP. Thailand has two other state-owned electric companies: the Metropolitan Electricity Authority (MEA) and the Provincial Electricity Authority (PEA).

Major Refineries (Crude oil refining capacity–bbl/d): Shell Company of Thailand (275,000); Thai Oil Co. Ltd (192,850); Esso Standard Thailand Ltd. (173,500); Petroleum Authority of Thailand (61,750)

Major Ports: Bangkok; Laem Chabang; Pattani; Phuket; Sattahip; Sri Racha; Songkla

85

Turkey

Turkey's strategic location makes it a natural "energy bridge" between major oil producing areas in the Middle East and Caspian Sea regions on the one hand, and consumer markets in Europe on the other. Turkey's port of Ceyhan is an important outlet both for current Iraqi oil exports as well as for potential future Caspian oil exports. Turkey's Bosporus Straits are a major shipping "choke point" between the Black and Mediterranean Seas. Finally, Turkey is a rapidly growing energy consumer in its own right.

Recent Developments

Turkey is currently attempting to recover from a severe economic contraction. After years of steady economic growth (3.7 per cent annually between 1991 and 2000, with a decline in 1999 due to two severe earthquakes), Turkey's economic situation deteriorated sharply in February 2001 as a devastating financial crisis forced the country to sharply devalue its currency, the lira. In addition, Turkey's inflation and unemployment soared, and real gross domestic product (GDP) fell sharply (down 7.4 per cent in 2001). Turkey's crisis was triggered in part by underlying structural weaknesses (*i.e.*, current account deficits, serious problems in the country's banking sector, political instability). The September 2001 terrorist attacks on the United States further exacerbated Turkey's problems, with the International Monetary Fund (IMF) pointing in particular to "lower export demand, loss of tourism receipts, reduced access to international financial markets, and weaker privatization and foreign direct investment prospects." For 2003, Turkey's real GDP is expected to grow by about 4.1 per cent, with an inflation rate of 24 per cent. On April 15, 2003, Turkey's Economy Minister, Ali Babacan, said that the war in Iraq had not caused major damage to Turkey's economy, despite the loss of a possible $30 billion U.S. assistance package after Turkey's parliament voted to reject deployment of U.S. troops in Turkey for the Iraq operation. In addition, the important tourism sector could be adversely affected by the war.

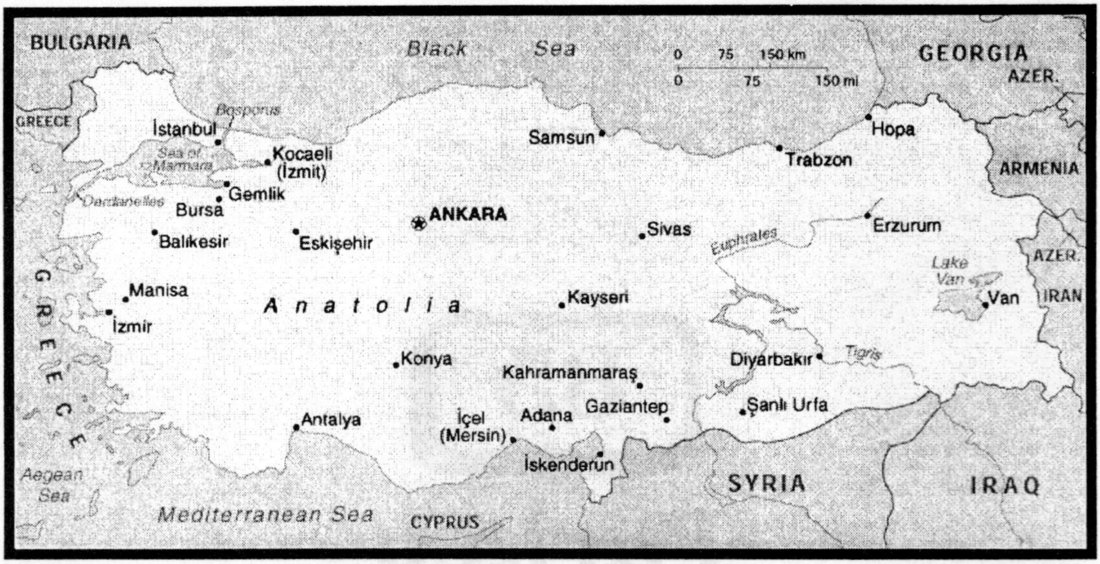

In response to Turkey's economic problems, the IMF since 2001 has been working closely with the country. An IMF assistance package begun in December 1999 ultimately disbursed around $15 billion to Turkey. This was followed by a further, $18 billion "Stand-By" assistance package, approved by the IMF on February 4, 2002 (as of April 2003, around $14 billion of this had been drawn). IMF assistance to Turkey is conditioned on implementation of a variety of reform measures aimed at addressing the root causes of the country's economic problems. Among other things, Turkey has pledged to cut state spending and subsidies, reform the country's banking sector, accelerate privatization of state-owned industries, lower the inflation rate, reduce the country's heavy debt burden, and in general create "a stable macroeconomic environment conducive to economic growth." On July 18, 2002, the IMF stated that Turkey's economic stabilization and reform program was "broadly on track," although this appears optimistic given the country's difficulties meeting its fiscal targets. In July 2002, Turkey's privatization head, Turgut Bozkurt, said that Turkey was aiming to accelerate the sell-off of energy assets in 2003.

Even prior to its recent economic crisis, Turkey faced numerous economic challenges, including: a large "underground" economy (estimated at 30 per cent -100 per cent the size of the reported economy); sharp income inequalities (between urban and rural areas in particular); low levels of private investment (Turkey hopes to increase this dramatically); a large, inefficient state sector; overly complicated legal and administrative procedures; a lack of foreign investment; and a failure to generate sufficient jobs for the country's rapidly growing population. Turkey also faces political instability, including a contentious coalition government and intense disagreement over key economic reforms required by the IMF. In addition, Turkey's desire to join the European Union (EU) has increased political debate over such issues as rights for ethnic Kurds, the death penalty and human rights, emergency rule in four eastern provinces, and democracy in general. On July 16, 2002, Turkey's governing coalition (led by Prime Minister Bulent Ecevit's Democratic Left Party–DSP) effectively lost its majority in Parliament, and new elections were held in November 2002. The outcome of these elections was a majority for the pro-islamic Justice and Development (AKP) party.

Energy

Prior to Turkey's recent severe economic difficulties, the country's energy consumption had been growing much faster than its production, making Turkey a rapidly growing energy importer. Assuming that the Turkish economy and energy demand return to a rapid growth path, Turkey will require billions of dollars worth of investments in coming years. On April 5, 2001, Turkey announced that it had ratified the Energy Charter Treaty, the international legal framework for energy investment. Also, in early 2001, the Turkish parliament passed an energy liberalization law aimed at ending the government's monopoly in the energy sector, and also geared towards attracting foreign energy investment.

Oil

In general, Turkish oil consumption has increased in recent years, although the country's recent economic recession plus price deregulation measures (which have raised the price of many oil products) since June 1999 appear to have interrupted this trend for the time being. During 2002, for instance, it appears that Turkish oil consumption and imports were down approximately 30,000 barrels per day (bbl/d) from 2000 levels. In the long-run, Turkish oil demand and imports are expected to resume steady growth. Oil provides around 43 per cent of Turkey's total energy requirements, but its share is declining (as the share of natural gas rises). Around 90 per cent of Turkey's oil supplies are imported, mainly from the Middle East (Saudi Arabia, Iran, Iraq, Syria) and Russia. Turkey's port of Ceyhan is a major outlet for Iraqi oil exports, with optimal pipeline capacity from Iraq of about 1.5-1.6 million bbl/d, but oil flows have been halted since late March 2003, following the outbreak of the Iraq war.

Turkey's oil production is accounted for primarily by three companies–the Turkish State Petroleum Company (TPAO), and foreign operators Royal Dutch/Shell (Shell) and ExxonMobil. Smaller companies include Petrom of Romania (produces around 2,600 bbl/d in the Selmo block) and Aladdin Middle East (480 bbl/d in Siirt and Gaziantep). TPAO alone accounts for about 80 per cent of the country's total oil output (currently around 51,000 bbl/d, down from 90,000 bbl/d in 1991). Turkish oil fields are generally small, and scattered around the country. Oil fields in the country's southeast (specifically the Hakkari Basin, Turkey's main oil producing area) are generally old and expensive to exploit. In addition to the Hakkari Basin, Turkey contains oil prospects in its European provinces, in the Black Sea shelf region, and in other oil basins in southern and southeastern Turkey. Potential oil reserves in the Aegean Sea have not been explored due to conflicting Greek claims over the area.

For several years, it has been reported that as much as 100,000 bbl/d of oil and oil products were being smuggled into Turkey via tanker truck, mainly from northern Iraq. This "border trade" costs the Turkish treasury millions of dollars in lost tax revenue. In March 2000, Turkey's National Security Council (MGK), concerned at lost tax revenues as well as harm to state companies Petrol Ofisis (POAS, the country's largest fuel retailer) and Tupras (which controls 85 per cent of Turkey's refining capacity), imposed controls on petroleum product smuggling from Kurdish areas of northern Iraq, Iran, Georgia, the Azeri enclave of Nakhchevan, Syria, and Bulgaria. A previous crackdown on smuggling in May 1999 reportedly had little effect. On September 18, 2001, Turkey reportedly stopped the diesel oil trade at the Habur border gate with Iraq, but allowed it to restart on January 7, 2002.

In March 2003, a major petroleum market reform bill was sent to Turkey's parliament. If enacted, the law would liberalize pricing of oil and oil products as well as integrate pipeline, refining, and distribution functions. Tupras and POAS are to be privatized as well. In 2002, POAS reportedly made a net profit of $141 million, up from $115 million in 2001. In February 2003, the Turkish government initiated a legal challenge to a proposed merger between POAS and IS Dogan Petrol Yatirimlari AS.

Pipelines

Oil and gas transportation is a crucial and contentious issue in the Caspian Sea/Central Asia regions. Turkey and the United States have pushed for a "Western route" pipeline that will carry oil from Azerbaijan's port of Baku through Azerbaijan and Georgia and then across Turkey to Ceyhan. The planned 1-million-bbl/d capacity, Baku-Tblisi-Ceyhan (BTC) "Main Export Pipeline" will stretch approximately 1,038 miles (281 miles through Azerbaijan, 135 miles through Georgia, and 622 miles through Turkey) and is expected to cost $2.8-$2.9 billion to construct. Despite initial opposition to the pipeline, which several oil companies criticized as too costly and uneconomical with the planned volumes from Azerbaijan, construction on the Turkish section of the pipeline began in June 2002. In April 2003, the BTC consortium criticized Turkey's state natural gas and pipeline company BOTAS for bureaucratic delays on the project, and called for BOTAS to be excluded from the project.

Russia, on the other hand, has promoted a "Northern route" across the Caucasus to the Russian Black Sea port of Novorossiisk. In March 2001, the Caspian Pipeline Consortium (CPC) commissioned the 990-mile, $2.5 billion, 1.34 million-bbl/d-capacity pipeline. From there, oil is transported through the Bosporus Straits. Preliminary plans are to increase exports via the CPC pipeline to 520,000 bbl/d in 2003, but the pipeline is not scheduled to reach its full capacity until about 2015. Turkey has raised concerns about the ability of the Bosporus Straits to handle additional tanker traffic that will be necessary to handle the planned volume of Kazakh oil to be exported via the CPC pipeline. Turkey has expressed its concern that the Straits, already a major chokepoint for oil tankers, cannot handle the strain of additional traffic, raising environmental concerns about a collision leading to an oil spill in the Straits. Although Kazakhstan has argued against limiting oil tanker traffic through the Straits, a number of "Bosporus bypass" options are under consideration or being developed in southeastern Europe. In addition, Ukraine already has constructed a new pipeline, the Odessa-Brody pipeline, specifically to transport oil from the Caspian Sea region to European markets.

One advantage which BTC has over other potential options for Caspian oil transport is that Ceyhan can handle Very Large Crude Carriers (VLCCs), while the ports of Supsa (Georgia) and Novorossiisk (Russia) are restricted to smaller LR-2 tankers which can transit the Bosporus. Another advantage for Ceyhan is that it can remain open all year, compared to Novorossiisk, which is closed up to two months per year due to bad weather. After failing to come to agreement with other energy companies to join the sponsor group, in March 2002 the Azerbaijan State Oil Company (SOCAR) reduced its stake in the pipeline project to 25 per cent, distributing 20 per cent among other group members. In June 2002, SOCAR sold an additional 5 per cent share to TotalFinaElf (France-Belgium), but rejected a proposal from ChevronTexaco to join the sponsor group. At the end of June 2002, the head of the sponsorship group, Michael Townshend of BP, said that the pipeline ownership group was complete. Shares in MEPCO are as follows: BP (38.21 per cent), SOCAR (20 per cent), Unocal (9.58 per cent), Statoil (8.9 per cent), TPAO (7.55 per cent), TotalFinaElf (5 per cent), ENI (5 per cent), Itochu (3.4 per cent), and Saudi Delta Hess (2.36 per cent).

Refining/Downstream

Turkey has refining capacity of 719,275 bbl/d at 6 refineries. Refining and other downstream operations in Turkey are dominated by partly-state-owned company Tupras, which has four main refining complexes: Batman in the southeast, Aliaga near Izmir, Izmit near Istanbul (the country's largest refinery, damaged during the August 1999 earthquake), and the Central Anatolian Refinery at Kirikkale near Ankara. In 2002, Tupras' share of the Turkish fuels and lubricants market was around 78 per cent, with other major retailers including BP, ExxonMobil, TotalFinaElf, Agip, and ConocoPhillips. Tupras is planning a fifth refinery–a $700-$800 million facility near Yarimca in western Turkey–to be completed by 2007. Tupras has a modernization program designed to switch output at its refineries towards lighter products. Turkey's sole private refinery is ATAS, near Mersin on the Mediterranean coast, a joint venture of Mobil (51 per cent), Shell (27 per cent), BP Amoco (17 per cent), and local company Marmara Petrol ve Rafineri Isleri AS (5 per cent).

In July 2002, Turkey's government announced that it would sell its 25.8 per cent share in POAS to the majority shareholder, Is Dogan Petrol Yatirimlari AS. The announcement came amidst calls by the IMF for an acceleration in Turkey's privatization process. In a related development, Turkey's privatization agency stated in early July 2002 that the government hoped to privatize most of the country's energy sector during 2003.

Natural Gas

Consumption and Production

Turkey consumed 520 billion cubic feet (Bcf) of natural gas (nearly all imported) in 2000, accounting for around 17 per cent of Turkey's total energy consumption (Turkish gas consumption in 2002 is estimated at around 700 Bcf). Prior to Turkey's recent severe economic problems (plus price deregulation moves), Turkish natural gas demand had been projected to increase extremely rapidly in coming years, with the prime consumers expected to be natural-gas-fired electric power plants and industrial users. Now, however, state natural gas and pipeline company BOTAS has revised its natural gas demand growth projections down sharply based on Turkey's economic problems, from about 1.6 trillion cubic feet (Tcf) in 2005 to only 1.1 Tcf in that year, a 37 per cent downward revision (but still higher than EIA's forecast–see graph). Many analysts now believe that, given lower Turkish natural gas consumption forecasts, only one of the main import options under development (*i.e.*, Blue Stream, Trans-Caspian Pipeline–TCP, Shah Deniz)–can be supported for some time.

This sharp downward revision in Turkey's projected natural gas demand could have significant repercussions, since Turkey already has signed contracts for far more natural gas than it is expected to need. To date, Turkey has signed deals for around 2 Tcf per year of natural gas imports beginning in 2005, more than double EIA's forecast for Turkish gas consumption in that year. Of this total, over 20 per cent is already coming from Russia via Bulgaria (studies on expanding the Russia-Bulgaria-Turkey "Main Line" are underway), 17 per cent from Iran, and 9 per cent from Algeria and Nigeria combined as liquefied natural gas (LNG). Turkey has one 112-bcf/year capacity LNG terminal, adjacent to the existing Marmara Ereglisi combined cycle gas turbine power station.

In the future, around one-fourth of Turkey's gas imports are to be supplied from Russia via the Black Sea (see "Blue Stream" below), another quarter from Turkmenistan (beginning in 2005), and about 10 per cent from Azerbaijan (beginning in 2005). Under the "take-or-pay" provisions of natural gas supply contracts with countries like Iran and Russia, Turkey theoretically could be forced to pay cash penalties of up to $1 billion per year if it fails to purchase contracted gas. In this context, Turkish energy officials have discussed the possibility of storing surplus natural gas in underwater depots beneath the Sea of Marmara.

Natural gas is Turkey's preferred fuel for new power plant capacity for several reasons: environmental (gas is less polluting than coal, lignite, or oil); geographic (Turkey is located near to huge amounts of gas in the Middle East and Central Asia); economic (Turkey could offset part of its energy import bill through transit fees it could charge for oil and gas shipments across its territory); and political (Turkey is seeking to strengthen relations with Caspian and Central Asian countries, several of which are potentially large gas exporters). The United States, among others, has been encouraging Turkey to utilize its unique geographical position to become a major transit center for natural gas from the Caspian/Central Asia to Europe. At the same time, however, Turkey's reliance on Russia for gas imports could reach 70 per cent or higher, seemingly undercutting Turkey's goal of diversifying its fuel suppliers. In April 2003, Turkish Energy Minister Hilmi Guler told Parliament that the government had a "strategic goal" of sharply reducing Turkey's reliance on Russian natural gas–from 70 per cent now to 30 per cent within five years. This goal appears to conflict, however, with the volume of Russian gas already contracted for via Bulgaria and "Blue Stream."

Turkish natural gas production in 2000 (23 billion cubic feet–Bcf) met around 4 per cent of domestic natural gas consumption requirements. Major natural gas producers in Turkey include Arco, TPAO and Shell. Marmara Kuzey (North Marmara), which came onstream in May 1997, is the country's largest non-associated gas field. Marmara Kuzey is located offshore in the Thrace-Gallipoli Basin of the Sea of Marmara. In March 2002, the Gocerler natural gas field was officially opened, 16 months after its discovery in the Thrace basin. Production potential is estimated to be as high as 6.6 Bcf per year. Also, in July 2001, TPAO announced that it had found gas in the Mersin and Iskenderun Bays in Turkish areas of the Mediterranean. Currently, most Turkish associated gas is reinjected into oilfields as part of an Enhanced Oil Recovery (EOR) system.

"Blue Stream" Pipeline

In October 2002, a twin 866-mile natural gas pipeline running from Russia under the Black Sea to Turkey was completed, with natural gas flows starting in February 2003, about one year behind the original schedule. The $3.2 billion "Blue Stream" pipeline runs from Izobilnoye in southern Russia, to Dzhugba on the Black Sea, then under the Black Sea for about 247 miles to the Turkish port of Samsun, and on to Ankara.

In March 2003, Turkey halted gas imports from Blue Stream for six months, with a BOTAS official stating simply that "we don't need the gas right now." According to the International Oil Daily,

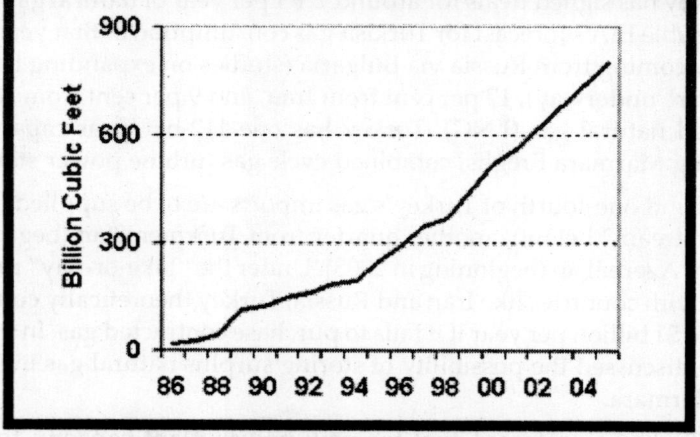

TURKEY'S NATURAL GAS DEMAND 1986-2004

Turkey's natural gas purchases from Blue Stream in 2003 will be only about 35 Bcf, just one-fourth the originally planned volume. By 2009, Blue Stream had been expected to reach peak capacity of 565 Bcf per year, but this is now somewhat doubtful given Turkey's lower gas demand forecasts. Over the course of the 25-year agreement signed in December 1997, Turkey was to import 14.1 Tcf of natural gas from Russia via Blue Stream, with the pipeline eventually extended to other Mediterranean countries, including Greece.

Other Natural Gas Import Deals

In late January 2002, Iran and Turkey officially inaugurated a much-delayed natural gas pipeline link between the two countries. This follows several years of delays due to economic, political, and technical factors. In 1996, Iran and Turkey had signed a $20 billion agreement that called for Iran to supply Turkey with more than 8 Tcf of natural gas over a period of 22 years beginning in late 1999. Officials in Turkey and Iran variously blamed U.S. sanctions, financing problems on the Turkish leg of the $1.9 billion pipeline, economic recession in Turkey, and delays by the Iranians in completing an important metering station for delaying the project. Exports of Iranian natural gas to Turkey are expected at about 105 Bcf in 2002, rising to 350 Bcf per year by 2007. There are questions, however, whether Turkish demand will grow rapidly enough to absorb this volume of natural gas from Iran, in addition to gas slated to be supplied by Russia, Algeria, and Nigeria. In June 2002, for instance, Turkey halted natural gas imports from Iran, citing problems with "gas quality," even though a lack of demand on the Turkish side appeared much more likely. On November 13, 2002, Turkey announced that it had resumed gas imports from Iran after reportedly securing a lower price and a reduction in the "take-or-pay" percentage.

If Turkish demand does not support the level of natural gas imports for which it has contracted (from Iran and others), Turkey could become an important transit center for natural gas exports to Greece and beyond. Along these lines, Greece and Turkey signed a binding agreement in late February 2003 which calls for extending the natural gas pipeline from Iran to Turkey into Greece. Reportedly, the 177-mile-long pipeline would connect Ankara to Alexandroupolis in northern Greece, would supply around 18 Bcf of gas per year, and would cost $250-$300 million. After that, natural gas could be transported to Europe via Bulgaria or via an undersea pipeline to Italy, where gas demand–especially for electric power generation–is expected to grow rapidly in coming years. A deep water option could be extremely expensive, however, making an overland route more likely. Also, in December 2002, Hungary's oil and gas company MOL signed a deal to study construction of a gas pipeline from Turkey to Austria, through Bulgaria, Romania and Hungary.

On May 21, 1999, Turkey's state natural gas and pipeline company BOTAS signed an agreement on building a $2-$2.4 billion, 1,050-mile, gas pipeline from Turkmenistan, underneath the Caspian Sea, across Azerbaijan and Georgia (both of which would collect transit fees), and on to Turkey. Gas deliveries of 565-1,060 Bcf per year are possible, with additional gas possibly being sent onwards to Europe. The consortium is led by U.S. company Bechtel and includes General Electric, Shell, and PSG International. In mid-July 1999, a top Turkish energy official stated that the Trans-Caspian Gas Pipeline (TCGP) from Turkmenistan was still the preferred option for Turkey despite the potentially huge (as high as 35 trillion cubic feet–Tcf) Shah Deniz gas field in Azerbaijan, which is located hundreds of miles closer (and on the western side of the Caspian Sea) to Turkey than Turkmenistan. Currently, however, progress on the TCGP appears stalled indefinitely, with the international consortium essentially having suspended operations, and with Turkey already oversupplied with gas from Iran and Russia ("Blue Stream").

After months of negotiation and delay, Azerbaijan and Turkey signed a long-term natural gas purchase and supply contract on March 12, 2001 (granted final approval on the Turkish side in February 2003). Starting in 2006, two years later than the original target date, Azerbaijan is to deliver 70 Bcf of natural gas to Turkey, rising to 177 Bcf in 2007 and around 230 Bcf per year from 2008 through 2020. Natural gas for the deal is to come mainly from Azerbaijan's $3.2 billion, BP-led Shah Deniz Phase I field development project (given the financial green light in February 2003). To transmit the gas, the $900 million South Caucasus (Baku-Tbilisi-Erzurum) pipeline would stretch some 630 miles, including 290 miles in Azerbaijan and approximately 170 miles in both Georgia and Turkey. Like Turkey's other possible gas deals, the pipeline from Shah Deniz has been called into question by lack of gas demand in Turkey, and now may not be built at all.

Egypt, with huge gas reserves of its own, represents yet another possible source of gas for Turkey, either by pipeline or via LNG tanker. This latter option would include construction of a $1.2 billion liquefaction terminal near Port Said on the Mediterranean coast, and a regasification facility at Aliaga (near Izmir) in Turkey. Egypt and Turkey signed a preliminary agreement for LNG exports in 1996, but analysts have raised serious questions about whether the project is economically feasible. Also, as mentioned above, Turkey already has committed to buying more gas than it probably needs for years to come.

Other Natural Gas Issues

In 2001, Turkey passed legislation which will abolish BOTAS' monopoly, separating the company into units for natural gas import, transport, storage, and distribution by 2009. At that point, the various components (except for transport) are to be privatized. In the meantime, BOTAS is to sell off 10 per cent of its market share every year, eventually getting down to 20 per cent. To date, however, progress in these areas has lagged, with–among other things–parliamentary delays in approving an amendment easing restrictions on foreign players in Turkey's gas sector. A bill introduced in March 2003 would have BOTAS focusing on gas transmission activities and moving away from purchasing gas.

In July 2002, TPAO said that it would begin negotiations with Germany's Lurgi Oel Gas Chemie on building Turkey's first natural gas storage unit. The facility would be located 50 miles west of Istanbul on the Marmara Sea coast, and will include equipment to for gas purification. Meanwhile, BOTAS reportedly plans to build another gas storage plant at Salt Lake (Tuz Golu) in central Anatolia.

In February 2002, faced with strong public opposition to a 200 per cent increase in natural gas prices, the Turkish government ordered a price cut of 6 per cent for households and 10 per cent for businesses.

Coal

Turkey has hard coal (anthracite and bituminous) reserves of around 1.1 billion short tons, plus lignite reserves as high as 8 billion short tons. Around 40 per cent of Turkey's lignite is located in the Afsin-Elbistan basin of southeastern Anatolia, while hard coal is mined only in one location–the Zonguldak basin of northwestern Turkey. Turkey produced 74 million short tons (Mmst) of coal (mainly lignite) in 2000, and consumed 91 Mmst. Between 1990 and 2000, the number of workers in Turkey's coal sector fell from 63,993 to 35,665. Turkish coal is generally of poor quality and highly polluting. It is used mainly for power generation

Electricity

With a young and growing population, low per capita electricity consumption, rapid urbanization and–until recently–strong economic growth, Turkey for nearly two decades has been one of the fastest growing power markets in the world. Prior to Turkey's recent economic difficulties, projections by Turkey's Electricity Generating and Transmission Corporation (TEAS), a public company which owns and operates 15 thermal and 30 hydroelectric plants generating 91 per cent of Turkey's electricity, had indicated that rapid growth in electricity consumption would continue over the next 15 years. Now, though, power demand growth looks much weaker, with demand hit hard by Turkey's 2001 economic crisis. Still, the government anticipates the need for significant increases in power generating capacity through 2020, requiring billions of dollars in foreign investment. The IMF and World Bank are pressing Turkey to move towards rapid privatization of the country's power sector in order to increase efficiency and to relieve pressure on the government's budget situation.

A major dilemma now faced by Turkey is how to invest in new electric power capacity while at the same time adhering to foreign debt ceilings mandated under lending rules set by the IMF. Conventional financing of major infrastructure projects would only increase the amount of foreign credit, thus MENR has conceived other options for financing projects. One option used until now has been the so-called Build, Operate and Transfer (BOT) model, under which private investors build and operate private sector generation facilities for a set number of years, at which point they transfer ownership to the state. First introduced in 1984 (under Law 3096) by then Prime Minister Turgat Ozal, BOT projects have been plagued by legal problems, which has slowed their implementation. In January 2001, the Turkish energy ministry announced 29 BOT power projects, worth $1.5 billion, that are to be the last benefiting from treasury guarantees. The projects are mainly wind and small hydro facilities, with a combined generating capacity of 1,379 MW.

In February 2001, Turkey passed the long-anticipated Electricity Market Law, which paves the way for a free market in power generation and distribution in the country. Among other things, the legislation (which President Sezer signed into law in July 2001) calls for: 1) TEAS to be broken up into separate generation, distribution, and trade companies; 2) trade and generation companies to be privatized, while transmission remains in state hands; and 3) a new regulatory board to be set up which will oversee the Turkish power market, set tariffs, issue licenses, and prevent uncompetitive practices. The new law throws into doubt the fate of dozens of BOT and TOR (transfer-of-operating-rights) power projects. In May 2002, the Energy Ministry put six power plants (Orhaneli, Hamitabat, Catalagzi, Soma A-B, and Ergil) and nine distribution grids on sale, transferring their assets to the country's privatization authority. Overall, however, progress has been slower than expected in implementing the 2001 Law.

Germany's Siemens AG is leading a consortium of companies in building a $1.45 billion, 1,300-MW, coal-fired power plant near Iskenderun, in southern Turkey. The plant is scheduled for completion in 2003 and is to burn imported coal. Aside from this large coal-fired facility, Turkey is mainly focused on increased natural gas use for thermal electric power production. In September 2002, two new combined cycle gas-fired power plants (the 770-MW Adapazari and 1,540-MW Gebze facilities) came online, ahead of schedule. A third, 1,525-MW, gas fired plant was scheduled to come online in December 2002 at Izmir. Finally, Tractabel is scheduled to complete a 763-MW gas-fired plant at Baymina, near Ankara, by late 2003. Several pipeline projects have been proposed to supply gas to these facilities, as well as several LNG terminals. In addition, BOTAS is expanding its natural gas transmission network along the Black Sea and the Aegean.

In addition to increasing domestically generated electricity through construction of new power plants, Turkey is looking outside its borders to help meet the country's growing power demand. In May 1999, for instance, Turkish and Turkmen officials reached agreement on power supplies from Turkmenistan. Turkey already is importing around 2.5-4 billion kilowatthours (bkwh) from Bulgaria annually, and has signed a memorandum with other Black Sea Economic Cooperation (BSEC) members to look into creation of a regional power grid. Turkey also imports power from Russia (via Georgia) and Iran. In September 2000, Turkey reportedly stated its desire to increase its power imports from Bulgaria to 5 bkwh by 2005. In April 2003, however, Turkey announced that it was unilaterally terminating power deliveries from Bulgaria, after declaring that Bulgaria had not met its obligations under a 1998 bilateral, 10-year energy agreement. Also, in March 2003, Turkey's energy minister said that Turkey could produce its own electricity more cheaply, and therefore would cease importing power from Bulgaria in 2006.

Turkey has significant hydroelectric power resources (more than 104 total plants, installed capacity over 10.2 GW), and is developing a great deal more, especially as part of the $32 billion Southeast Anatolia–GAP–hydropower and irrigation project. When completed, GAP, which is considered one of the most ambitious water development projects ever undertaken, will include 21 dams, 19 hydro plants (with around 7.5 GW of power generating capacity), and a network of tunnels and irrigation canals. Major Turkish hydro dams as part of the GAP include: Ataturk (2,400 MW capacity); Karakaya (1,800 MW); Ilisu (1,200 MW; the largest hydro project on the Tigris River, but highly controversial due to environmental concerns); Cizre (240 MW); Silvan/Kayser (240 MW); Hakkari (208 MW); Alpaslan II (200 MW); Batman (198 MW); Konaktepe (180 MW); and Karkamis (180 MW).

In July 2000, the Turkish government decided to abandon a planned, but oft-delayed, $4 billion, 1,300-MW nuclear power plant. Three international consortia (AECL of Canada, Westinghouse-Mitsubishi of the United States and Japan, and NPI of France and Germany) had submitted bids to build the plant, which would have been Turkey's first nuclear plant. The project was to have been turnkey and would have been located at Akkuyu, on the southern Mediterranean coast. Reportedly, the plant was killed for financial reasons, although there also had been opposition from environmental and anti-nuclear groups, as well as neighboring countries like Greece. Prime Minister Ecevit said that Turkey was not abandoning nuclear power completely, and would consider building the plant in 10-20 years, particularly if nuclear technology improves.

Turkey is considered to have a large amount of wind, geothermal, and solar power potential. In January 2001, Turkey announced approval for 17 wind and one geothermal BOT power plants. Currently, wind power capacity in Turkey is around 19 MW, with units located all over the country. Potential for wind power may be as high as 120,000 MW, with particularly attractive areas for wind located along Turkey's west coast and in southeastern Anatolia. Solar energy is mainly used for roof-top hot water. Geothermal energy potential is estimated at around 35 GW.

Environment

Turkey's explosive economic growth in the mid-1990s had significant repercussions on the country's environment. Economic growth and energy consumption have gone hand-in-hand, and the effect has been an increasing air pollution in cities that are already suffering from high pollution levels. Although Turkey is beginning to take steps to improve air quality (including a switch towards unleaded gasoline by 2005), the increased number of automobiles on Turkish streets is hampering this effort.

Of special concern to Turkey is the threat of marine pollution, especially from oil transport through the narrow Bosporus Straits. The 12-mile passage is already one of the most difficult in the world to navigate, and increased shipping–from oil and gas imports flowing into Turkey, as well as increased Russian shipping from the Black Sea through the Straits to world markets–raise the possibility of an accident. Collisions in the Straits have resulted in large oil spills, and additional oil shipping from the Caspian Sea region via the Black Sea and the Bosporus could put the Istanbul area at further environmental risk.

Industrial production has meant that Turkey's carbon emissions are on the rise, and Turkey is not a party to the U.N. Framework Convention on Climate Change. Compared to other International Energy Agency countries, Turkey's energy and carbon intensities are low, but per capita energy consumption and per capita carbon emissions are trending upwards.

Turkey has substantial renewable energy resources–especially hydroelectric power–and it is currently constructing a series of dams and hydroelectric power plants. As Turkey looks towards possible European Union membership, it will need to continue utilizing this cleaner energy as a means to achieve sustainable economic development. Turkey also has a great degree of potential for energy efficiency improvements.

Country Overview

Independence: October 29, 1923 (successor state to the Ottoman Empire)

Population (July 2002E): 67.3 million

Location/Size: Southwest Asia/780,580 sq. km (301,930 sq. mi.), slightly larger than Texas

Major Cities: Ankara (capital), Istanbul, Izmir, Adana

Languages: Turkish (official), Kurdish, Arabic

Ethnic Groups: Turkish (80 per cent), Kurdish (20 per cent)

Religions: Muslim (99.8 per cent, mostly Sunni), other 0.2 per cent

Defense (8/1/98): Army (525,000), Navy (51,000), Air Force (63,000), Coast Guard (2,200), Reserves (378,700)

Economic Overview

Currency: Turkish lira (TL)

Market Exchange Rate (5/1/03): US$1=1,588,256 TL (compared to US1$=640,260 TL on 8/3/00)

Gross Domestic Product (GDP) (2002E, market exchange rates): $182.9 billion

Real GDP Growth Rate (2001E): -7.4 per cent *(2002E):* 4.9 per cent *(2003F):* 4.1 per cent

Consumer Price Inflation Rate (2001E): 54.4 per cent *(2002E):* 45.0 per cent *(2003F):* 24.4 per cent

Unemployment Rate (2002E): 7.6 per cent *(2003F):* 6.9 per cent

Current Account Balance (2001E): $3.2 billion *(2002E):* -$3.0 billion

Major Trading Partners: Germany, United States, Italy, France, United Kingdom, Russia

Merchandise Exports (2002E): $39.3 billion (around half going to the EU)

Merchandise Imports (2002E): $48.3 billion

Merchandise Trade Balance (2002E): -$9.0 billion

Major Export Products: Agricultural, textiles, iron, steel

Major Import Products: Oil, machinery, chemicals, iron, steel

Foreign Debt (1/03): $58.6 billion

Energy Overview

Proven Oil Reserves (1/1/03E): 300 million barrels

Oil Production (2002E): 50,674 barrels per day (bbl/d) of which 46,674 bbl/d was crude oil

Oil Consumption (2002E): 635,000 bbl/d

Net Oil Imports (2002E): 584,326 bbl/d

Crude Oil Refining Capacity (1/1/03E): 719,275 bbl/d

Natural Gas Reserves (1/1/03E): 300 billion cubic feet (Bcf)

Natural Gas Production (2001E): 11 Bcf

Natural Gas Consumption (2001E): 563 Bcf (more than triple the 150 Bcf consumed in 1991)

Net Natural Gas Imports (2001E): 552 Bcf

Coal Production (2001E): 72.6 million short tons (Mmst)

Coal Consumption (2001E): 81.1 Mmst

Net Coal Imports (2001E): 8.5 Mmst

Estimated Recoverable Coal (2001E): 4.1 billion short tons

Electric Generation Capacity (2001E): 27.3 gigawatts (59 per cent thermal, 41 per cent hydroelectric)

Net Electricity Generation (2001E): 116.6 terawatthours (Twh)

Electricity Consumption (2001E): 112.6 Twh (up sharply from 54.0 Twh in 1991, but down from 114.0 in 2000)

Environmental Overview

Total Energy Consumption (2001E): 2.9 quadrillion Btu* (0.7 per cent of world total energy consumption)

Energy-Related Carbon Emissions (2001E): 50.1 million metric tons of carbon (0.8 per cent of world total carbon emissions)

Per Capita Energy Consumption (2001E): 42.1 million Btu (vs. U.S. value of 341.8 million Btu)

Per Capita Carbon Emissions (2001E): 0.7 metric tons of carbon (vs. U.S. value of 5.5 metric tons of carbon)

Energy Intensity (2001E): 15,126 Btu/$1995 (vs U.S. value of 10,736 Btu/$1995)**

Carbon Intensity (2001E): 0.26 metric tons of carbon/thousand $1995 (vs U.S. value of 0.17 metric tons/thousand $1995)**

Fuel Share of Energy Consumption (2001E): Oil (43.0 per cent), Coal (26.2 per cent), Natural Gas (20.3 per cent)

Fuel Share of Carbon Emissions (2001E): Oil (44.2 per cent), Coal (38.8 per cent), Natural Gas (16.9 per cent)

Number of People per Motor Vehicle (1998): 12.3 (vs. U.S. value of 1.3)

Status in Climate Change Negotiations: Turkey is not a signatory to the United Nations Framework Convention on Climate Change or to the Kyoto Protocol.

Major Environmental Issues: Water pollution from dumping of chemicals and detergents; air pollution, particularly in urban areas; deforestation; concern for oil spills from increasing Bosporus ship traffic.

Major International Environmental Agreements: A party to Conventions on Air Pollution, Antarctic Treaty, Biodiversity, Desertification, Hazardous Wastes, Nuclear Test Ban, Ozone Layer Protection, Ship Pollution, Wetlands and Whaling. Has signed, but not ratified, Antarctic-Environmental Protocol and Environmental Modification. Has not signed the UN Framework Convention on Climate Change.

The total energy consumption statistic includes petroleum, dry natural gas, coal, net hydro, nuclear, geothermal, solar, wind, wood and waste electric power. The renewable energy consumption statistic is based on International Energy Agency (IEA) data and includes hydropower, solar, wind, tide, geothermal, solid biomass and animal products, biomass gas and liquids, industrial and municipal wastes.

Oil and Gas Industries

State Oil Company: Turkish State Petroleum Company (TPAO)

State Refining Company: Turkish Petroleum Refineries Corporation (Tupras)

State Pipelines and Gas Agency: BOTAS

State Oil Products Retailer: Petrol Ofisi AS (POAS)

Major Ports: Iskenderun, Istanbul, Mersin, Izmir

Major Oil and Gas Fields: Bati Raman, Karakus, K. Karakus

Major Pipelines: Turkey-Iraq; Turkey contains 1,078 miles of crude oil pipelines, 1,439 miles of oil product pipelines, and 439 miles of natural gas pipelines

Major Refineries (crude oil capacity): Izmit (251,600 bbl/d), Aliaga-Izmir (226,440 bbl/d), Kirikkale (113,200 bbl/d), Mersin (100,000 bbl/d), Batman-Siirt (22,015 bbl/d); Kahramanmaras (6,000 bbl/d)

86

Ukraine

Ukraine is important to world energy markets because it is a critical transit center for exports of Russian oil and natural gas to Europe, as well as a significant energy consumer in its own right.

General Background

The Ukrainian economy grew by roughly 5 per cent in 2002 to $39.8 billion, marking the country's third consecutive year of economic growth following an eight-year post-Soviet recession. Economic growth has been fueled primarily by growing industrial and agricultural output- exported both eastwards to Russia and westwards to Europe. This unique geographic position, linking East and West while also holding critical warm water ports on the Black Sea, has made Ukraine a trade link of growing importance between the former Soviet Union and Europe.

Accordingly, Ukraine has undertaken efforts to synchronize its trade policies with its neighbors and trade partners in both directions. The country has been engaged in a partnership agreement with the European Union (EU) since 1998, and the two sides are currently considering a draft action plan which Ukraine hopes will pave the way for eventual integration with the EU. Meanwhile, in September 2003, Ukraine joined Russia, Belarus and Kazakhstan in creating a "single economic area" designed to coordinate the countries' trade regulations and reduce tariffs. While some observers have criticized this dual-track economic strategy as contradictory, Kiev insists that the newly formed post-Soviet economic bloc will compliment its EU integration strategy and thereby serve the national interest.

Leonid Kuchma was elected President of Ukraine in July 1994, and was re-elected to a second term in November 1999. In recent years the Kuchma government has been the subject of increasing scrutiny from the U.S. State Department and international observers such as the Organization for Security and Cooperation in Europe (OSCE) and Amnesty International, owing to perceived threats to free speech and human rights in the country. In August 2002, President Kuchma proposed political

reforms designed to transfer more political power from the presidency to the country's unicameral legislative body, the Supreme Council, or "Verkhovnaya Rada". The country's next presidential elections are scheduled for October 2004, during which President Kuchma will be ineligible for re-election under the current Ukrainian constitution.

Oil

Ukraine has 395 million barrels of proven oil reserves, the majority of which are located in the eastern Dnieper-Donetsk basin. Although Ukraine has made efforts at exploration, particularly in its sector of the Sea of Azov, oil production has remained relatively flat since independence (see graph). Consumption, on the other hand, has fallen dramatically, from 813,000 barrels per day (bbl/d) in 1992 to 296,000 bbl/d in 2002. Despite this decline in consumption, Ukraine remains highly dependent on imported oil, most of which comes from Russia with lesser amounts coming from Kazakhstan. In 2002, net oil imports totaled roughly 212,000 bbl/d, representing 70 per cent of consumption.

Oil Transit

Ukraine's geographic location makes it an ideal corridor for oil and natural gas to transit from Russia and the Caspian Sea region to European markets. Most of the oil transited via Ukraine is Russian oil, and is sent through the 1.2-million-bbl/d Druzhba pipeline, the southern fork of which runs through Ukraine. In July 2003, the Ukrainian government approved a new agreement with Russia calling for the transit of Russian oil through Ukraine to increase by 60 per cent, to 1.6 million bbl/d, for 15 years. However, the Russian side, in a surprise negotiating tactic, has refused to sign the agreement, owing to a dispute over Ukraine's newest oil pipeline, Odessa-Brody.

Odessa-Brody Reversal Project

Ukraine also hopes to become a transit center for oil from the Caspian Sea region, which is expected to increase significantly over the decade. The leading potential conduit for this oil in Ukraine

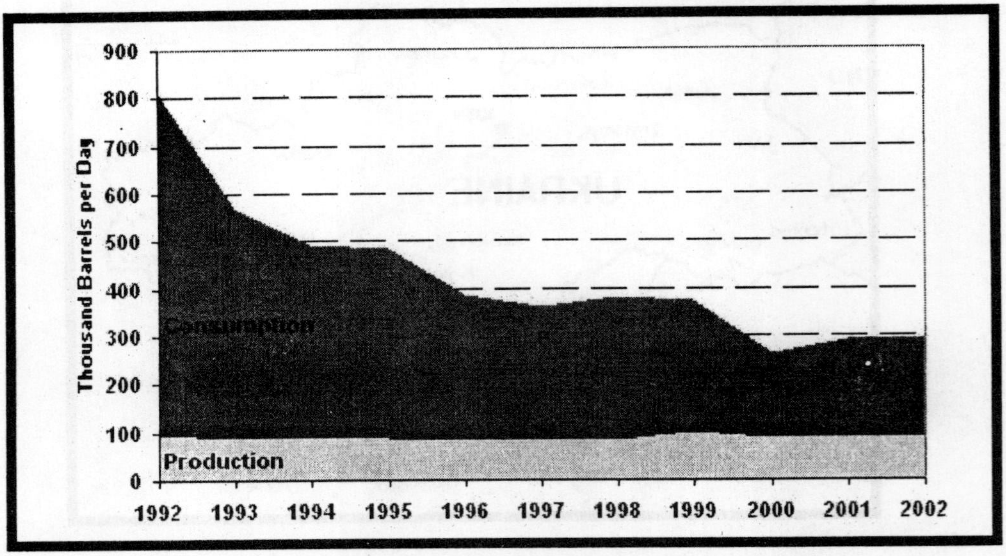

UKRANIAN OIL PRODUCTION AND CONSUMPTION, 1992-2002

is the Odessa-Brody pipeline, which was completed in 2001 and extends from Ukraine's Black Sea port of Odessa northward to the city of Brody. The pipeline was initially intended to load Caspian Sea oil from the newly completed Black Sea marine terminal, Pivdenniy (or Yuzhniy), and carry it northward through the Ukrainian system and on to Europe with an initial capacity of roughly 300,000 bbl/d. However, for roughly two years the pipeline has been mostly dormant. Russia has suggested that the pipeline be used in reverse, to move oil from Russia southwards to tankers in the Black Sea and shipped onwards to world markets. Since January 2003, the last 32-mile leg of the pipeline has been used (in reverse) by Russian oil companies for these purposes.

Faced with the possibility of losing direct access to Caspian Sea region oil, European governments have voiced their opposition to the reversal project in newspaper articles and public statements. Leading Caspian sea region producer, Kazakhstan, has also taken counter-measures. In July 2003, for instance, the country agreed to construct a 32-mile pipeline parallel to the segment currently being used in reverse to transit Russian oil. However, in late September 2003, the Ukrainian government announced that in 2004, the pipeline will be used in its originally intended north-south direction to carry 180,000 bbl/d of Caspian sea region crude to Europe. Later, however, on October 3, 2003, the Ukrainian state oil company UkrTransNafta effectively reversed that decision, declaring that it had accepted an offer from the Russian-British company TNK-BP to ship 9,000 bbl/d from Brody to Odessa (in reverse). Whether the Odesssa-Brody pipeline will be permanently and entirely reversed, per the Russian proposal, remains unknown at the present time. Some analysts have suggested that the pipeline be reversed temporarily, to allow Russian oil to flow through the pipeline in the meantime until major Caspian Sea region projects come online.

Refining/Downstream

Ukraine has six crude oil refineries, with a combined throughput capacity of approximately 1 million bbl/d. However, with domestic demand at just over 30 per cent of the country's refining

PROPOSED PIPELINES REVERSALS

capacity, Ukraine's refineries are operating significantly below capacity. Until recently, Ukraine's refineries did not even receive enough crude oil supplies to supply the country's domestic petroleum product demand.

Ukraine has begun to achieve better results in securing sufficient crude oil supplies for its refineries by offering oil exporters in Russia and Kazakhstan a stake in the country's refineries. Ukraine's recent success in privatizing its refineries has allowed the country to secure additional oil supplies to meet domestic demand, as well as to attract funds for necessary renovation work and to boost utilization rates at its refineries.

Natural Gas

Ukraine has 39.6 trillion cubic feet (Tcf) of natural gas reserves, from which roughly 0.64 Tcf was produced in 2001. That year, the country consumed 2.6 Tcf of natural gas, making it the former Soviet Union's largest net natural gas importer (1.97 Tcf, or 75 per cent of consumption). Historically, Russia has met this demand, partially through natural gas offered as payment in-kind for transiting its gas on to Europe, and partially through annual sales contracts. In the past few years, Ukraine has begun to import additional natural gas from Turkmenistan as well. In May 2001, Ukraine and Turkmenistan signed an agreement calling for Turkmenistan to supply Ukraine with 8.8 Tcf of natural gas been 2002 and 2006, and the agreement is expected to be extended through 2015.

Transit Gas to Europe

To world energy markets, Ukraine's real significance is as a intermediary connecting Russia, the world's largest natural gas producer, with growing European markets. In 2002, approximately 4 Tcf of Russian and Turkmen natural gas transited Ukraine en route to European markets. This represented roughly 24 per cent of OECD Europe's natural gas consumption, and 38 per cent of imports.

Accordingly, Ukraine's natural gas infrastructure is of growing importance both to European consumers and Russian producers. In June 2002, heads of state from Ukraine, Russia, and Germany

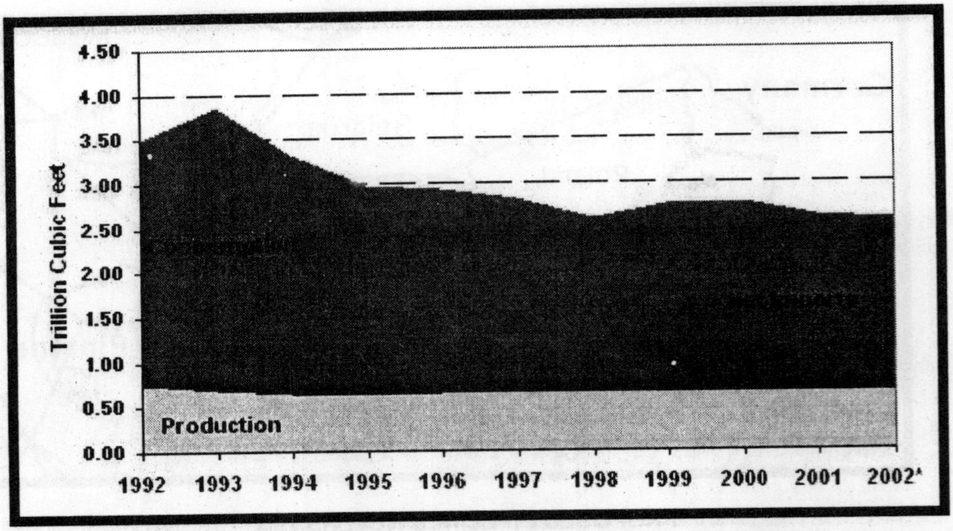

UKRANIAN NATURAL GAS AND CONSUMPTION, 1992-2002*

*2002 figures are estimates

agreed to begin developing an international consortium to manage and upgrade Ukraine's natural gas distribution infrastructure. In October 2002, Ukrainian and Russian state-owned oil and gas concerns, Gazprom, and Naftohaz Ukrainy, signed preliminary agreements, and in January 2003, the new company was registered in Kiev, with each company holding 50 per cent.

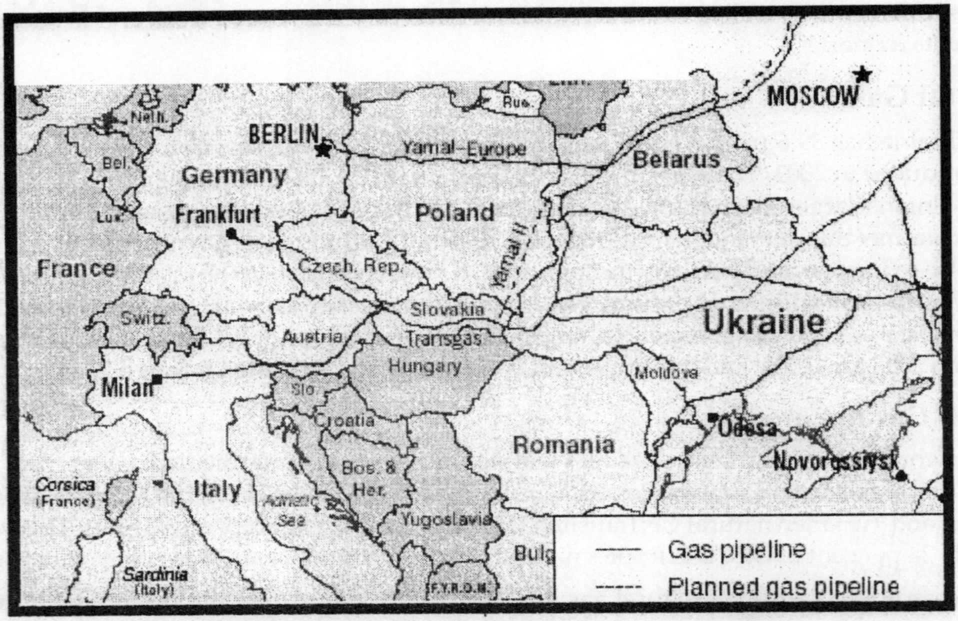

EXISTING AND PLANNED GAS ROUTES TO EUROPE

The partners are still considering several proposals for the structure and membership of the consortium. Germany's Ruhrgas has been present at the consortium's negotiations, but its role is as of yet undecided. Several other parties have shown interest in the consortium, including Gaz de France and the European Bank for Reconstruction and Development (EBRD). Ukraine has also suggested that Caspian Sea region producers Turkmenistan, Kazakhstan, and Azerbaijan be invited to participate.

In August 2003, the consortium agreed to its first infrastructure development project, the construction of a new 930-mile natural gas pipeline between eastern and western Ukraine with a capacity of roughly 1 Tcf per year. As conceived by the partners, the pipeline would allow Russian natural gas exports to Europe through Ukraine to increase by roughly 25 per cent. The consortium is currently conducting feasibility studies, and expect that once commenced, construction will take approximately two years.

Coal

Ukraine has 37.6 billion short tons in proven coal reserves (17.9 billion short tons of which is antracitie and bituminous coal, and 19.7 billion short tons of which is lignite and sub-bituminous), accounting for about 15 per cent of the former Soviet Union's total reserves. Production and consumption of coal in Ukraine have been relatively flat since 1996. In 2001 the country produced 90 million short tons, while consuming roughly 94 million short tons, making Ukraine a net importer, despite its sizeable resources.

Most of Ukraine's coal is mined in the Donetsk/Donbas basin in the eastern region of the country. The country's coal industry, which counts slightly less than 200 mines and employs around 450,000 people, is managed by a hierarchy of state organizations and suffers from numerous problems, including: labor strikes, hazardous working conditions, inefficiency and low productivity. The government has made restructuring the coal industry a priority, and in December 2002, Ukraine's Fuel and Energy Ministry announced plans to hand the industry over to 21 open joint-stock companies designed for eventual privatization. The handover was completed in March 2003, but privatization has not yet been completed.

Electricity

Ukraine's power sector is the twelfth largest in the world in terms of installed capacity, with 54 gigawatts (GW), slightly more than South Korea (52 GW). However, generation and consumption have fallen sharply since independence (see graph), and in 2001 the country generated 165 billion kilowatt hours (bkwh) of electricty–representing only 60 per cent of South Korea's generation. The country is currently in the process of revamping its electricity sector, by privatization, increased utilization at existing facilities, and the completion of two new nuclear plants.

In Ukraine, thermal power plants (oil, natural gas, coal) account for nearly 50 per cent of generation, with nuclear power generating another 40 per cent, and hydroelectric accounting for approximately 10 per cent (see graph). Ukraine has more than enough generating capacity to produce twice its electricity needs. However, the country's distribution system is in need of investment and maintenance as significant quantities of generation are wasted via line losses and several of the country's nuclear facilities are often intermittently shut down throughout the year owing to technical problems. In April 2003, President Kuchma announced that all of Ukraine's 27 regional electricity distribution companies should be privatized in 2003-2004 in an effort to encourage investment. Currently, only six Ukrainian distribution companies have been fully privatized.

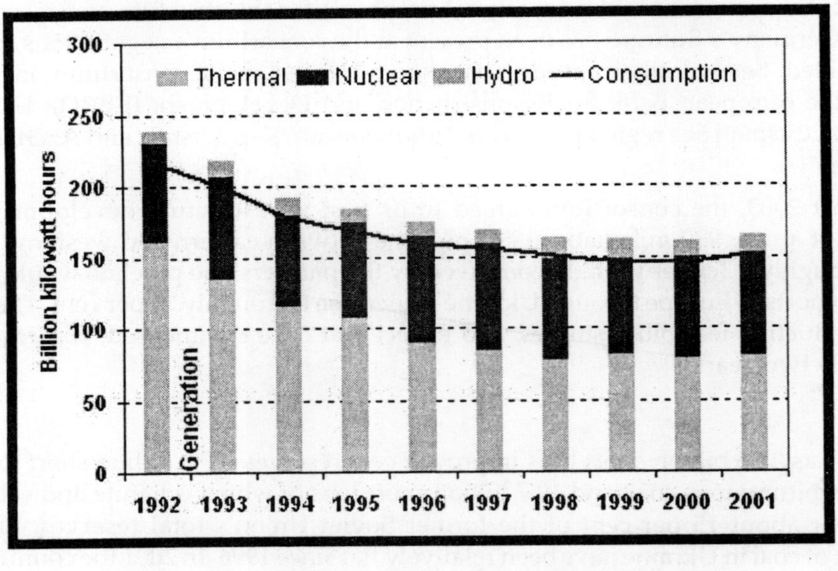

ELECTRICITY GENERATION IN UKARINE, 1992-2001

Nuclear

Ukraine currently has four operating nuclear power plants. These power plants have a combined capacity of 12.8 gigawatts, which accounts for approximately 24 per cent of the country's total power-generating capacity. Ukraine's nuclear power plants produce 40 per cent of the country's power, despite frequent malfunctions and lengthy repairs and maintenance.

On December 15, 2000, Ukraine permanently shut down the 925-MW, Unit 3 at the Chornobyl power plant, disabling the last remaining working reactor at the ill-fated power plant. To replace the power generated by Chornobyl, which Ukrainian officials say produced approximately 5 per cent of the country's total, Ukraine has resumed construction of two 1-GW reactors at the Khmelnitsky and Rivne power plants.

Construction of Khmelnitsky-2 and Rivne-4 was begun under the Soviet Union, and both were more than 80 per cent finished when Ukraine received its independence and ran out of money to complete them. Ukraine had been hoping to finish construction of both reactors with the help of financing from the European Bank for Reconstruction and Development (EBRD), but an EBRD loan for the project was put on hold in December 2001. In September 2003, after lengthy deliberations, the Ukrainian government announced its intentions to compete the reactors on its own, without the involvement of the EBRD. However, in late September 2003, talks between the two sides resumed. The plan has been criticized by environmentalists and others as unnecessary given Ukraine's existing overcapacity.

Environment

The 1986 Chornobyl nuclear meltdown exposed the Soviet Union's negligent environmental record and triggered alarm across the globe. The world's worst nuclear accident created disastrous consequences for the environment, both in Ukraine and in neighboring countries. As a result, Soviet policies that encouraged industrial development at the expense of the environment came under harsh international criticism, and Chornobyl became a rallying cry for environmentalists around the world.

While Chornobyl remains the lasting symbol of environmental degradation in Ukraine, today air pollution in the major cities is a major problem. Yet, despite increased vehicle traffic, energy use is significantly lower now than in the mid-1990s. Although policies encouraging energy conservation and energy efficiency can take some of the credit, Ukraine's economic woes account for much of the reduction: as the economy contracted through the 1990s, industrial production and consumer demand dropped as well, resulting in lower carbon emissions. Ukraine's recent economic growth has led to increases in both carbon emissions and energy consumption.

In terms of energy consumption per dollar, Ukraine suffers from one of the highest levels of energy intensity in the world. The country's heavy dependence on coal makes it correspondingly high in carbon intensity, and the continued reliance on nuclear power-as well as a lack of financial resources or economic incentives–has stifled the country's use of renewable energies. In order to protect its environment better in the coming years, Ukraine will need to shift away from fossil fuels and break the link of economic output from environmental pollution.

Country Overview

Independence: December 1, 1991 (from Soviet Union); National holiday: Independence Day, August 24, 1991

Population (7/01E): 48.7 million

Location: Eastern Europe, bordering the Black Sea between Poland and Russia

Size: 233,090 square miles, slightly smaller than Texas

Major Cities: Kiev (capital), Kharkiv, Donetsk, Dnipropetrovsk, Odesa, L'viv

Languages: Ukrainian (official), Russian, Romanian, Polish, Hungarian

Ethnic Groups: Ukrainian 73 per cent, Russian 22 per cent, Jewish 1 per cent, other 4 per cent

Religions: Ukrainian Orthodox–Moscow Patriarchate, Ukrainian Orthodox–Kiev Patriarchate, Ukrainian Autocephalous Orthodox, Ukrainian Catholic (Uniate), Protestant, Jewish

Economic Overview

Currency: Hryvnia

Market Exchange Rate (9/29/03): US $1=5.49 hryvnia

Nominal Gross Domestic Product (GDP) (2002E): $39.8 billion; *(2003F):* $43.3 billion

*Real GDP Growth Rate (2002E):*4.8 per cent; *(2003E):* 5.3 per cent *(2004F):* 4.6 per cent

Inflation Rate (Change in Consumer Prices year-on-year, 2002): 0.8 per cent; *(2003E):* 6.1 per cent *(2004F):* 8.9 per cent

Official Unemployment Rate (2002E): 3.6 per cent; *(2003E):* 4.0 per cent

Current Account Surpluss (2002E): $3.17 billion; *(2003E):* $2.85 billion

Major Trading Partners: Russia, EU, U.S., Turkey

Merchandise Exports (2002E): $18.8 billion; *(2003E):* $19.8 billion

Merchandise Imports (2002E): $18.1 billion; *(2003E)*: $19.0 billion

Merchandise Trade Surpluss (2002E): $700 million; *(2003E)*: $800 million

Major Exports: ferrous and nonferrous metals, fuel and petroleum products, machinery and transport equipment, food products

Major Imports: energy, machinery and parts, transportation equipment, chemicals

External Debt (12/02E): $10.2 billion

Energy Overview

Proven Oil Reserves (1/1/03E): 395 million barrels

Oil Production (2002E): 84,400 barrels per day (bbl/d)

Oil Consumption (2002E): 296,000 bbl/d

Net Oil Imports (2002E): 211,600 bbl/d

Crude Refining Capacity (1/1/03E): 1 million bbl/d

Natural Gas Reserves (1/1/03E): 39.6 trillion cubic feet (Tcf)

Natural Gas Production (2001E): 0.64 Tcf

Natural Gas Consumption (2001E): 2.62 Tcf

Net Natural Gas Imports (2001E): 1.98 Tcf

Coal Reserves (2001E): 37.6 billion short tons

Coal Production (2001E): 90.6 million short tons (Mmst)

Coal Consumption (2001E): 93.7 Mmst

Electricity Generation Capacity (2001E): 54 gigawatts (GW)

Electricity Production (2001E): 165 billion kilowatt hours (Bkwh)

Electricity Consumption (2001E): 152 Bkwh

Environmental Overview

Total Energy Consumption (2001E): 6.08 quadrillion Btu* (1.5 per cent of world total energy consumption)

Energy-Related Carbon Emissions (2001E): 96.58 million metric tons of carbon (1.5 per cent of world total carbon emissions)

Per Capita Energy Consumption (2001E): 123.7 million Btu (vs. U.S. value of 341.8 million Btu)

Per Capita Carbon Emissions (2001E): 2.0 metric tons of carbon (vs. U.S. value of 5.5 metric tons of carbon)

Energy Intensity (2001E): 31,001 Btu/$1995 (vs U.S. value of 10,918 Btu/$1995)CHECK

Carbon Intensity (2000E): 3.13 metric tons of carbon/thousand $1995 (vs U.S. value of 0.17 metric tons/thousand $1995)

Fuel Share of Energy Consumption (2001E): Natural Gas (45 per cent), Coal (30 per cent), Nuclear (13 per cent), Oil (10 per cent)

Fuel Share of Carbon Emissions (2001E): Coal (48 per cent), Natural Gas (41 per cent), Oil (11 per cent)

Status in Climate Change Negotiations: Non-Annex I country under the United Nations Framework Convention on Climate Change (ratified May 13th, 1997). Signatory to the Kyoto Protocol (signed March 15th, 1999, not yet ratified)

Major Environmental Issues: Inadequate supplies of potable water; air and water pollution; deforestation; radiation contamination in the northeast from 1986 accident at Chornobyl Nuclear Power Plant.

Major International Environmental Agreements: A party to Conventions on Air Pollution, Air Pollution-Nitrogen Oxides, Air Pollution-Sulphur 85, Antarctic Treaty, Biodiversity, Endangered Species, Environmental Modification, Hazardous Wastes, Law of the Sea, Marine Dumping, Nuclear Test Ban, Ozone Layer Protection, Ship Pollution, Wetlands. *Has signed, but not ratified*: Air Pollution-Persistent Organic Pollutants, Air Pollution-Sulphur 94, Air Pollution-Volatile Organic Compounds, Antarctic-Environmental Protocol.

The total energy consumption statistic includes petroleum, dry natural gas, coal, net hydro, nuclear, geothermal, solar and wind electric power. The renewable energy consumption statistic is based on International Energy Agency (IEA) data and includes hydropower, solar, wind, tide, geothermal, solid biomass and animal products, biomass gas and liquids, industrial and municipal wastes.

87

United Arab Emirates

The United Arab Emirates (UAE) is important to world energy markets because it contains 98 billion barrels, or nearly 10 per cent, of the world's proven oil reserves. The UAE also holds the world's fifth-largest natural gas reserves and exports significant amounts of liquefied natural gas.

General Background

The overall performance of the UAE's economy is heavily dependent on oil exports, which account for about 30 per cent of total gross domestic product (GDP). Growth in real GDP had slowed to 1.8 per cent in 2001, largely as a result of cuts in oil export revenues, but it is projected to recover to 2.5 per cent for 2002 and 3.3 per cent for 2003. Growth in the non-oil sectors of the economy is expected to outpace growth in the oil sector over the next several years, as a result of increasing capital investment and the government's expansionary monetary policy.

Government Structure

The UAE is a federation of seven emirates–Abu Dhabi, Dubai, Sharjah, Ajman, Fujairah, Ras al-Khaimah, and Umm al-Qaiwain. Political power is concentrated in Abu Dhabi, which controls the vast majority of the UAE's economic and resource wealth. The two largest emirates–Abu Dhabi and Dubai–provide over 80 per cent of the UAE's income. In June 1996, the UAE's Federal National Council approved a permanent constitution for the country. This replaced a provisional document which had been renewed every five years since the country's creation in 1971. The establishment of Abu Dhabi as the UAE's permanent capital was one of the new framework's main provisions.

Other Industry

In recent years, the UAE has undertaken several projects to diversify its economy and to reduce its dependence on oil and natural gas revenues. The non-oil sectors of the UAE's economy presently

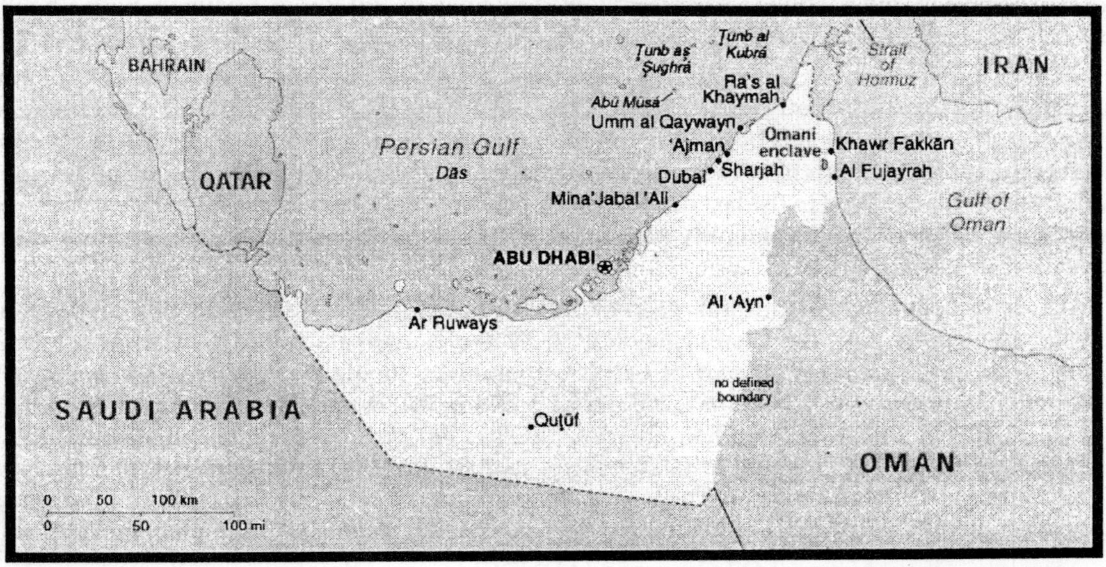

contribute more than two-thirds of the UAE's total GDP, and about 30 per cent of its total exports. The federal government has invested heavily in sectors such as aluminum production, tourism, aviation, re-export commerce, and telecommunications. As part of its strategy to further expand its tourism industry, the UAE is building new hotels, restaurants and shopping centers, and expanding airports and duty-free zones. Dubai has become a central Middle East hub for trade and finance, accounting for about 70 per cent of the Emirates' non-oil trade. The UAE has been a member of the World Trade Organization (WTO) since 1995, and has one of the most open economies in the region.

Foreign Affairs

The UAE and Iran continue to dispute the ownership of three islands, Abu Musa and the Greater and Lesser Tunb Islands, which are strategically located in the Strait of Hormuz. All three islands were effectively occupied by Iranian troops in 1992. The Mubarak field, which is located six miles off Abu Musa, has been producing oil and associated natural gas since 1974. In 1995, the Iranian Foreign Ministry claimed that the islands are "an inseparable part of Iran." Iran rejected a 1996 proposal by the Gulf Cooperation Council (GCC) for the dispute to be resolved by the International Court of Justice, an option supported by the UAE. In early 1996, Iran took further moves to strengthen its hold on the disputed islands. These actions included starting up a power plant on Greater Tunb, opening an airport on Abu Musa, and announcing plans for construction of a new port on Abu Musa. In the dispute, the UAE has received strong support from the GCC, the United Nations, and the United States. Although Iran remains a continuing concern for officials in Abu Dhabi, they have chosen not to escalate the territorial dispute. The two governments have recently held high-level discussions about the territorial dispute. Iran is one of Dubai's major trading partners, accounting for 20 per cent to 30 per cent of Dubai's business.

Oil

The UAE contains proven crude oil reserves of 97.8 billion barrels, or slightly less than 10 per cent of the world total. Abu Dhabi holds 94 per cent of this amount, or about 92.2 billion barrels. Dubai contains an estimated 4.0 billion barrels, followed by Sharjah and Ras al-Khaimah, with 1.5 billion and 100 million barrels of oil, respectively.

The majority of the UAE's crude oil is considered light, with gravities in the 32° to 44° API range. Abu Dhabi's Murban 39° and Dubai's Fateh 32° blends are the UAE's primary export crude streams, though Dubai's production is been falling in recent years due to the decline of its modest reserves. Most of the UAE's oil fields have been producing since the 1960s or early 1970s. Proven oil reserves in Abu Dhabi have doubled in the last decade, mainly due to significant increases in rates of recovery. Abu Dhabi has continued to identify new finds, especially offshore, and to discover new oil-rich structures in existing fields.

Under the UAE's constitution, each emirate controls its own oil production and resource development. Although Abu Dhabi joined OPEC in 1967 (four years before the UAE was formed), Dubai does not consider itself part of OPEC or bound by its quotas. The UAE's current OPEC production quota (effective January 1, 2002) is 1.89 million bbl/d, and its crude oil production in the fourth quarter of 2002 is 2.01 million bbl/d. OPEC has left prduction quotas unchanged in 2002, after several rounds of cuts due to falling prices in 2001, but overall output by OPEC members, including the UAE, has been rising. The UAE's total production capacity is 2.60 million bbl/d, making it second only to Saudi Arabia for excess production capacity among OPEC member states.

The Abu Dhabi National Oil Company (ADNOC) is currently planning a limited opening of UAE upstream oil production to foreign firms. The initial asset sale will involve 28 per cent of the offshore Upper Zakhum field. Bids have been solicited from BP, TotalFinaElf, ChevronTexaco, ExxonMobil, and the Japan Oil Development Company (JODCO), and an award decision is expected in the near future. JODCO already holds a 12 per cent stake in thew field from a previous investment in 1972, when the field was first developed.

Several projects to upgrade infrastructure at existing oilfields are planned or underway. A $300 million project to increase the capacity of the onshore Bu Hasa field is planned, including construction of natural gas separation units and drilling of natural gas reinjection wells. The goal is to increase production capacity to 480,000 bbl/d from the present 100,000 bbl/d. A natural gas reinjection project also is planned for the onshore Bab field. These projects are part of an overall goal of raising the UAE's production capacity to 3 million bbl/d within the next four years.

Refining

The UAE has two refineries operated by ADNOC. The Ruwais refinery has a capacity of 145,000 bbl/d. It produces light products mainly for export to Japan and India. Fuel oil from Ruwais is sold as bunkers by ADNOC and also used for domestic electric power generation. A $480-million contract was awarded to the Italian engineering firm Technip in June 2002 for an expansion of the Ruwais complex to a capacity of 500,000 bbl/d, including refits of existing units and expansion of units for production of unleaded gasoline and low-sulfur fuel oil. Work under this contract is to be completed by 2005.

UAE has five smaller refineries. Umm al-Nar, in Abu Dhabi, has a capacity of 88,500 bbl/d. Since its construction in 1976, the Umm al-Nar plant has undergone debottlenecking as well as a recent expansion. The refinery primarily supplies the domestic market. Metro Oil has a 90,000-bbl/d refinery in Fujairah. The Emirates National Oil Company (ENOC) Jebal Ali condensate refinery, with a capacity of 120,000 bbl/d, began operations in Dubai in May 1999. A 40,000 bbl/d second-hand gasoline unit, owned by the private firm ISO Octane, opened near Jebal Ali in May 2000. Another 71,250 bbl/d second-hand unit was set up by the Sharjah Oil Refining Company in 2001.

Foreign Downstream Operations

In October 1998, the International Petroleum Investment Company (IPIC), the UAE's downstream investment outfit, purchased 50 per cent of the Hyundai Oil Refinery Company of South Korea for $500 million. The UAE is the second-largest crude oil supplier to South Korea after Saudi Arabia. IPIC's overseas holdings also include a 10 per cent stake in Spain's CEPSA and a 19.6 per cent share of Austria's OMV.

Natural Gas

The UAE's natural gas reserves of 212 trillion cubic feet (Tcf) are the world's fifth largest after Russia, Iran, Qatar, and Saudi Arabia. The largest reserves of 196.1 Tcf are located in Abu Dhabi. Sharjah, Dubai, and Ras al-Khaimah contain smaller reserves of 10.7 Tcf, 4.1 Tcf, and 1.1 Tcf, respectively. In Abu Dhabi, the non-associated Khuff natural gas reservoirs beneath the Umm Shaif and Abu al-Bukhush oil fields rank among the world's largest. Current natural gas reserves are projected to last for about 150-170 years.

Increased domestic consumption of electricity and growing demand from the petrochemical industry have provided incentives for the UAE to increase its use of natural gas. Over the last decade, natural gas consumption in Abu Dhabi has doubled, and is projected to reach 4 billion cubic feet per day (bcf/d) by 2005. The development of natural gas fields also results in increased production and exports of condensates, which are not subject to OPEC quotas.

Projects

The past few years have seen the UAE embark on a massive, multi-billion dollar program of investment in its natural gas sector including a shift toward natural gas-fired power plants and the transformation of the Taweelah commercial district into a natural gas-based industrial zone. An ambitious plan, the Dolphin Project, to interconnect the natural gas grids of Qatar, the UAE, and Oman, also is planned. Most of the UAE's increased nautral gas needs in the next decade are to be satisified with imported natural gas from Qatar. Much of the natural gas development in the UAE itself involves the extraction of natural gas liquids (NGLs) and reinjection of the gas to maintain pressure in oilfields.

The second phase of the UAE's $1-billion onshore natural gas development program (OGD-2) at the Habshan complex located directly over the Bab oil and natural gas field was completed in early 2001. This second phase included the construction of four trains to process 1 bcf/d of natural gas, 300-500 tons per day (t/d) of natural gas liquids (NGLs), 35,000-55,000 t/d of condensate and up to 2,100 t/d of sulphur. Additional capacity expansion is planned in the third phase, OGD-3, and will involve the construction of two additional natural gas processing trains. Bechtel was awarded the initial engineering and design work for OGD-3 in May 2002, which they are to complete in 2003.

Another project closely linked with OGD-2 is the Asab natural gas development project, which was completed in 1999. The Asab development processes around 830 million cubic feet per day (Mmcf/d) of associated wet natural gas from the Thamama F and G reservoirs and produces up to 100,000 bbl/d of condensate for processing at the Ruwais refinery. The natural gas also will support other industries in Ruwais and be re-injected into Asab reservoirs to maintain field pressure. A second phase AGD-2, is planned, which will add more NGL processing capacity, but will not increase natural gas production capacity.

Supplying Dubai

Dubai's natural gas consumption has been gowing by nearly 7 per cent annually due to expansion of the emirate's industrial sector, a switch to natural gas by its power stations, and the need for an enhanced oil recovery (EOR) system based on natural gas injections for its mature oilfields. Dubai projects future demand will average 810 Mmcf/d in 2005, with major swings between summer and winter consumption patterns. Until mid-2001, Dubai's entire natural gas supply came entirely from fellow UAE member Sharjah. BP operates three fields and the 800-Mmcf/d Sajaa processing facility in conjunction with the Sharjah government. In May 2001, a pipeline from the Maqta area of Abu Dhabi to Dubai commenced operations, delivering 200 Mmcf/d of natural gas. Pipeline throughput is expected to reach 900 Mmcf/d in 2003 once additional compressor stations are completed.

The Dolphin Project

The Dolphin Project aims to develop links between the natural gas infrastructures of Qatar, the UAE, and Oman, with a possible future link to the Indian Subcontinent. It will allow the export of non-associated natural gas from Qatar's massive offshore North Dome field. A Statement of Principles for the project was signed in March 1999 between the UAE Offsets Group (UOG) and the Qatar General Petroleum Corporation (QGPC). The two firms signed a natural gas sales agreement in March 2001, with natural gas supplies to start in late 2005. Estimated to cost $8-10 billion over the next decade, the project will begin as a subsea pipeline from Ras Laffan in Qatar to a landfall in Abu Dhabi, which will then be extended to Dubai and northern Oman. It will start at 48 inches in diameter, narrowing to 30 inches by the time it reaches Oman. In its initial phase, the pipeline is to carry 3 Bcf/d of Qatari natural gas to the UAE and Oman, accounting for nearly 10 per cent of total world natural gas supplies shipped by pipeline.

In October 1999, UOG and ADNOC issued a joint declaration dividing up natural gas distribution between them. Natural gas from the Dolphin Project will be the exclusive supply for natural gas-fired power plants, except in the Western Region of Abu Dhabi, and will also supply natural gas for ADNOC contracts with Dubai. Natural gas from the Dolphin Project will use the ADNOC distribution network until the project develops its own network. In March 2000, UOG signed a contract with two foreign firms, TotalFinaElf and Enron, after securing purchase agreements with Abu Dhabi, Dubai, and Oman. Originally, the two main foreign firms participating in the project were Enron and TotalFinaElf. In May 2001, however, Enron announced that it was backing out of the project, and selling its 24.5 per cent stake back to UOG. UOG sold this share to Occidental Petroleum in June 2002 after receiving bids from several foreign companies. Upstream development in Qatar and construction of the subsea pipeline are expected to commence in 2003, with natural gas deliveries to the UAE beginning in 2005. The proposed extension from Oman to Pakistan might be built later. This phase of the project is dependent on Pakistan's ability to pay for the natural gas, which is questionable given the current weakness of its economy.

Electricity

The UAE's soaring demand for electric power, coupled with volatile swings in peak loads, led the Emirates in 1997 to form a Privatization Committee for the Water and Electricity Sector. In early 1998, the committee called for a comprehensive restructuring, including the elimination of the state-owned Abu Dhabi Water and Electricity Department (ADWED). ADWED was transformed into a regulatory body, the Abu Dhabi Water and Electricity Authority (ADWEA). The government plans to take a majority holding in the new ventures with minority interests held by foreign firms. The government may gradually privatize its shares through initial public offerings (IPOs), allowing UAE nationals to become shareholders, though this is still uncertain.

TotalFinaElf and Tractebel were awarded a contract by ADWEA in August 2000 for an upgrade to the Taweelah A-1 plant, which gives a 20 per cent ownership stake to each of the foreign partners, with the rest remaining with ADWEA. The upgrade will bring the capacity of the plant to 1,350 megawatts (MW). The consortium, Gulf Total Tractebel Power Company (GGTPC), currently is preparing to receive bids for the construction of the upgrades.

Another step in the reorganization is the expansion of the Taweelah cogeneration facility. The expansion, known as Taweelah A-2, is the UAE's first independent water and power project (IWPP), and reached financial close in April 1999. It is the second independent power project in the Gulf after Oman's al-Manah facility. With a price tag of some $800 million, the expansion is to add about 763 megawatts (MW) of power and 50 million gallons of desalinated water to the UAE's supplies. The first 370-MW came online in July 2000. The rest of the generating units became operational in August 2001. The Taweelah A-2 project is run by Emirates CMS Power, a joint venture between CMS Energy (40 per cent ownership interest) and the newly-formed Emirates Power Company (EPC) (60 per cent).

The al-Taweelah Power Company manages the Taweelah B facility. The plant, which currently has six 122-MW steam turbines and six 13 million gallon-per-day (g/d) multi-stage flash units, is now undergoing a $360 million expansion. The addition of two new natural gas-turbine units will bring the plant's capacity to 1,220 MW and 103 million gallons per day (g/d) of water. The Umm al-Nar Power Company operates the plant by the same name with a 1,215-MW, 97-million-g/d facility, which will be upgraded with two new 3.5 million g/d desalination units. The new units run on steam already available at the site. The company also will operate the 120-MW Baniyas station. ADWEA received bids in November 2002 for the partial privatization of the company, which will be structured similarly to the two Taweelah IWPPs.

The Abu Dhabi Water and Electricity Authority (ADWEA) signed a contract for the Shuweihat IWPP project in August 2001 with a consortium of CMS Energy and International Power PLC. The $1.6 billion deal provides for the construction and operation of a 1,500-MW combined cycle plant with a desalination capacity of 100 million gallons per day. Construction began in early 2002, with commercial operation expected by mid-2004.

The UAE also is taking part in a $1-billion plan to build a regional power grid throughout the countries of the Gulf Cooperation Council (GCC). The first phase of the plan would link Saudi Arabia, Kuwait, Bahrain and Qatar; the UAE and Oman would join the grid in the second phase of the plan. GCC electricity ministers signed a final agreement on the project in June 1999. The plan is based on the assumption that each country will have its own unified power grid, and the UAE is doing its part by connecting all the power stations along its western coast with the central region. The process of building interconnections between all seven Emirates' transmission grids is expected to be completed by early 2003.

Country Overview

Independence: December 2, 1971 (from United Kingdom)

Population (2002E): 2.4 million

Location/Size: Persian Gulf between Oman and Saudi Arabia/30,000 square miles

Major Cities: Abu Dhabi (capital), Dubai, Sharjah, al-Ain

Languages: Arabic (official), Persian, English, Hindi, Urdu

Ethnic Groups: Arab (UAE citizens) (19 per cent), other Arab and Iranian (23 per cent), South Asian (50 per cent), other expatriate (Western and East Asian) 8 per cent .

Religion: Muslim 96 per cent (Shi'a 16 per cent), Christian, Hindu, Other 4 per cent

Defense (1998): Total manpower 64,500 (Army 59,000; Air Force 4,000; Navy 1,500)

Economic Overview

Currency: Dirham (AED)

Market Exchange Rate (12/9/02): US$1 = 3.67 Dirhams

Gross Domestic Product (2002E): $63.1 billion

Real GDP Growth Rate (2002E): 2.5 per cent

Inflation Rate (consumer prices)(2002E): 2.1 per cent

Major Trading Partners: Japan, United Kingdom, United States, Singapore, Germany, South Korea, Iran, India

Current Account Balance (2002E): $7.4 billion

Merchandise Exports (2002E): $43.8 billion

Merchandise Imports (2002E): $34.9 billion

Merchandise Trade Balance (2002E): $8.9 billion

Major Export Products: Crude oil, natural gas, re-exports, aluminum, dried fish, dates

Major Import Products: Manufactured goods, machinery, and transportation equipment, food

International Reserves (2002E): $11.7 billion

Energy Overview

Proven Oil Reserves (1/1/02E): 97.8 billion barrels

Oil Production (3rd Quarter of 2002E): 2.27 million bbl/d, of which 1.99 million bbl/d is crude oil

OPEC Crude Oil Production Quota (effective 1/1/02): 1.89 million bbl/d

Crude Oil Production Capacity (4th Quarter of 2002): 2.60 million bbl/d

Oil Consumption (2002E): 353,000 bbl/d

Net Oil Exports (3rd Quarter of 2002E): 1.92 million bbl/d

Major Crude Oil Customers (2002E): Japan (about 60 per cent), other Far East (about 20 per cent)

Crude Oil Refining Capacity (1/1/02E): 514,750 bbl/d

Natural Gas Reserves (1/1/02E): 212 trillion cubic feet (Tcf)

Natural Gas Production (2000E): 1.41 Tcf

Natural Gas Consumption (2000E): 0.97 Tcf

Net Natural Gas Exports (2000E): 0.44 Tcf

Electric Generation Capacity (1/1/00E): 5.6 gigawatts

Electricity Production (2000E): 38.7 billion kilowatthours

Environmental Overview

Total Energy Consumption (2000E): 1.7 quadrillion Btu* (0.4 per cent of world total energy consumption)

Energy-Related Carbon Emissions (2000E): 30.2 million metric tons of carbon (0.5 per cent of world total carbon emissions)

Per Capita Energy Consumption (2000E): 560.1 million Btu (vs. U.S. value of 351.0 million Btu)

Per Capita Carbon Emissions (2000E): 9.7 metric tons of carbon (vs. U.S. value of 5.6 metric tons of carbon)

Energy Intensity (2000E): 36,463 Btu/$1995 (vs U.S. value of 10,918 Btu/$1995)**

Carbon Intensity (2000E): 0.6 metric tons of carbon/thousand $1995 (vs U.S. value of 0.17 metric tons/thousand $1995)

Sectoral Share of Energy Consumption (1998E): Transportation (10.3 per cent), Industrial (58.4 per cent), Residential (16.2 per cent), Commercial (15.1 per cent)

Sectoral Share of Carbon Emissions (1998E): Industrial (56.7 per cent), Transportation (13.2 per cent), Residential (15.6 per cent), Commercial (14.5 per cent)

Fuel Share of Energy Consumption (2000E): Oil (41.5 per cent), Natural gas (58.5 per cent), Coal (0.0 per cent)

Fuel Share of Carbon Emissions (2000E): Natural gas (50.6 per cent), Oil (49.4 per cent), Coal (0.0 per cent)

Renewable Energy Consumption (1998E): 0.71 trillion Btu* (0 per cent increase from 1997)

Number of People per Motor Vehicle (1998): 71.4 (vs. U.S. value of 1.3)

Status in Climate Change Negotiations: Non-Annex I country under the United Nations Framework Convention on Climate Change (ratified December 29th, 1995). Not a signatory to the Kyoto Protocol.

Major Environmental Issues: Lack of natural freshwater resources being overcome by desalination plants; desertification; beach pollution from oil spills.

Major International Environmental Agreements: A party to Conventions on Climate Change, Desertification, Endangered Species, Hazardous Wastes, Marine Dumping and Ozone Layer Protection. Has signed, but not ratified, Biodiversity and Law of the Sea.

The total energy consumption statistic includes petroleum, dry natural gas, coal, net hydro, nuclear, geothermal, solar, wind, wood and waste electric power. The renewable energy consumption statistic is based on International Energy Agency (IEA) data and includes hydropower, solar, wind, tide, geothermal, solid biomass and animal products, biomass natural gas and liquids, industrial and municipal wastes.

Oil and Gas Industries

Organizations: Abu Dhabi National Oil Company (ADNOC); Operates three main oil and natural gas operating companies, five Service companies, three joint ventures to fully utilize the produced natural gas, two maritime transport companies for crude oil, refined product and LNG and one refined product distribution company.

Major Refineries: Ruwais (145,000 bbl/d), Emirates National Oil Company (ENOC) - Dubai (120,000), Umm al-Nar (88,500 bbl/d), Metro Oil (Fujairah)(90,000 bbl/d), Sharjah Oil Refining Company (71,250)

Major Natural gas Processing Plants: Bab, Bu Hasa, Das Island, Habshan (2), Jebel Ali, Ruwais

Major Oil Fields: Abu Dhabi: 'Asab, Bab, Bu Hasa, Al-Zakum *Dubai:* Fallah, Fateh, Southwest Fateh, Margham, Rashid *Sharjah:* Mubarak (near Abu Musa Island)

Major Associated Natural Gas Fields: Abu Dhabi: Abu al-Bukhush, Bab, Bu Hasa, Umm Shaif, Zakum

Ports: Abu Dhabi: Das Island, Delma Island, Jebel as Dhanna, Ruwais, Abu al Bukhush, Al Mubarraz, Zirku Island, Port Zayed, Umm al Nar *Dubai:* Jebel Ali, Fateh, Port Rashid *Sharjah:* Mubarak

88

United Kingdom

With its significant North Sea reserves, the United Kingdom is a major European oil and natural gas producer. It is also one of the largest energy consumers in Europe.

General Background

The United Kingdom (official name: United Kingdom of Great Britain and Northern Ireland, abbreviated: UK) is a major political power, and also the world's fourth-largest economy. The country joined the European Union (EU) in 1973 (confirmed by referendum in 1975), but has not yet decided whether to adopt the European Monetary Union's single currency, called the euro. The UK government is in the process of assessing the national economy to determine whether there is sustainable convergence and sufficient flexibility with the economies of the euro zone to make joining feasible. The government is also studying the possible effects on employment, investment, and the financial services industry of adopting the euro. Results from these studies are expected to be available in June 2003, after which the UK government will decide whether to recommend joining the euro zone. If the recommendation is made, a vote will be taken in Parliament and a referendum of the British people will follow.

Despite the UK's lack of participation in the euro, the country has continued to attract foreign direct investment (FDI)–about $586 billion total at the end of 2001, second in the world after the United States. The UK is an even larger exporter of capital–outward FDI at the end of 2001 totaled $992 billion, making it the third largest target for foreign investment after China and the United States. The UK maintains a smaller public sector than many of its EU counterparts.

The UK, like most of the OECD, saw growth rates remain low in 2002. Real gross domestic product (GDP) is estimated at only 1.6 per cent for 2002, but is projected to increase to 2.7 per cent in 2003. Despite this slow rate of economic growth, unemployment continues to fall, reaching a new 27-year low of 3.1 per cent in December 2002. Although it recognized the risks of a continuing

sluggish economy in the euro zone, as well as the potential for U.S. consumer spending to falter, the Bank of England chose to leave interest rates untouched at 4 per cent in November 2002. Housing price inflation has become another concern, having increased at the unsustainable rate of 26 per cent in 2002.

The UK is by far the largest petroleum producer and exporter in the EU (Norway is not a member of the EU). It also is the largest producer and an important exporter of natural gas in the EU. Most of the UK's oil and gas reserves and production are located off the coast of Scotland, with the Scottish city of Aberdeen considered to be the nation's oil and gas capital. The International Petroleum Exchange (IPE), the second-largest energy futures exchange in the world, is located in London. The second and third-largest publicly traded energy companies in the world in terms of market value–Royal Dutch/Shell and BP, respectively–are based in the UK (Royal Dutch/Shell is also based in the Netherlands). Because major UK energy companies are private, the eventual decline in British oil and gas production could translate to an increase in UK companies' involvement abroad, mitigating the effect in the overall UK economy, though Scottish employment is particularly sensitive to North Sea production levels. The UK has high taxes on petroleum products, making for among the highest prices in the EU, though its electricity prices have been steadily declining.

In July 1999, a Scottish Parliament met for the first time in almost 300 years. "Devolution" gives the Scottish Parliament the ability to tax its own citizens, plus jurisdiction over local issues such as education, health, transport, and agriculture. It has no effect on the economic and industrial structure of the United Kingdom, which remains a single market. Devolution has had no effect on North Sea oil and gas production.

North Sea Oil and Gas

North Sea oil and gas reserves were first discovered in the 1960s. The North Sea did not emerge immediately as a key non-OPEC oil producing area, but North Sea production grew as major discoveries continued throughout the 1980s and into the 1990s. Although the region is a relatively high cost producer, its high quality crude oil, political stability, and proximity to major European consumer markets have allowed it to play a major role in world oil and gas markets.

Many of the world's major crude oil prices are linked to the price of the North Sea's Brent crude oil. (Brent crude is a blend of North Sea crude oils and does not come exclusively from the Brent field.) Because Brent crude is traded on the International Petroleum Exchange in London, fluctuations in the market are reflected in the price of Brent. Therefore, all other crude oils linked to Brent can be priced according to the latest market conditions.

The North Sea is considered a "mature" area, with few large discoveries likely to be made. Only a few frontier areas hold the possibility of further discoveries of large oil and natural gas fields. In both of the major North Sea producing nations, Norway and the UK, government and industry are taking steps to restructure their oil and natural gas sectors to make them more internationally competitive. In November 2002, Chancellor of the Exchequer Gordon Brown announced plans to abolish all royalty taxes on North Sea oil and natural gas, which apply to fields that were granted development approval before April 1982. Measures have also been taken to make the area more attractive to smaller firms.

Oil

The UK holds about 5 billion barrels of proven oil reserves, almost all of which is located in the North Sea. Most of the country's production comes from basins east of Scotland in the central North Sea. The northern North Sea (east of the Shetland Islands) also holds considerable reserves, and smaller deposits are located in the North Atlantic Ocean, west of the Shetland Islands. There are over 100 oil and gas fields currently onstream, and several hundred companies are active in the area. In 2002, the United Kingdom's production declined to an estimated 2.53 million barrels per day (bbl/d), down from a historical high of 2.95 million bbl/d in 1999. Production is expected to decline further in 2003. Most of the UK's crude oil production ranges in gravity from 30 ° to 40 ° API. Most high quality crude is exported, while cheaper, lower quality (mainly from the Middle East) crude oils are imported for refining. According to the UK's Department of Trade and Industry (DTI), unit costs for UK oilfields averaged just over $17 per barrel in 2001, though fields that started production in the 1990s have lower costs.

The domestic UK oil and gas industry is expected to decline as reserves are depleted in the coming decade. The British Oil and Gas Industry Task Force (OGITF) was set up in 1998 to bring

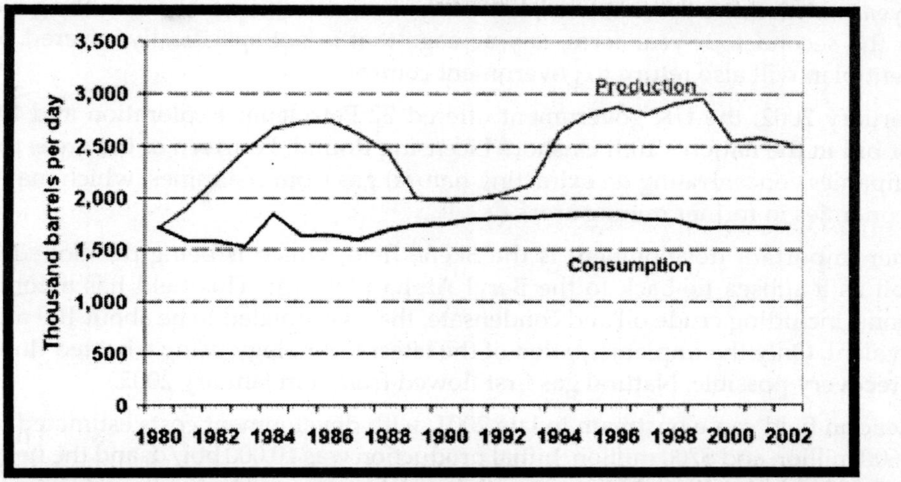

OIL PRODUCTION AND CONSUMPTION IN THE UNITED KINGDOM, 1980-2002

together government departments and oil and gas industry representatives (the oil and gas industry is 100 per cent in the hands of the private sector) to discuss the future of the industry. A successor body to the Task Force, known as "PILOT", now has been created to oversee the execution of Task Force recommendations and future developments. PILOT's goals for 2010 are to maintain oil production at or above 3 million bbl/d, have $4.8 billion annual investment in the industry, create 100,000 more jobs, create $1.6 billion in additional revenue for new business, and make the UK the safest place to work in the world oil and gas industry. By 2005 PILOT hopes to increase oil exports by 50 per cent.

Production

As the UK's oil fields mature, the industry's focus has been shifting from searching for new oil discoveries to finding ways to continue productivity of mature fields. The number of fields in production in the UK at the end of 2002 was just over 260, while the number of fields under development was 84. There were 144 new projects being considered on mature oil fields, an increase from 96 in 2001.

The eleventh onshore and twenty-first offshore licensing rounds were announced by Energy Minister Brian Wilson in February 2003. These rounds will mark the beginning of a new type of license for the UK, in which the traditional licensing fee will be cut by 90 per cent for the first two years, thereby providing more opportunity for smaller companies to enter the North Sea market. The two-year "promote" license will expire after two years unless a substantive work plan has been agreed upon, *e.g.* the drilling of a well. This two-year period is intended to give companies a chance to explore for oil and gas before promoting the licensed area to investers to acquire funding for drilling and other work.

Talisman Energy, Limited announced in October 2002 the discovery of oil in an area of the UK Central North Sea adjacent to the Buchan field. A test well discovered a gross oil column of 164 feet. The 40° API oil flowed at a rate of 6,600 bbl/d. Talisman already produces 145,000 bbl/d from its North Sea concessions.

In July 2002, the UK government awarded 25 offshore production licences to groups comprised of 33 different companies. Six new companies have acquired their first license in the area, and licenses were awarded to small niche players as well as the larger, more familiar firms. Licences will cover two, four-year terms. After the first term, half of the license will revert back to the government. At the end of the second four-year term, any acreage that is not specifically covered by a specific development plan will also return to government control.

In February 2002, the UK government offered 22 Petroleum Exploration and Development licenses for bid in the nation's 10th Onshore Licensing Round. Fourteen of the areas to be licensed are for companies concentrating on extracting natural gas from coalmines, which may boost hard-hit local economies in former mining areas.

Another important development is the Skene field, which is being developed by operator ExxonMobil as a subsea tie-back to the Beryl Alpha platform. This field has a complex mix of hydrocarbons, including crude oil and condensate, that is estimated to be about 100 million barrels of oil equivalent. Only the implementation of the latest technology using a heated flowline bundle has made recovery possible. Natural gas first flowed from it in January 2002.

The Leadon field came onstream in late 2001, with development costs estimated to have been between $600 million and $700 million. Initial production was 10,000 bbl/d, and the field is expected to peak at 45,000 bbl/d, with field life estimated at sixteen years. The field's water depth of 370 feet required the use of subsea horizontal wells that are tied to a floating, production, storage, and offloading vessel, in this case called the Kerr-McGee Global Producer III, which has a storage capacity of 500,000 barrels.

Europe's largest on-shore oilfield is Wytch Farm. Estimated reserves are 500 million barrels. Egdon Exploration is active in the area, and it is hoped that even smaller fields can be economically viable as they are on-shore. Other smaller on-shore fields are clustered in east-central England.

Industry Structure

Industry reorganization that started with BP's 1998 merger with Amoco continues. The merged BP Amoco, (now BP) already one of the world's largest petroleum companies, announced in April 1999 its intentions to take over Los Angeles-based Atlantic Richfield (Arco), which was completed in April 2000. The merged company is truly global and is the world's third-largest publicly traded oil and gas company. Most of the majors have a share of UK North Sea production, including BP, ChevronTexaco, Conoco, ENI, ExxonMobil, Royal Dutch/Shell, Texaco, and TotalFinaElf. Amerada Hess, Enterprise, and Statoil also have large shares.

BP Exploration is managed from Aberdeen, Scotland (as are most other companies that are active in the British North Sea). BP produces oil and gas and brings ashore about 40 per cent of the UK's total oil production through the Forties Pipeline System to Grangemouth, Scotland. BP has producing fields in the North Sea and, since the end of 1997, in the North Atlantic, west of the Shetland Islands. It operates the Sullom Voe oil terminal in the Shetlands, which is Europe's largest oil terminal. The 206,000-bbl/d oil refinery and petrochemical complex at Grangemouth represents one of Scotland's largest industrial complexes.

British independent oil companies, important in the North Sea oil scene, were particularly hard hit by the oil price collapse of 1998. However, as oil prices rise and the UK government makes investment more attractive for smaller, independent firms through special licenses, they are beginning to show renewed interest. The Energy Ministry has shown some concern that British independents are in danger of being taken over by the large corporations, as evidenced by the Energy Minister's support of Enterprise Oil's management when it resisted an unsolicited takeover bid by ENI. Enterprise, the largest British independent, unsuccessfully attempted to take over the second largest, Lasmo, which was later acquired by ENI in 2001. Italian oil and gas giant ENI began its acquisition of British independents when it bought British-Borneo in March 2000. Small British companies such as Tullow, Highland, Consort, Intrepid, Paladin, and Venture have all done well in the North Sea market, often by concentrating their efforts on areas with small reserves that are not considered materially viable by major operators.

Downstream

The UK's crude oil refining capacity is approximately 1.79 million barrels per day, just slightly more than the country's consumption. However, the UK imports and exports refined products because British refineries produce an excess of some grades and products and insufficient quantities of others for local demand. Additionally, demand for gasoline varies seasonally. The largest refinery is ExxonMobil's (Esso's) 311,240-bbl/d Fawley refinery in Southhampton, one of the largest in Europe and marine tanker accessible. It also has a pipeline to the on-shore Wytch Farm field. The 100,000-bbl/d Port Clarence Phillips-Imperial Petroleum refinery at North Tees is connected by pipeline to the Phillips Consortium Ekofisk Oil Terminal at Seal Sands, giving it a direct feed from the North Sea. The Grangemouth refinery is also directly connected to the North Sea through the Forties Pipeline System.

In January 2003, BP announced that it would invest the $41 million necessary to reinstate the fluidized catalytic cracking unit (FCCU) at its Grangemouth refinery. The FCCU has not been operational since 2000, when it suffered fire damage. BP aims to have the unit operational in 2004, with output expected to be maintained at 19,000 bbl/d of high grade motor gasoline components.

The retail gasoline market is dominated by Esso (ExxonMobil), BP, Shell, TotalFinaElf, Texaco, and Conoco, which together account for 58 per cent of gasoline sales. Supermarkets now account for 8 per cent of retail sales. Total retail sales were 28 billion liters (7.4 billion gallons) in 2000. The transport sector consumed 74 per cent of petroleum products in 2000, whereas the energy industry consumed just 7 per cent. Fuel oil use has declined 30 per cent since 1998, as industrial and home-heating demand has dropped in favor of natural gas.

Natural Gas

The UK contains an estimated 26.0 trillion cubic feet (Tcf) of natural gas reserves, most of which are in non-associated gas fields located off the English coast in the Southern Gas Basin, adjacent to the Dutch North Sea sector. The UK shares the declining Frigg field with Norway (39.18 per cent to the UK), which produced only 17 million cubic feet in 2002, and will shut down in 2004. It also has a small share of the 0.44-Tcf Statfjord field (14.53 per cent). There are a few small fields on-shore, while the Irish Sea contains the large Morecambe and Hamilton fields. Morecambe alone accounts for up to 20 per cent of British natural gas production. Due to a decline in production and the maturation of its gas fields, the UK could become a natural gas importer in the next three years. Key producing gas fields in the North Sea include BP's 5.7-Tcf Leman, Chevron and Conoco's 3-Tcf Brittania, Shell's 1.7-Tcf Indefatigable and 0.8-Tcf Clipper, and TotalFinaElf's 0.85 Tcf Elgin. Key pipelines are the Scottish Area Gas Evacuation (SAGE) system to the St Fergus Terminal, which handles gas produced from a number of North Sea fields, including Britannia, the Beryl and Brae areas, and others in the central/northern North Sea, the Central Area Transmission System (CATS) that also goes to the Central North Sea, and takes gas from several fields, including Everest, Judy, and Jade, and others, and the Far North Liquids and Associated Gas System (FLAGS) that takes gas from the northern North Sea, including the Brent, Magnus, Cormorant, Ninian, and Hutton fields.

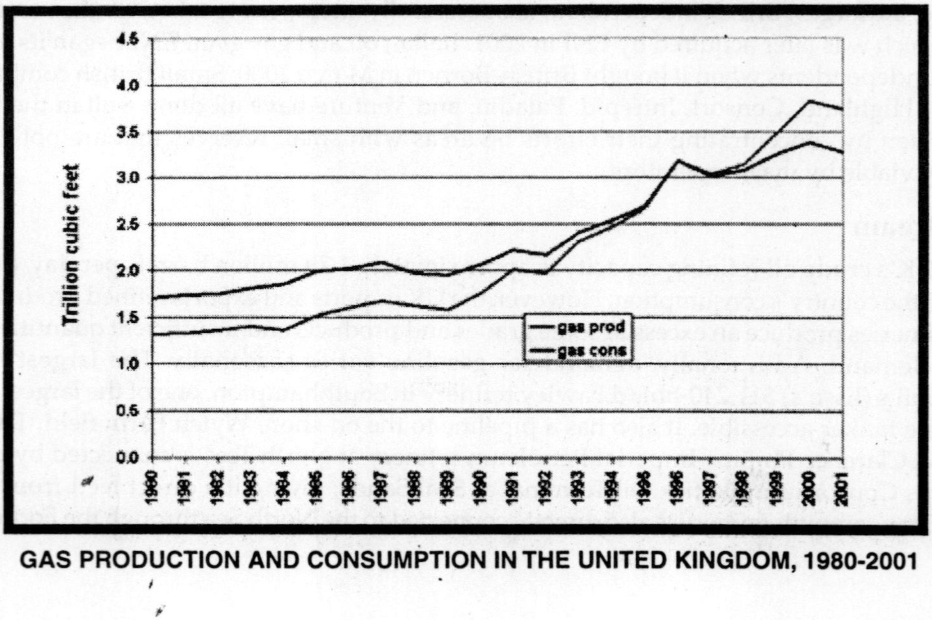

GAS PRODUCTION AND CONSUMPTION IN THE UNITED KINGDOM, 1980-2001

ATP Oil & Gas Corporation announced the successful testing of the Helvellyn well, which is located in the North Sea's Southern Gas Basin, in January 2003. The well tested at 60 million cubic feet per day (Mmcf/d). The initial discovery of natural gas in this field was made by BP in 1985, though the company did not develop it. At present, ATP is making the final preparations for production, which include modifications to the BP-operated Amethyst platform to support a 9.3 mile pipeline running from the platform to the well. Production is scheduled to begin during 2003.

A division of National Grid Transco (NGT) has submitted a proposal to build the UK's first LNG import terminal in 40 years. If approved in early 2003, the terminal would be built on the Thames estuary 37 miles east of London. NGT is also preparing to change the way in which in sells capacity in its transportation system, and hopes to draw in more Norwegian supplies.

The largest project to come online in 2001 (in March) in the British North Sea is the TotalFinaElf-operated Elgin/Franklin platform, which might prove to be the last big North Sea production platform. It is the world's largest high-pressure, high temperature development. The Elgin/Franklin platform has extensive processing facilities, unlike most North Sea platforms. The $2.3 billion platform is expected to last for 22 years in its location in the central North Sea, in the Graben area, off the coast of Scotland. It is to produce 700 million barrels of oil equivalent, about half condensate and half natural gas. This equates to peak production of 350 million cubic feet per day (Mmcf/d) of natural gas. The export pipelines are shared with the Shearwater field, and include a 294-mile gas pipeline to Bacton and a 24-mile condensate pipeline to the Marnock platform. The Shell-operated Shearwater field in the central North Sea was inaugurated in September 2000, and has reserves of 0.71 Tcf natural gas and 110 million barrels of condensate. Gas production is expected to peak at 375 Mmcf/d.

Industry Structure

British Gas was the monopoly supplier to the interruptible market until the passage of the 1995 Gas Act, which split the company into supply and shipping (British Gas Trading Limited) and while other functions remained with British Gas, including transport subsidiary Transco. In 1997, Centrica was demerged from from British Gas, and British Gas was renamed BG. Centrica is the holding company for British Gas Trading, British Gas Services, the Retail Energy Centers, and is the producer in the Morecambe fields. BG retained Transco, along with exploration and production, international

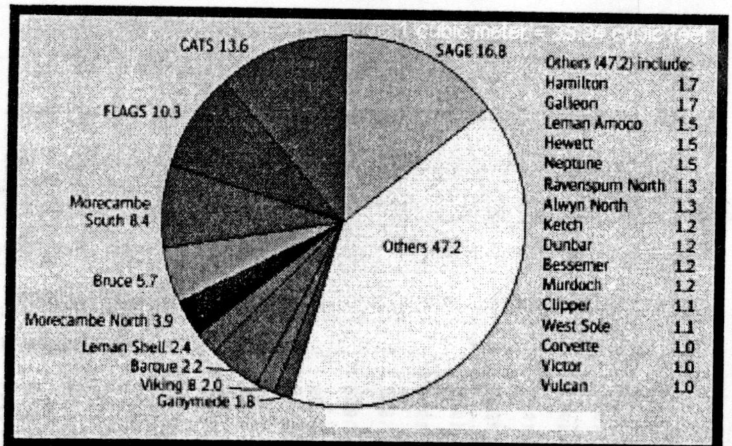

GAS PRODUCTINO (BILLION CUBIC METRES)

downstream, R&D and properties. In October 2000, BG again split, with Transco becoming part of a separate holding company Lattice Group. Independent Gas suppliers entered the firm (non-tariff) market in 1990, but the larger interruptible market (smaller customers) brought in competition in 1995. The consumer gas market was deregulated by region from October 1997 to June 1998, such that all residential and commercial customers could choose their supplier at the end of this process. At the end of 2000, suppliers other than British Gas Trading had captured 20-30 per cent of the market in many regions of the UK. In July 2001, Houston-based Dynegy purchased BG Storage from what remains of BG for $590 million, acquiring gas production wells and platforms, salt caverns, pipelines, and a natural gas processing terminal.

The UK's gas and electricity regulatory body is the Office of Gas and Electricity Markets (Ofgem). Ofgem has proposed reforming price controls on Transco's pipeline usage fees. The privatization of the UK's gas industry, leading to an increased gas supply and reduced prices, has helped gas to replace much of the UK's reliance on coal as a source for electricity generation. The natural gas share of utility fuels was 1 per cent in 1988 and is expected to increase to almost 50 per cent by 2010. Privatization in the UK has progressed well in advance of EU requirements.

Coal

Coal production and consumption in the United Kingdom have decreased dramatically since 1986. UK coal production fell from 119 million short tons (Mmst) in 1986 to 35.3 Mmst in 2000. In 2000, steam coal accounted for 80 per cent of coal demand, coking coal for 15 per cent, and anthracite for 5 per cent. Electricity demand accounts for 75 per cent of coal demand in the UK. In the late 1980s, coal accounted for about two-thirds of the United Kingdom's thermal electricity production. By 2001, only 37.2 per cent of UK electricity was coal-fired, and the figure is expected to drop further by the end of the decade. Coal mines are located primarily in central and northern England and southern Wales, with some coal mines also found in southern Scotland. The UK produced 34.9 million tons of bituminous coal and 353 thousand tons of anthracite coal in 2000. The UK also produces some metallurgical coke, although not in sufficient quantities to meet the demands of its steel industry. Net imports of coal have been rising each year, and in 2000 were 30.8 million Mmst.

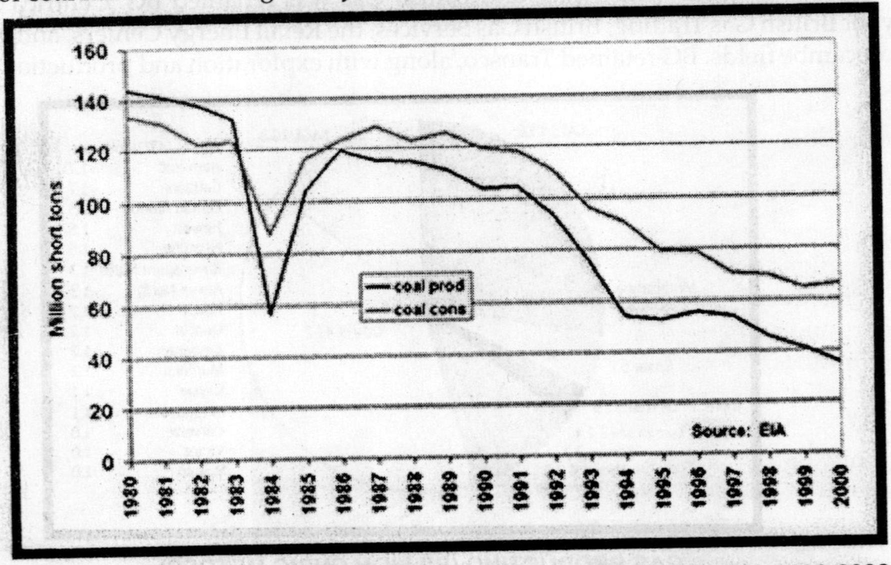

COAL PRODUCTION AND CONSUMPTION IN THE UNITED KINGDOM, 1980-2000

Between 1984 and 1985, the British coal miners' union staged a year-long strike. The strike dramatically altered energy production and consumption patterns in the United Kingdom for that year and precipitated the longer term decline of the industry (see graph). Employment in the industry has plummeted since the late 1980s. The United Kingdom began liberalizing its electricity market in 1989, and this liberalization is one of the major reasons for the decline of the country's coal industry. Prior to the privatization of electricity, the cost of domestic coal to electric utilities had far exceeded the cost of coal traded in international markets. The Central Electricity Generation Board (CEGB) had been the primary purchaser of British coal. The CEGB largely subsidized the British coal industry, purchasing domestic coal at above world market prices and then passing on those costs to consumers. This ended when National Power and PowerGen, two private electricity generation companies, were formed in the early 1990s, weakening the bargaining power of British Coal, the national coal company.

In 1992, the British coal industry reached a turning point. Growing competition from increasingly available natural gas, the imminent removal of the regional electricity companies' captive franchise supply markets, and newly-enacted pollution abatement goals all worked to initiate the steady decline of the industry. The industry was privatized in 1994, at which point RJB Mining bought the major British Coal assets and become the country's major producer. Mining Scotland and Celtic Energy are the other two remaining companies. The UK coal industry received its first subsidies since 1995 after the European Commission approved a modernization plan and aid scheme in November 2000. The aid, distributed in three Tranches, was originally slated to be used toward mines/production units with long-term economic viability on the world market, but which were having temporary difficulties as they restructured in an effort to reduce production costs. Though scheduled to expire in July 2002, a fourth Tranche was added to extend the scheme through to the end of the year. Only units that received subsidy payments in Tranche 3 were eligible to receive continued assistance in Tranche 4. The total amount of aid was not to exceed £110 million, although Tranche 4 added £10 million to provide miners at the Selby Coalfield (which will close by the end of 2004) with favorable redundancy terms; a multi-agency taskforce to provide regeneration and retraining for the Selby mine; £5,000 per person for ex-British Coal miners at Longannet mine; and consultation on an investment aid plan for the coal industry. Production costs over the period 1992 to 1999 fell 35 per cent, and the expectation is that these costs can fall further still, thereby allowing coal to remain a continually viable source of energy.

Some estimates predict that UK coal burn could fall to half its current level over the next eight to ten years, and that all coal-fueled electricity generation could end by the year 2016. New EU environmental directives, which are expected to further increase British coal production costs, are cited as a primary cause of this decline. RJB Mining is more optimistic about the future of British coal. RJB maintains that foreign coal prices will increase, making British coal more competitive, and that clean coal technology will allow power stations to burn increased amounts of coal without increased greenhouse gas emissions. Higher natural gas prices, gas-fired power plant outages for maintenance and repair, and reduced nuclear power led to a 14 per cent increase in coal consumption by power producers in 2000.

Electricity

The United Kingdom has 72.4 million kilowatts of installed electric capacity, about 80 per cent of which is thermal, 18 per cent nuclear, and 2 per cent hydropower. The country generated 355.8 billion kilowatt hours (bkwh) of electricity in 2000, making it the third-largest electricity market in Europe (behind Germany and France). In 2001, only 37.2 per cent of UK electricity was coal-fired. The remainder was accounted for by natural gas (31.5 per cent), and primary electricity sources such as nuclear and hydroelectricity (25.8 per cent).

Electricity privatization began in the early 1990s, and the final phase of transition ended in May 1999. Initially, all non-nuclear state-owned power stations were privatized and four major generating companies–PowerGen and National Power in England and Wales, and Scottish Power and Hydro-Electric in Scotland–were formed to operate the stations. The grid distribution system in England and Wales became the property of the National Grid Company. Regional Electricity Boards were privatized as separate distribution companies. Large customers were the first to be able to choose their suppliers, with all small customers (below 100 kW peak load) being able to choose by May 1999.

The number of electric generation companies in the United Kingdom has increased to 27 as a result of the liberalization process, according to DTI, such that 40 per cent of the UK's electricity was generated by these new companies in 2000. In March 2001, the structure of the electricity industry changed yet again. Under the former system, generators and suppliers in England and Wales traded electricity through the electricity pool, which was regulated by the National Grid Company, owner of the transmission network. The New Electricity Trading Arrangements (NETA) changed this to a system based on bilateral trading between generators, suppliers, traders, and customers. The system includes forwards and futures markets, a balancing mechanism to enable the National Grid Company to balance the system, and a settlement process. Dallas-based TXU purchased United Utilities' retail electricity and natural gas business, Norweb Energi, for $465 million in August 2000. This, added to TXU's European retail business Eastern Energy, created the UK's largest electricity retailer, with over 5.6 million customers.

The UK electricity sector encountered several problems in 2002, most significantly the financial troubles of some large companies. TXU Corp. sought to bolster its finances by selling its assets in the UK to Powergen, a subsidiary of E.ON. However, the deal left TXU with five long-term contracts to purchase large amounts of wholesale electricity in the UK at fixed prices, which now exceed market levels by about $15/MWh. While restructuring of the electricity industry was intended to end market abuses by large electricity generators and increase the market's efficiency, the system has been so successful that in the last four years, prices have decreased by about 40 per cent.

TXU's difficulties have caused a domino effect in the UK energy market. In October 2002, UK Coal announced that it would suspend shipments of coal to the Drax power plant, the nation's largest, because the firm AES was uanble to pay the balance of its multi-million dollar bill. AES was unable to pay the bill because it was owed a large amount of money by TXU. AES had managed to pay half of the outstanding balance, but UK Coal was still reluctant to resume coal shipments to Drax due to concern that the remainder of the balance might not be paid soon enough. A deal was reached in late October and shipments were resumed.

Powergen, with 2.6 million retail customers as well as 14 per cent of electricity generation in England and Wales, merged with Louisville-based LG&E Energy in December 2000. Powergen announced in October 2002 that it will close two of its own plants and several of TXU's plants in order to reduce the UK's 25-30 per cent overcapacity in electricity generation. Powergen will also take over TXU's 5.6 million customers when the deal is finalized.

In Scotland, the two main companies, Scottish Power and Scottish and Southern Energy, cover the full range of electricity provision. Ofgem has made proposals to further reform the Scottish power market. Northern Ireland, part of the United Kingdom but not part of Great Britain, is served by Northern Ireland Electricity, one of the largest companies in Northern Ireland and part of the Viridian Group. Northern Ireland has a separate electricity and gas regulatory body, Ofreg. The grids of Northern Ireland and the Republic of Ireland are connected for electricity import/export.

Nuclear

In 1995, the government announced that it would privatize its more modern nuclear stations while retaining ownership of older stations. In 1996, more modern stations were privatized and British Energy became the holding company of Nuclear Electric and Scottish Nuclear, which merged in 1998 to form British Energy Generation, the nation's largest private nuclear generator and the world's first wholly privatized nuclear utility. British Energy operates eight nuclear power stations in the UK (as well as several in the U.S. through its AmerGen subsidiary that is jointly owned with PECO). Each station consists of two advanced gas-cooled reactors, except Sizewell B, which is a modern pressurized-water reactor. Nuclear power stations were not privatized simultaneously with non-nuclear stations.

No new nuclear plants have been built since 1995, but because of limited domestic coal and gas reserves in the long run, new construction is under discussion, at least to maintain nuclear's market share as older nuclear plants are retired. Of the UK's 33 reactors, 26 are of the old Magnox design. Six of the Magnox reactors are being decommissioned, as well as the Dounreay prototype fast reactor. The remaining Magnox plants are run by the state-owned British Nuclear Fuels. British Nuclear Fuels operates the Sellafield reprocessing plant, and is one of only two companies in the world that provides reprocessing and recycling technologies. The British nuclear industry is regulated by the Department of Trade and Industry's Nuclear Directorate.

British Energy PLC encountered serious financial problems in 2002, which prompted the UK government to provide a £650 million ($1 billion) loan and £10 billion ($16 billion) in subsidies for the company. Nuclear energy provides nearly 20 per cent of the UK's electricity at present, and is reported to be losing approximately $3 million a day, due to an increase in the cost of producing electricity combined with lower wholesale market prices. In an effort to improve its finances and avoid government administration, British Energy agreed to sell its Canadian subsidiary, Bruce Power, to a consortium led by Cameco Corporation for £300 million. Other members of the consortium include TransCanada Pipelines and Borealis Capital. In February 2003, BE agreed to a four-year plan to supply 15 per cent of its output to Centrica, the nation's largest natural gas and electricity supplier, in a deal that is hoped will cushion the ailing nuclear energy company from further declines in the wholesale energy market.

Environment

With a reduction in sulfur dioxide and carbon dioxide emissions, environmental conditions in the United Kingdom have improved over the past couple of decades. Some of these environmental improvements, such as a reduction in air pollution, can be attributed to the United Kingdom's energy use choices. Partially as a result of deregulation and the elimination of coal subsidies, coal's share of total primary energy consumption is gradually being replaced by natural gas.

Improvements in energy efficiency have led to a gradual reduction in both energy and carbon intensity. In 1999, energy intensity in the United Kingdom registered 7.8 thousand Btu per $1995, decreasing to 7.3 thousand Btu per $1995 in 2000, a 6 per cent decline. However, carbon intensity in 2000 registered 0.11 metric tons of carbon per thousand $1995, maintaining 1999 levels. Per capita energy consumption, at 166.1 million Btu in 2000, is down slightly from 1999.

As the United Kingdom enters the 21st century, many energy and environment-related policies reflect the country's awareness of climate change issues. With introduction of the Climate Change Levy in 2001, and its exemption for renewable energy resources like solar and wind, these alternative sources of energy are beginning to gain more attention. For example, the United Kingdom hopes to increase the share of electricity generated by renewables from the current 2.3 per cent, to 10 per cent by 2010.

Country Overview

Population (2002E): 59.8 million

Location/Size: Western Europe, islands including the northern one-sixth of the island of Ireland between the North Atlantic Ocean and the North Sea, northwest of France/244,820 sq km (slightly smaller than Oregon)

Capital City: London

Language: English

Ethnic groups: English 81.5 per cent, Scottish 9.6 per cent, Irish 2.4 per cent, Welsh 1.9 per cent, Ulster 1.8 per cent, West Indian, Indian, Pakistani, and other 2.8 per cent

Religions: Anglican and Roman Catholic 40 million, Muslim 1.5 million, Presbyterian 800,000, Methodist 760,000, Sikh 500,000, Hindu 500,000, Jewish 350,000 (1991 est.)

Defense (8/98): Army, 113,900; Navy, 44,500; Air Force, 52,540

Economic Overview

Currency: Pound sterling

Exchange Rate (1/30/03): 1 US Dollar = 0.61 pounds

Gross Domestic Product (2002E): $1,527 billion

Real GDP Growth Rate (2002E): 1.6 per cent *(2003E):* 2.7 per cent

Inflation Rate (consumer prices, 2002E): 2.0 per cent *(2001E):* 2.3 per cent

Unemployment Rate (2002E): 3.1 per cent

Merchandise Exports (2002E): $283 billion

Merchandise Imports (2002E): $324 billion

Merchandise Trade Deficit (2002E): $41 billion

Current Account Balance (2002E): -$22.8 billion

Major Trading Partners: United States, Germany, France, Netherlands

Major Exports: Food, beverages, and tobacco; crude materials, fuels, chemicals, machinery, transport equipment

Major Imports: Food, beverages, and tobacco; crude materials, fuels, chemicals, machinery, transport equipment

Energy Overview

Proven Oil Reserves (1/1/03): 4.9 billion barrels

Oil Production (2002E): 2.53 million bbl/d, of which 2.25 million bbl/d was crude oil

Oil Consumption (2002E): 1.7 million bbl/d

Net Oil Exports (2002E): 0.83 million bbl/d

Crude Oil Refining Capacity (1/1/03E): 1.78 million bbl/d

Natural Gas Reserves (1/1/03E): 26.0 trillion cubic feet (Tcf)

Natural Gas Production (2000E): 3.8 Tcf

Natural Gas Consumption (2000E): 3.4 Tcf

Natural Gas Net Exports (2000E): 0.4 Tcf

Major Systems: Brent, Ninian, Forties, Flotta, Fulmar

Major Fields: E. Brae, Brent, Forties, Magnus, Miller, Scott

Oil and Gas Companies: Amerada Hess, BP Amoco, BHP, Chevron, ExxonMobil, Kerr-McGee, Phillips, Ranger Oil, Shell, Texaco

Recoverable Coal Reserves (12/31/99E): 1.65 billion short tons

Coal Production (2000E): 35.3 million short tons (Mmst)

Coal Consumption (2000E): 66.1 Mmst

Electrical Generation Capacity (1/1/00): 72.4 gigawatts (79.8 per cent thermal, 17.9 per cent nuclear, 2.0 per cent hydro, 0.2 per cent other)

Electricity Generation (2000E): 355.8 billion kilowatt hours (bkwh)

Electricity Consumption (2000E): 345.0 bkwh

Environmental Overview

Total Energy Consumption (2000E): 9.9 quadrillion Btu* (2.6 per cent of world total energy consumption)

Energy-Related Carbon Emissions (2000E): 147.8 million metric tons of carbon (2.5 per cent of world carbon emissions)

Per Capita Energy Consumption (2000E): 166.1million Btu (vs. U.S. value of 355.8 million Btu)

Per Capita Carbon Emissions (2000E): 2.5 metric tons of carbon (vs. U.S. value of 5.6 metric tons of carbon)

Energy Intensity (2000E): 7,632 Btu/$1995 (vs U.S. value of 10,918 Btu/$1995)

Carbon Intensity (2000E): 0.11 metric tons of carbon/thousand $1995 (vs U.S. value of 0.17 metric tons/thousand $1995)

Fuel Share of Energy Consumption (2000E): Oil (35.1 per cent), Natural Gas (36.3 per cent), Coal (15.4 per cent)

Fuel Share of Carbon Emissions (2000E): Oil (38.9 per cent), Natural Gas (35.5 per cent), Coal (25.5 per cent)

Status in Climate Change Negotiations: Annex I country under the United Nations Framework Convention on Climate Change. Under the negotiated Kyoto Protocol (signed on April 29th, 1998–not yet ratified), the UK has agreed to reduce greenhouse gases 8 per cent below 1990 levels by the 2008-2012 commitment period.

Major Environmental Issues: Sulfur dioxide emissions from power plants contribute to air pollution; some rivers polluted by agricultural wastes and coastal waters polluted because of large-scale disposal of sewage at sea.

Major International Environmental Agreements: A party to Conventions on Air Pollution, Air Pollution-Nitrogen Oxides, Air Pollution-Sulphur 94, Air Pollution-Volatile Organic Compounds, Antarctic-Environmental Protocol, Antarctic Treaty, Biodiversity, Climate Change, Desertification, Endangered Species, Environmental Modification, Hazardous Wastes, Law of the Sea, Marine Dumping, Marine Life Conservation, Nuclear Test Ban, Ozone Layer Protection, Ship Pollution, Tropical Timber 83, Tropical Timber 94, Wetlands and Whaling.

The total energy consumption statistic includes petroleum, dry natural gas, coal, net hydro, nuclear, geothermal, solar, wind, wood and waste electric power. The renewable energy consumption statistic is based on International Energy Agency (IEA) data and includes hydropower, solar, wind, tide, geothermal, solid biomass and animal products, biomass gas and liquids, industrial and municipal wastes.

89

United States of America

The United States of America is the world's largest energy producer, consumer, and net importer. It also ranks twelfth worldwide in reserves of oil, sixth in natural gas, and first in coal.

General Background

As of mid-May 2003, the U.S. economy continued to grow slowly and uncertainly, with first quarter 2003 real growth in Gross Domestic Product (GDP) of only 1.6 per cent. On November 6, 2002, the U.S. Federal Reserve had cut its interest rate target by a half-per centage point, to 1.25 per cent, in an effort to stimulate an economic recovery. The move came following data in September and October indicating an increase in the unemployment rate (from 5.6 per cent to 5.7 per cent), slower-than-expected economic growth (3.1 per cent in the third quarter, following a 1.3 per cent gain in the second quarter), a decline in job creation, a sharp decrease in consumer confidence, and a reduction in consumption spending. Since then, however, unemployment has risen to 6.0 per cent in April, while industrial output fell and factory capacity utilization hit a two-decade low. Following the removal of uncertainty surrounding the months-long buildup to war with Iraq, however, some analysts believe that the US economy is poised for a stronger recovery in coming months.

The recent difficulties experienced by the U.S. economy follow a period during the mid- and late-1990s of strong growth, low inflation, low unemployment, rapid productivity growth, and a booming stock market. Real (inflation adjusted) U.S. gross domestic product (GDP) growth for 2003 now is expected at 2.6 per cent, up from 2.4 per cent growth in 2002.

For Fiscal Year (FY) 2002, the federal unified budget ran an estimated deficit of $180 billion after running a surplus of around $127 billion in 2001. For FY 2003, a budget deficit of more than $300

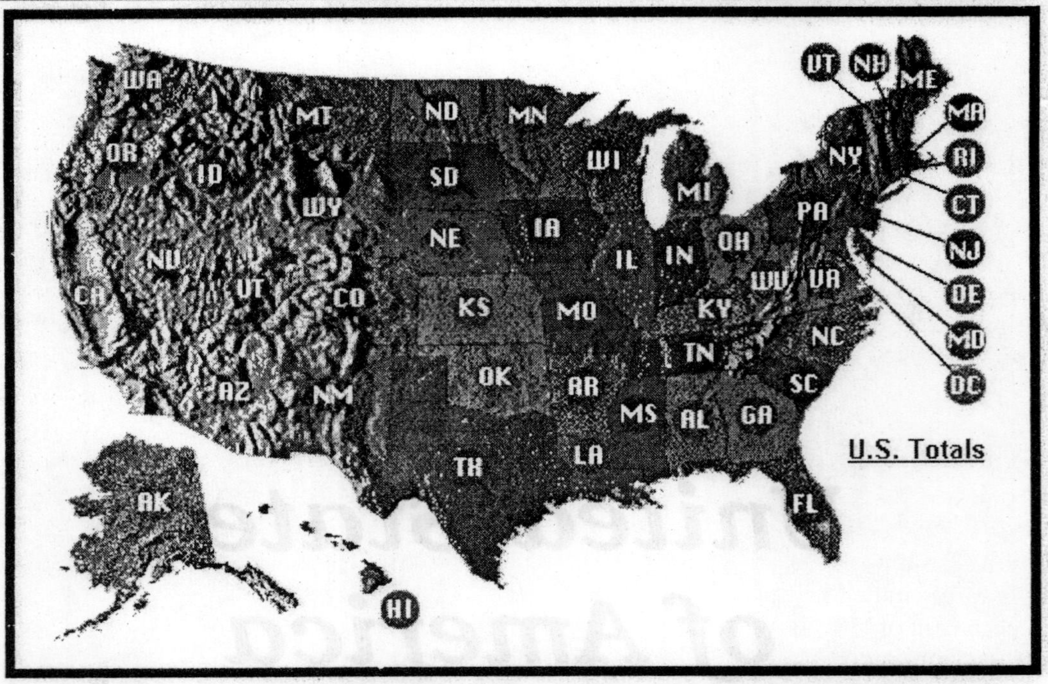

billion is possible. The turn from surplus to deficit has come about as a consequence of several factors, including economic slowdown, tax rebates and cuts, and increased government spending on homeland security and war following the Setpember 11, 2001 terrorist attacks. Meanwhile, the U.S. merchandise trade deficit is estimated at $484 billion for 2002. This deficit mainly reflects the relative strength of the U.S. economy compared to major U.S. trading partners. The current account deficit now is running at over 4 per cent of GDP, compared to 1.7 per cent in 1997.

In mid-May 2001, the Bush administration issued a series of energy policy recommendations as part of its new National Energy Policy Report, developed by a task force led by Vice President Dick Cheney. In August 2001, the U.S. House of Representatives passed an energy bill (the "Securing America's Future Energy"–SAFE–Act of 2001) which contained many of the energy plan's recommendations. In April 2002, the U.S. Senate passed its own version of an energy bill, which must be reconciled with the House version.

Oil

According to the Oil and Gas Journal, the United States had 22.4 billion barrels of proved oil reserves as of January 1, 2003, twelfth highest in the world. These reserves are concentrated overwhelmingly (over 80 per cent) in four states–Texas (24 per cent including the state's reserves in the Gulf of Mexico), Alaska (22 per cent), Louisiana (20 per cent including the state's reserves in the Gulf of Mexico), and California (19 per cent, including the state's Federal Offshore reserves). U.S. proven oil reserves have declined by around 20 per cent since 1990, with the largest single-year decline (1.6 billion barrels) occurring in 1991.

During 2002, the United States produced around 8.1 million barrels per day (MMBD) of oil, of which 5.8 MMBD was crude oil, and the rest natural gas liquids and other liquids. U.S. total oil production in 2002 was down sharply (around 2.5 MMBD, or 24 per cent) from the 10.6 MMBD averaged in 1985. U.S. crude oil production, which declined following the oil price collapse of late 1985/early 1986, leveled off in the mid-1990s, and began falling again following the sharp decline in oil prices of late 1997/early 1998. With the rebound in world oil prices since March 1999, U.S. crude production basically leveled off once again in 2000 and 2001, rising slightly in 2002. Despite this increase, U.S. crude production remains near 50-year lows.

The United States contains over 500,000 producing oil wells, the vast majority of which are considered "marginal" or "stripper" wells, generally producing only a few barrels per day of oil. During 2002, top oil producing areas included the Gulf of Mexico (1.6 million bbl/d), Texas onshore (1.1 million bbl/d), Alaska's North Slope (954,000 bbl/d), California (707,00C bbl/d), Louisiana onshore (274,000 bbl/d), Oklahoma (181,000 bbl/d), and Wyoming (150,000 bbl/d).

Domestic oil exploration and development spending by U.S. major oil companies rebounded during 2001 from the deep cuts made during the oil price collapse of 1997/1998. Improved technology and new or increased offshore production in the Gulf of Mexico (including at deepwater areas beyond the continental shelf) also could help matters. In 2000, deepwater production in the Gulf of Mexico for the first time surpassed shallow water production. In January 2000, Chevron and Shell–the largest producer in the Gulf of Mexico–signed an agreement to share drilling rigs and to drill exploratory wells jointly in the deep-water Gulf of Mexico. In August 2002, a U.S. government lease sale for the western Gulf of Mexico produced bids totaling $182 million. Bidders included Amerada Hess, Kerr-McGee, Dominion Exploration and Production, Shell, and Nexen.

In April 2003, TotalFinaElf announced that it was the highest bidder on four deepwater Gulf of Mexico exploration blocks, located in the Mississippi Canyon and Atwater Valley areas of the Gulf. The blocks were among 561 tracts offered by the Minerals Management Service in Lease Sale 185.

Overall, production from deepwater areas of the Gulf of Mexico has been increasing rapidly, with deepwater wells accounting for about two-thirds of total U.S. Gulf output. Large fields include ExxonMobil's $1.1 billion Hoover-Diana development (which started up in May 2000 and is producing 80,000 bbl/d), plus several by BP: the $2 billion Atlantis project (scheduled to come online in 2005); Crazy Horse (the largest single field every discovered in the Gulf of Mexico), Crosby, Holstein, King,

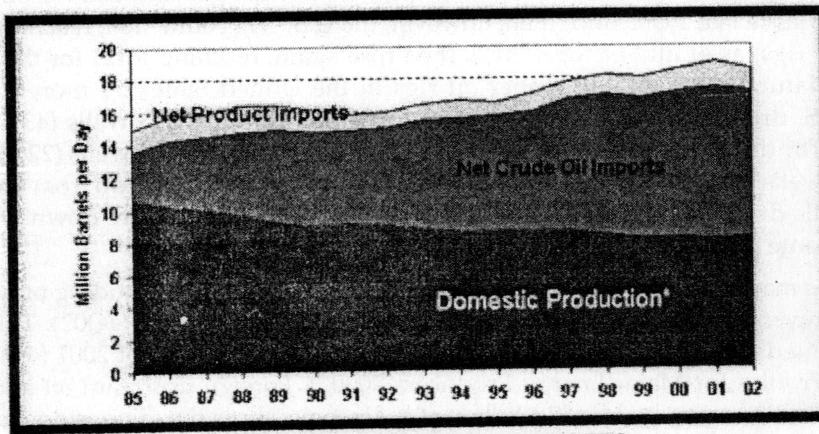

U.S. OIL PRODUCTION AND IMPORTS

Note: Domestic Production includes crude oil, natural gas liquids and other hydrocarbons and alcohol production, but does not include refinery gain.

King's Peak, Mad Dog, Marlin, and Nakika fields. BP has stated that it plans to accelerate its deepwater Gulf of Mexico production plans, possibly including construction of a $1 billion deep-sea pipeline, and to increase its production from around 200,000 bbl/d currently to 750,000 bbl/d in 2007. This will require billions of dollars worth of investment.

Crude oil production in the lower 48 states is expected to increase slightly in 2003, to 4.85 million bbl/d, while Alaskan crude production falls slightly, to 0.95 million bbl/d. Alaskan production, which accounts for around 16 per cent of the U.S. total, is down more than 50 per cent from the 2.0 MMBD reached during the peak year of 1988. Most of Alaska's oil output comes from the giant Prudhoe Bay Field, and is transported via the Alyeska pipeline. A new oilfield, known as Alpine (owned 78 per cent by Phillips Petroleum, 22 per cent by Anadarko), began production in November 2000. Alpine represents the largest North American onshore oil discovery in a decade, and currently is producing aroudn 100,000 bbl/d of high quality, light crude oil. Production at Alpine is to be maintained using tie-ins to the Nanuq and Fiord satellite fields beginning in 2006. Phillips has been the largest oil producer in Alaska since acquiring Arco's Alaska fields in early 2000. In November 2000, two oil and natural gas lease sales conducted by the State of Alaska drew bids worth $11 million for offshore tracts in the Beaufort Sea and onshore in the North Slope.

In other news from Alaska, the critical Trans-Alaska Pipeline System (TAPS) shut down briefly in early November 2002 due to an earthquake. In October 2001, TAPS also was shut down for a short time after being punctured by a gunshot. In early 2000, the Energy Information Administration (EIA), in response to a Congressional request, issued a report on potential oil reserves and production from the Arctic National Wildlife Refuge (ANWR). The report, which cited a 1998 U.S. Geological Survey study of ANWR oil resources, projected that for the mean resource case (10.3 billion barrels technically recoverable), ANWR peak production rates could range from 1.0 to 1.35 MMBD, with initial ANWR production possibly beginning around 2010, and peak production 20-30 years after that.

According to Baker Hughes Inc., which has tallied weekly U.S. drilling activity since 1940, domestic oil and natural gas drilling has rebounded sharply since the low point of 488 reached in late April 1999 following the oil price collapse of late 1997. In mid-October 2001, for instance, the U.S. weekly "rig count" reached the 1,141 mark (933 for natural gas and 208 for oil), close to the highest number since late 1990. Since then, however, the U.S. "rig count" fell, reaching 843 (703 gas rigs and 137 oil rigs) as of mid-October 2002, then rose again, reaching 1,021 for the week ending May 9, 2003. Natural gas rigs outnumber oil rigs in the United States by more than five-fold. Historically, U.S. drilling activity peaked in 1981, with a total of 91,553 wells (43,598 oil, 20,166 natural gas, 27,789 dry wells) drilled in that year. For 2001, a total of 34,179 wells (22,083 natural gas wells, 8,060 oil wells, and 4,036 dry wells) were drilled in the United States, up from the low point of 18,377 total wells drilled in 1999. For 2002, total wells drilled (24,540) were down sharply–28 per cent–from the same period a year earlier.

Twenty-two major energy companies reported overall net income (excluding unusual items) of $6.6 billion on revenues of $149 billion during the fourth quarter of 2002 (4Q02). This level of net income represented a 296 per cent increase relative to the fourth quarter of 2001 (4Q01) (see EIA's "Performance Profiles for Major Energy Producers 2000"). Foreign upstream oil and natural gas production operations accounted for $3.6 billion of net income, about the same as domestic upstream oil and natural gas production operations ($3.6 billion) and domestic downstream natural gas and power operations ($0.6 billion).

Independent oil and natural gas producers, oil field companies and refiner/marketers reported a decrease in net income (down 39 per cent) during 4Q02 compared to 4Q01. This decline in net income was due to a sharp decline in oil and gas drilling activity. Refining margins, on the other hand, increased due in part to higher oil product prices.

Consumption/Marketing

The United States consumed an average of about 19.7 MMBD of oil in 2002. Of this, 8.9 MMBD (or 45 per cent of the total) was motor gasoline, 4.8 MMBD (24 per cent) "other oils," 3.8 MMBD (19 per cent) distillate fuel oil, 1.6 MMBD (8 per cent) jet fuel, and 0.7 million bbl/d (3 per cent) residual fuel oil. U.S. oil demand is expected to increase by about 3 per cent (540,000 bbl/d) in 2003. Following the September 11 terrorist attacks, U.S. jet fuel demand fell sharplv. For all of 2002, U.S. jet fuel consumption was down 7 per cent compared to 2000, and still had not recovered to levels seen in August 2001, prior to 9/11.

Imports/Exports

The United States averaged *total gross oil* (crude and products) imports of an estimated 11.4 MMBD during 2002, representing around 58 per cent of total U.S. oil demand. Around two-fifths of this oil came from OPEC nations, with Persian Gulf sources accounting for about one-fifth of total U.S. oil imports. Overall, the top suppliers of oil to the United States during 2002 were Canada (1.9 MMBD), Saudi Arabia (1.6 MMBD), Mexico (1.5 MMBD), and Venezuela (1.4 MMBD).

U.S. Energy Sanctions Issues

The United States maintains energy sanctions against several countries, including Iran and Libya (the status of Iraq, which was placed under comprehensive sanctions after its invasion of Kuwait in August 1990, remains in flux as of late May 2003). Iran and Libya are affected by the Iran-Libya Sanctions Act (ILSA), passed unanimously by the U.S. Congress and signed into law by President Clinton in August 1996. ILSA imposes mandatory and discretionary sanctions on non-U.S. companies which invest more than $20 million annually (lowered in August 1997 from $40 million) in the Iranian oil and natural gas sectors. The passage of ILSA was not the first U.S. sanction

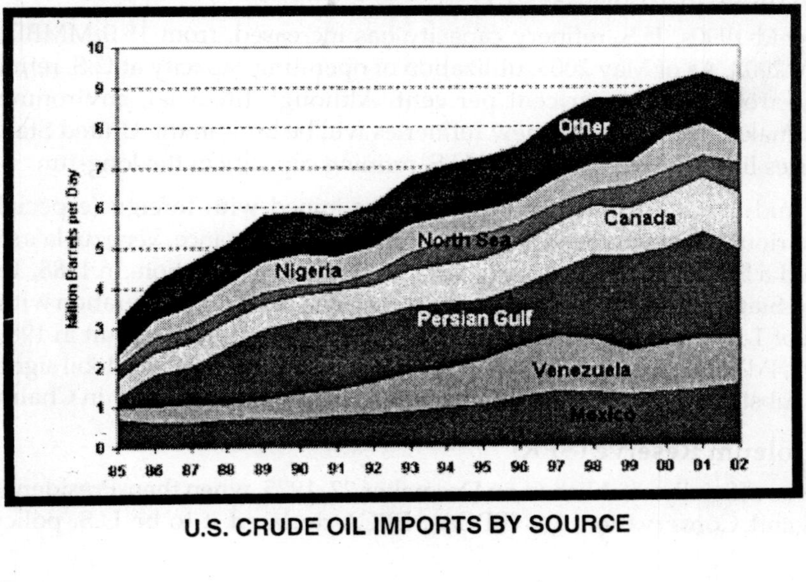

U.S. CRUDE OIL IMPORTS BY SOURCE

against Iran. In early 1995, President Clinton signed two Executive Orders which prohibited U.S. companies and their foreign subsidiaries from conducting business with Iran. The Orders also banned any "contract for the financing of the development of petroleum resources located in Iran." On March 13, 2001, President Bush, citing threats posed by Iran to U.S. national security, extended Clinton's two Executive Orders on Iran for another 6 months. On August 3, 2001, President Bush signed into law the ILSA Extension Act of 2001. This Act provides for a 5-year extension of ILSA with amendments that affect certain of the investment provisions.

Attempts by the United States to implement ILSA have run into opposition from a number of foreign governments. The European Union (EU) opposes the enforcement of ILSA sanctions on its members, and on November 22, 1996 passed resolution 2271 directing EU members to not comply with ILSA. On May 18, 1998, the EU and the U.S. reached an agreement on a package of measures to resolve the ILSA dispute at the EU/U.S. Summit in London, but the Summit deal is contingent upon acceptance by the U.S. Congress before full implementation may take place.

On April 5, 1999, following the Libyan handover of two suspects in the 1988 bombing of Pan Am flight 103 to stand trial before a Scottish Court in the Netherlands, the United States modified its Libya sanctions on April 28, 1999 to allow shipments of donated clothing, food and medicine for humanitarian reasons (trade in informational materials such as books and movies is also allowed). However, all other U.S. sanctions against Libya remain in force. On February 1, 2001, one suspect was convicted by the Scottish court, while another was acquitted. The U.S. and British governments both said that they still expected Libya to accept responsibility for the murders, which Libya has said it would not do.

Refining/Downstream

The United States experienced a steep decline in refining capacity between 1981 and the mid-1990s. Between 1981 and 1989, for instance, the number of U.S. refineries fell from 324 to 204, representing a loss of 3 MMBD in operable capacity, while refining capacity utilization increased from 69 per cent to 86 per cent. Much of the decline in U.S. refining capacity resulted from the 1981 deregulation (elimination of price controls and allocations), which effectively removed the major prop from underneath many marginally profitable, often smaller, refineries.

Since the mid-1990s, U.S. refinery capacity has increased, from 15.0 MMBD in 1994 to 16.6 MMBD in early 2003. As of May 2003, utilization of operating capacity at U.S. refineries reportedly was averaging around 92-94 per cent per cent. Although financial, environmental, and legal considerations make it unlikely that new refineries will be built in the United States, expansion at existing refineries likely will increase total U.S. refining capacity in the long-run.

Since the mid-1980s, several U.S. refiners have joined with foreign (especially Venezuelan) companies in various joint venture arrangements. In 1986, for instance, Venezuela's state oil company PdVSA acquired a 50 per cent interest in Citgo's U.S. refining operation. In 1988, Texaco and Saudi Aramco created Star Enterprise, an integrated refining and marketing operation with three refineries and a network of Texaco gasoline stations. Unocal and PdVSA followed suit in 1989, forming Uno-Ven Co. (in 1997, PdVSA bought out Unocal's share). In late October 1997, Mobil signed an agreement with a PdVSA subsidiary on joint ownership of the 170,000-bbl/d refinery in Chalmette, Louisiana.

Strategic Petroleum Reserve (SPR)

The SPR was officially established on December 22, 1975, when then-President Ford signed the Energy Policy and Conservation Act (EPCA). EPCA declared it to be U.S. policy to establish a petroleum reserve of up to 1 billion barrels. In order to store the reserve oil, the U.S. government in April 1977 acquired several salt caverns along the Gulf of Mexico coastline. The first crude oil was delivered to the SPR on July 21, 1977, and stored at the West Hackberry storage site near Lake Charles, LA. Other major storage sites include: Bryan Mound and Big Hill in Texas; and Bayou Choctaw, the St. James Terminal in Louisiana, with a total storage capacity of 700 million barrels.

Approximately $327 million worth of SPR oil was sold off in 1996, and an additional $220 million in 1997. On September 22, 2000, President Clinton authorized the release of 30 million barrels of oil from the SPR over 30 days in an attempt to bolster U.S. oil supplies and to alleviate possible shortages of heating oil during the upcoming winter. The release took the form of a "swap" (bidding results were announced on October 4) in which crude oil volumes drawn from the SPR is to be replaced by the recipients at a later date.

In mid-November 2001, President Bush directed the Department of Energy (DOE) to fill the SPR to its capacity of 700 million barrels in order to "maximize long-term protection against oil supply disruptions." Under the DOE plan, the SPR is to be filled with "royalty in kind" (RIK) oil. As of May 14, 2003, the SPR contained around 600 million barrels of oil–the largest emergency oil stockpile in the world. The SPR has a maximum drawdown capability of 4.3 million bbl/d for 90 days, with oil beginning to arrive in the marketplace 15 days after a presidential decision to initiate a drawdown. The SPR drawdown rate declines to 3.2 million bbl/d from days 91-120, to 2.2 million bbl/d for days 121-150, and to 1.3 million bbl/d for days 151-180.

Under EPCA, there is no preset "trigger" for withdrawing oil from the SPR. Instead, the President determines that drawdown is required by "a severe energy supply interruption or by obligations of the United States" under the International Energy Agency. EPCA defines a "severe energy supply interruption" as one which: 1) "is, or is likely to be, of significant scope and duration, and of an emergency nature;" 2) "may cause major adverse impact on national safety or the national economy" (including an oil price spike); and 3) "results, or is likely to result, from an interruption in the supply of imported petroleum products, or from sabotage or an act of God."

Should the President decide to order an emergency drawdown of the SPR, oil would be distributed mainly by competitive sale to the highest bidder(s). This would be accomplished in a 4-step process, including a "Notice of Sale," receipt of bids, selection of bidders, and finally delivery of oil.

Oil Mergers and Acquisitions

Pushed in part by low oil prices during 1998 and into early 1999, but also by the desire for oil reserves, cost cutting, and higher refining/marketing shares, merger activity in the oil business accelerated sharply during 2000 and 2001 (before slowing considerably in 2002 and early 2003). The largest merger/acquisition announcements came from BP and Amoco, Exxon and Mobil, BP Amoco and ARCO, and, most recently, Chevron and Texaco. BP and Amoco completed their $53 billion merger on December 31, 1998, a day after the deal received regulatory approval from the U.S. Federal Trade Commission (FTC), subject to certain conditions.

In September 2002, U.S. regulators approved the purchase of Pennzoil-Quaker State Co. by Shell Oil Co. The deal, first reported in March 2002, was for $1.8 billion (with Shell also assuming $1.1 billion of Pennzoil-Quaker State debt). The transaction combines Shell's 3 per cent share of the U.S. market for passenger car motor oil with Pennzoil-Quaker State's 35 per cent share, making Shell the No. 1 U.S. lubricants company. Shell also adds Pennzoil-Quaker State's 46,200 barrels-per-day Shreveport, Louisiana refinery and more than 2,000 Jiffy Lube outlets. In October 2002, Shell announced that it would close or sell seven U.S. blending and packaging plants as part of its ongoing merger.

In November 2001, Phillips Petroleum and Conoco Inc. agreed to merge in a $15.2 billion transaction. The merger was completed in August 2002, creating a new company called ConocoPhillips. ConocoPhillips ranks as the sixth-largest oil and gas company in the world, the largest U.S. refiner, and the third-largest U.S.-based energy company.

Another major oil industry merger/acquisition was announced in October 2000, this time between Chevron and Texaco. According to the announcement, Chevron is to buy Texaco for $35 billion in stock, creating the world's fourth largest energy company (behind ExxonMobil, Shell, and BP). The deal received regulatory approval in early October 2001, and was approved by shareholders of the two companies on October 9, 2001, creating ChevronTexaco.

In November 2000, Russia's Lukoil announced that it intended to purchase Getty Petroleum Marketing for $71 million. Lukoil eventually intends to switch Getty's 1,300 retail outlets in the Northeastern and Middle Atlantic states to the Lukoil brand name. The purchase represents the first takeover of a publicly traded U.S. company by a Russian firm. In late January 2001, Getty shareholders approved the the buyout.

On April 13, 2000, the FTC approved the $27.6 billion BP Amoco-ARCO deal. This followed the March 15, 2000 announcement by Phillips Petroleum that it had agreed to purchase ARCO's assets in Alaska for $6.5 billion. The sale was made as part of an effort to secure approval from the FTC. On the same day, the FTC announced that it had suspended its antitrust lawsuit seeking to block the merger, citing progress in talks with the companies involved. Among other issues, the FTC was concerned that the BP Amoco-ARCO merger would control about 75 per cent of Alaskan North Slope crude oil output and over 70 per cent of the critically important TAPS line, potentially hurting consumers on the U.S. west coast. BP Amoco agreed to sell some pipeline and oil storage holdings in Cushing, Oklahoma. The new company (now called BP) will rank in the top three private oil companies in the world, along with ExxonMobil and Royal Dutch/Shell.

Meanwhile, the $81 billion merger between Exxon and Mobil, which formed the world's largest privately owned petroleum company (in terms of revenues), was approved by the FTC on December 1, 1999, subject to the divestiture of 2,400 service stations and other assets (on December 3, 1999, 1,740 of these stations were sold to Tosco, the largest U.S. independent oil refiner). In a related development, in April 2000, Duke Energy said that it had agreed to acquire Mobil's European natural gas trading and marketing business. The sale of Mobil's natural gas operations in Europe was required by the European Commission as part of its approval of the ExxonMobil merger.

Besides these large mergers, several defensive mergers among smaller, independent oil companies also have been unveiled recently, including Kerr-McGee Corp.'s (KMG) $1.86 billion takeover of Oryx Energy Co. (ORX), and an agreement between Seagull Energy Corp. (SGO) and Ocean Energy Inc. (OEI) to merge in a $1.1 billion deal (approved by shareholders in March 1999). On July 14, 2000, Anadarko Petroleum announced the closing of its merger transaction with the Union Pacific Resources Group. Union Pacific became a wholly owned subsidiary of Anadarko, creating one of the largest U.S. independent oil and natural gas companies. In January 2001, Amerada Hess announced that it was withdrawing a $3.5 billion offer to purchase Britain's Lasmo P.L.C., a move which would have created a "super-independent" oil company. Instead, Lasmo was purchased by Italy's ENI for $4 billion.

Due to low profitability in the refining/marketing line of business, U.S. integrated major energy companies began a process during the 1990s of selective refining/marketing divestiture, and numerous U.S. refineries were shut down. Among independent refiners, growth largely has been concentrated in the following group of companies: Citgo/PDV America, Clark Refining and Marketing, Diamond Shamrock (merged with Ultramar during 1996, creating Ultramar Diamond Shamrock), Koch Industries, Tesoro Petroleum, Ultramar, and Valero Energy. In May 2001, Valero agreed to acquire Ultramar Diamond Shamrock for $6 billion. Another company, Tosco Corporation, was purchased by Phillips Petroleum for $7.5 billion in September 2001, creating the second largest refining group in the United States, behind ExxonMobil.

In late 2002 and early 2003, several smaller companies announced or closed deals to purchase refining and/or related assets. For instance, Valero said in March 2003 that it would pay $289 million to buy El Paso's 102,500-bbl/d Corpus Christi refinery. This follow's Valero's sale of its 168,000-bbl/d Golden Eagle refinery and related assets in northern California to Tesoro Petroleum for $1.1 billion in May 2002. In other news, Canada's Suncor announced in April 2003 that it would buy ConcoPhillips' 58,000-bbbl/d Commerce City, Colorado reinfery and related assets for $150 million. And on April 30, 2003, Sunoco announced that it had signed a letter of intent to buy El Paso's 150,000-bbl/d Eagle Point refinery and related assets in New Jersey for $130 million.

Natural Gas

As of January 1, 2003, the United States had estimated proven natural gas reserves of 183 trillion cubic feet (Tcf), or 3.3 per cent of world reserves (6th in the world). For all of 2002, U.S. production of dry natural gas is estimated at 19.0 Tcf. Natural gas consumption is estimated at 23.2 Tcf, and net imports at around 3.5 Tcf (15 per cent of total demand). Around 94 per cent of U.S. natural gas imports come Canada, mainly the western provinces of Alberta, British Columbia, and Saskatchewan. Overall, the United States depends on natural gas for about 24 per cent of its total primary energy requirements (oil accounts for around 39 per cent and coal for 23 per cent).

Natural gas wellhead prices reached record highs of nearly $10.00 per thousand cubic feet (mcf) in late 2000/early 2001, but fell sharply soon thereafter to around $2.50 per mcf. Cold weather in the U.S. Northeast and Midwest during the winter of 2002/2003 raised prices once again, particularly in late February, as gas storage levels hit unusually low levels and cold weather limited pipeline operations.

For all of 2002, the average natural gas wellhead price averaged $2.96 per mcf, compared to over $4.00 per mcf in 2001. In 2003, wellhead prices are projected to average $4.79 per mcf. At the end of April 2003, working gas in storage stood about 52 per cent below end-of-April 2002 levels and 41 per cent below the previous 5-year average. The exceptionally large current shortfall in natural gas storage relative to normal levels continues to place unusually strong upward pressure on near-term gas prices because companies will need to obtain large amounts of natural gas from other uses in order to refill storage for the next heating season. Moreover, if abnormally warm weather prevails this summer the current market may become highly sensitive to demand, particularly in the Western and South Central United States, where natural gas is heavily used for power generation. Such conditions could cause a mid-year spike in prices.

Natural Gas Production and Storage

Domestic dry natural gas production is projected to increase by about 1.4 per cent in 2003 after falling by about 1.8 per cent in 2002. Strong increases in U.S. natural gas production and net imports are needed over the next two decades to meet demand. Improved unconventional and offshore recovery technology will be needed to effectively tap abundant natural gas reserves. Increased natural gas production is expected to come mainly from onshore sources, although offshore Gulf of Mexico production also is forecast to grow significantly. In August 2001, for instance, ExxonMobil began production at its $330 million Mica natural gas project in the deepwater Gulf of Mexico. Alaska's North Slope fields also represent a large potential natural gas source, with an estimated 30-35 Tcf of natural gas reserves. Alaska's Governor Tony Knowles has stated that he supports a $17.2 billion natural gas pipeline running from the North Slope along the Alaska Highway into Alberta and on to markets in the U.S. Midwest (another option would be to route the pipeline via the MacKenzie Delta in northern Canada).

Future increases in natural gas production likely will come mainly from lower 48 sources, with increased use of cost-saving technologies expected to result in continuing large natural gas finds, including in the deep waters of the Gulf of Mexico but also in conventional onshore fields. Currently, top natural-gas-producing states (in descending order) include Texas, Oklahoma, New Mexico, Louisiana, Wyoming, Colorado, Alaska, Kansas, California, and Alabama.

September 2002 hurricanes in the Gulf of Mexico temporarily shut in some natural gas production, causing spot prices at the Henry Hub and elsewhere to rise above the $4.00 per million btu mark for most of October 2002. Early cold weather in October 2002, particularly in the Midwest, also helped raise prices even as storage levels remained relatively high. With storage levels well above 3 Tcf at the end of October, further large price increases were not expected in the near term. However, cold weather, particularly in the Northeast for much of the 2002-2003 heating season, and high prices for competing fuel oil increased natural gas demand and rapidly depleted gas in storage. Spot natural gas prices surged, averaging near $7 per million Btu in February and March 2003. The shift from adequate to below-normal storage levels and the burden of exceptionally high prospective injection demand has pushed spring and expected summer natural gas prices to nearly $6 per million Btu for all of 2003.

Natural Gas Demand

From 1990 through 2002, natural gas consumption in the United States increased by about 21 per cent, although consumption has fallen slightly since 2000 in part as a result of an economic slowdown. Still, growth in U.S. natural gas demand is likely in coming years. Greater use of natural gas as an industrial and electricity generating fuel can be attributed, in part, to its relatively clean-burning qualities in comparison with other fossil fuels. An expanding transmission and distribution network have also helped expand its acceptance and use. Natural gas is consumed in the United States mainly in the industrial (34 per cent), electric power (24 per cent), residential (21 per cent), and commercial (14 per cent) sectors.

With high natural gas prices, natural gas demand is expected to fall slightly in 2003, despite sharply higher weather-related gas demand during the first quarter of 2003. Natural gas demand in 2004 is expected to rise about 1 per cent as industrial demand recovers from its 2002/2003 lows. Sluggish near-term natural gas demand is expected to delay the startup of some planned natural gas pipeline extension projects, at least until economic conditions pick up. Still, around 3,571 miles in natural gas pipelines were added during 2002, at a cost of $4.4 billion. Five major new natural gas pipeline systems were completed and placed in operation during 2002, accounting for 34 per cent of total new gas pipeline mileage in that year. These pipelines were: Gulfstream (1,130 MMcf/d capacity; Gulf of Mexico to Florida); North Baja (500 MMcf/d, southeastern California to Baja California);

Questar Southern Trails (87 MMcf/d); New Mexico/Utah to California/Arizona); Guardian (750 MMcf/d; Chicago hub to northern Illinois and southern Wisconsin); and Horizon (380 MMcf/d; same as Guardian).

U.S. natural gas consumption and imports, overwhelmingly from Canada–and to a far lesser extent from liquefied natural gas, or LNG, from Trinidad, Algeria, Qatar, and others–are expected to expand substantially in coming decades, with the fastest volumetric growth resulting from additional natural-gas-fired electric power plants. Increased U.S. natural gas consumption will require significant investments in new pipelines and other natural gas infrastructure. The largest natural gas pipeline project currently under construction is the $1.2 billion Gulf Stream pipeline, which will run 564 miles from Alabama to Florida.

Domestic and Import Pipelines

On November 1, 1993, FERC issued Order No. 636, which decoupled the various stages of the natural gas industry between wellhead and end-user. This order has led to significant restructuring of the interstate natural gas pipeline industry, including moves towards unbundled services, diversification into other energy sectors, and development of mega-pipeline systems.

During the past decade, interstate natural gas pipeline capacity has increased substantially. From January 1996 through August 1998 alone, at least 78 projects were completed adding approximately 11.7 billion cubic feet per day of capacity, and much more will be needed in coming years. Recently completed pipelines include the Pony Express project and the Trailblazer system expansion, providing access from the Wyoming and Montana production regions. Also, the Transwestern and El Paso natural gas pipeline expansions have increased capacity from New Mexico's San Juan Basin.

On December 1, 2000, the $2.9 billion, 1.3-Bcf/day Alliance Pipeline from western Canada (Fort St. John, British Columbia) to the Chicago area entered service. Another pipeline, the Independence Pipeline ($678 million) received FERC approval in July 2000, but was cancelled in June 2002 due to lack of customer interest.

Columbia Gas System's Millennium project ($700 million), which is to connect Canadian natural gas sources to New York and Pennsylvania, received FERC go-ahead on September 19, 2002. When complete, Millennium will transport up to 700 million cubic feet of natural gas per day, providing an environmentally preferred option for generating electricity. According to the Millennium Pipeline consortium's Web site, more than 90 per cent of the pipeline's 425-mile overland route uses existing utility corridors, with about 224 miles of the project replacing and upgrading a 50-year-old pipeline system owned and operated by Columbia Gas Transmission Corp. That existing system serves several major gas end-users, utilities and their customers in New York's Southern Tier region.

Growing U.S. demand for Canadian natural gas has been a dominant factor underlying many of the pipeline expansion projects this decade. The U.S. and Canadian natural gas grids are highly interconnected and Canadian natural gas has become an increasingly important component of the total natural gas supply for the United States. This is especially true for certain U.S. regions such as the Northeast, Midwest, and Pacific, which depend on Canadian natural gas for significant amounts of their supply. Overall, the United States received about 3.8 Tcf of natural gas (net) from Canada during 2002, up slightly from 3.7 Tcf in 2001. Mexico is a small net importer of natural gas from the United States.

There has been considerable progress in recent years on natural gas interconnections between Canada and the United States. The Northern Border Pipeline, an extension of the Nova Pipeline, came onstream in late 1999 and connects to Chicago through the upper Midwest. A further extension to Indiana entered service in 2001.

The Maritimes and Northeast Pipeline came onstream in January 2000, running from Sable Island to New England, with further extensions into the Boston area to be completed during 2003. The pipeline has a capacity of 400 MMcf/d. The $2.5 billion Alliance Pipeline, at 1,875 miles, is the longest pipeline ever built in North America, and is designed to carry about 1.3 billion cubic feet per day (Bcf/d) of gas from western Canada (Fort St. John, British Columbia) to the Chicago area. The pipeline began commercial service on December 1, 2000. The U.S. utility Pacific Gas & Electric imports natural gas from British Columbia via the Alliance pipeline. To date, the Alliance system has been operating at close to its capacity of 1,630 MMcf/d.

The long-delayed 714 MMcf/d Millennium Pipeline project could be placed in service sometime in late 2004. Millenium is slated to connect Canadian sources to southern New York and Pennsylvania. Indecision over the final route of the pipeline in New York has been one factor stalling progress.

Another possibility for future U.S. natural gas supplies lies in northern Canada, which contains around one third of that country's recoverable gas reserves. The Mackenzie Valley pipeline, being developed by Shell, ConocoPhillips, ExxonMobil and Imperial Oil, would carry about 1.9 Bcf/d of gas from Canada's far north to southern Canada and the United States.

On October 12, 2001, the U.S. Coast Guard lifted the ban on LNG tankers from Boston harbor. The ban, in effect since September 26 (two weeks after the terrorist attacks in New York and Washington, DC), was established in response to security and safety concerns about the ships that bring LNG to the import facility of Distrigas of Massachusetts (a Division of Tractebel, Inc.). The decision enabled the reopening of the Distrigas facility in Everett, Massachusetts, which received 45 shipments containing 99 Bcf of natural gas in 2000, mostly from Trinidad, accounting for 44 per cent of total LNG imports into the United States that year. The Distrigas facility is one of three currently active LNG facilities in the United States. The other two active facilities are located in Lake Charles, Louisiana, and the recently reopened facility in Elba Island, Georgia. An additional LNG facility, in Cove Point, Maryland, is currently used as a storage and peak shaving facility, but is being upgraded into the nation's largest LNG facility, scheduled to open in mid-2003. In August 2002, Williams announced that it was selling Cove Point (including an 87-mile pipeline) for $217 million to a subsidiary of Dominion Resources.

Overall, there is growing interest in LNG to supply natural gas for U.S. electric power generation and provide supply flexibility. EIA expects that LNG imports to the United States will increase at an average 11 per cent annual rate, to 2.14 Tcf by 2025 from 0.14 Tcf in 2002. Overall, during 2002, the United States received 169 Bcf of LNG in 2002, mainly from Trinidad and Tobago, Qatar, and Algeria.

Natural Gas Mergers, Acquisitions, Bankruptcies

As with oil, a number of major natural gas market participants are engaging in various forms of corporate combinations, such as mergers, acquisitions, and strategic alliances. The value of mergers and acquisitions within the natural gas industry quadrupled from $10.4 billion in 1990 to $39 billion in 1997. This increase parallels an enormous surge in corporate combinations (mergers, acquisitions, joint ventures and strategic alliances) across the energy sector. In August 2001, Devon Energy announced the acquisition of Mitchell Energy for $3.1 billion, forming the second largest independent natural gas producing company in the United States, behind Anadarko Petroleum Corp. In late January 2001, El Paso Energy completed its $24-billion merger with Coastal, creating the fourth-largest U.S. energy company by market capitalization (after BP, ChevronTexaco, and Enron at the time). The October 1999 merger between El Paso Energy Corporation and Sonat had created the largest transporter of natural gas in the country.

On December 2, 2001, Enron, formerly the world's largest electricity and natural gas trading company, filed for Chapter 11 bankruptcy in the Southern District of New York for 14 affiliated entities, including Enron, Enron North America, Enron Energy Services, Enron Transportation Services, Enron Broadband Services, and Enron Metals & Commodity Corporation. Enron had been the seventh-largest publicly-traded energy company in the world. Also in early December 2001, Enron filed a $10 billion lawsuit against Dynegy, alleging breach of contract, in connection with Dynegy's November 28 termination of its proposed merger with Enron. On November 9, 2001, Enron had agreed to an all-stock takeover by former competitor Dynegy. ChevronTexaco, a 27 per cent stakeholder in Dynegy, was to inject $1.5 billion of cash immediately into Enron, and an additional $1 billion into the combined entity. The merged company was to be called Dynegy Inc., and Dynegy executives were to occupy all top positions. On November 28, 2001, however, Dynegy withdrew from the merger deal.

On January 2, 2002, the U.S. Department of Justice confirmed that a criminal probe of Enron had been launched. A task force was formed to investigate whether the former giant energy company defrauded investors by deliberately withholding or falsifying crucial financial information. The U.S. Securities and Exchange Commission has been investigating Enron since October 2001. A number of civil suits already have been filed against Enron. In October 2002, the Justice Department filed a criminal complaint against Enron's former CFO, Andrew Fastow, alleging multiple counts of financial fraud.

Coal

The United States is forecast to produce 1,089 million short tons (Mmst) of coal in 2003, down from 1,094 Mmst in 2002. Also in 2003, the United States is expected to consume 1,081 Mmst (up from 1,065 Mmst in 2002). Led by Wyoming, the West is the leading U.S. coal-producing region (with about half of the U.S. total), overwhelmingly from surface mines. Appalachia (led by West Virginia and Kentucky) accounts for about 36 per cent of total U.S. coal production, mainly from underground mines.

Around three-fifths of U.S. coal production is bituminous, one-third subbituminous, and about one-tenth lignite (brown coal). Around 80,000 miners work in the $20 billion U.S. coal industry, down from a peak of 700,000 in 1923, when U.S. coal production was half what it is today. Major U.S. coal companies include Peabody Energy (the largest in terms of production), Arch Coal (the second largest coal producer); and Kennecott Energy.

During 2003, coal production is expected to fall in Appalachia and the U.S. Interior, and to rise in the West. In 1998, low-sulfur western coal production surpassed relatively higher-cost, higher-sulfur, Appalachian coal for the first time, following strong increases since 1994, prompted largely by Phase 1 of the Clean Air Act Amendments of 1990 (CAAA). CAAA originally took effect during 1995, and required lower sulfur emissions from coal combustion. In response, Wyoming increased its coal production sharply, particularly low-sulfur, low-ash (and low cost) coal from the Powder River Basin, where coal is strip-mined.

The electric power sector (made up of electricity producers whose primary business is producing power for public distribution) accounts for the vast majority (over 90 per cent) of U.S. coal consumption, with independent power producers (IPPs) and manufacturing taking nearly all the rest. This pattern is expected to continue through 2020 at least, with coal maintaining a fuel cost advantage over oil and natural gas. As sulfur dioxide emissions standards are tightened (in 2000, for instance, Phase 2 of CAAA took effect), the share of low-sulfur coal in the U.S. coal consumption mix is expected to increase. In 1999, low and medium-sulfur coals had approximately the same share of the U.S. coal market, with high-sulfur coal far behind.

U.S. coal exports have fallen precipitously since 1995 due mainly to lower world coal prices and increased competition from other coal-producing nations (*i.e.*, Australia, South Africa, China, Venezuela, Colombia), plus natural gas–especially in Europe. In 2001, total U.S. coal exports dropped to the lowest level since 1978, largely due to 1) a strong U.S. dollar, which gave an edge to other coal-exporting countries; and 2) the tight supply market in the United States, which resulted in increased spot prices of coal, influencing some producers to shift their output to the domestic market. Metallurgical coal exports experienced the greatest decline in 2001, accounting for 75 per cent of the total decline. Export markets for metallurgical coal have been declining over the past few years because of the expansion of new steel-making technologies requiring less high-grade coking coal. Consequently many U.S. metallurgical coal operations have closed, and increased amounts of metallurgical coal have been sold into the domestic utility steam coal market. The U.S. coal industry is expected to continue to face strong competition from other coal-exporting countries, with limited or negative growth in import demand in Europe and the Americas. Given this, it is likely that the U.S. share of world coal exports will decline in coming years.

U.S. gross coal imports, at 39.9 Mmst, are expected to be about about the same in 2003 than they were in 2002. Coal imports continue to rise, in part attributable to the heightened demand for low-sulfur coal, and in part to the need to meet stricter sulfur emission requirements of Phase II of the CAAA.

On January 29, the Fourth Circuit Court of Appeals ruled in favor of the coal industry and the Department of Justice by overturning Judge Charles Haden's May 2002 ban on new valley fill permits at coal mines in West Virginia and eastern Kentucky. The three-judge panel ruled that the 2002 ruling had been "over broad" and essentially supported the existing policies that the Army Corps of Engineers has followed for many years in issuing fill permits under the Clean Water Act. This action is not, however, expected to result in an immediate increase in new permits. The Corps has not accepted most new applications during the appeal period and the normal 45-day processing time could go to several months if a flood of applications is received.

Further, a new issue has developed during the off time. Last year, the Corps introduced a new impact mitigation policy that applies to both existing and new permits. Under the Nationwide 21 program, the new Corps policy applies to the Central Appalachian coalfields of West Virginia and eastern Kentucky. Operators will be required to save a stream or wetland at another location to compensate for those it fills during coal mine operations, or to pay "in lieu" fees if compensating wetlands cannot be preserved. The most pressing issue related to this is a February 11 deadline, after which no new fill can be done. Under Congressional pressure from the affected States, and now that the Haden ruling has been overturned, the Corps is free to move ahead and may extend the February 11 deadline.

Electricity

In 2002, the United States generated 3,836 billion kilowatthours (Kwh) of electricity, including 3,673 billion Kwh from the electric power sector plus an additional 162 billion Kwh coming mainly from combined heat and power (CHP) facilities in the commercial and industrial sectors. For the electric power sector, coal-fired plants accounted for 52 per cent of generation, nuclear 21 per cent, natural gas 16 per cent, hydroelectricity 7 per cent, oil 2 per cent, geothermal and "other" 1 per cent.

Natural gas-fired power plants have been gaining share rapidly over the past few years. Coal-fired power plants generally have been less attractive than natural-gas-fired plants due to relatively high capital costs, longer construction periods, and lower efficiencies than natural gas combined-cycle plants, and has been losing share. Nuclear power has been growing only slowly, far behind the rate of natural gas-fired power.

On a national level, the average retail price of electricity during 2002 averaged 7.25 cents per Kwh, down slightly from 7.32 cents per Kwh in 2001. Electricity prices in the United States fell every year between 1993 and 1999, but this trend reversed in 2000 and 2001.

As of 2001, U.S. total installed electric generating capacity was 813 gigawatts (GW). Of this total, 74 per cent was thermal (mainly coal and natural gas), 12 per cent nuclear, 12 per cent hydro, and 2 per cent "renewables" (geothermal, solar, wind). The amount and geographical distribution of capacity by energy source is a function of availability and price of fuels and/or regulations. Capacity by energy source generally shows a geographical pattern such as: significant nuclear capacity in New England, coal in the central U.S., hydroelectric in the Pacific West, and natural-gas-fired capacity in the Coastal South.

Total annual electricity demand (retail sales plus industrial and commercial generation for own use and other direct sales) grew 3.9 per cent for all of 2002. Abnormally high summer temperatures and high cooling demand increased electricity demand sharply in the third quarter of 2002. Total U.S. electricity demand is estimated to have been 5.8 per cent higher this past winter than during the 2001-2002 winter, due mostly to weather related demand. In 2003, electricity demand is expected to grow by a relatively subdued rate of about 1.8 per cent. Due to the sluggish economy and a huge increase in electric power generating capacity, the southeastern United States is even expecting a large-scale power oversupply problem in the near term.

As a result of the Federal and State initiatives, the electric power industry is transitioning from highly regulated, local monopolies which provided their customers with a total package of all electric services and moving towards competitive companies that provide the electricity while utilities continue to provide transmission or distribution services. States are moving away from regulations that set rates for electricity and toward oversight of an increasingly deregulated industry in which prices are determined by competitive markets. Almost half of the States have passed major legislation and/or regulations to restructure their electric power industry. States that historically had higher than average U.S. prices, such as California, Pennsylvania, New York, and most of New England, have opened their retail electricity markets to competition allowing customers to choose their electricity supplier, while other States are beginning with a limited number of consumers. One of the major goals in restructuring is to lower the price for electricity. Industry analysts have cited Pennsylvania as the most successful State in achieving its goals in restructuring.

In March 2001, the Energy Secretaries of Canada, Mexico, and the United States met to discuss a common energy strategy for the three countries, including integration of the three countries' power grids and creation of a US-Mexican working group to focus on promoting cross-border electricity trade. At present, power trade between Mexico and the United States is severely limited by infrastructure constraints, including inadequate power transmission capability (there are only two cross-border transmission lines: San Diego-Tijuana and El Paso-Matamoros). In January 2001, a small (50-MW), natural-gas-fired power plant in Baja California began exporting power to California. Canada exported about 36 bkwh of electricity to the United States in 2002, mostly from Quebec, Ontario, and New Brunswick to New England and New York. Smaller volumes are exported from British Columbia and Manitoba to Washington state, Minnesota, California, and Oregon. There is considerable reciprocity between the Canadian and U.S. power markets, as the United States also exports smaller volumes of electricity to Canada.

Nuclear

In 2002, U.S. nuclear power generation reached a record 780 billion kWh, or about 20 per cent of total U.S. electricity generation, second only to coal in the U.S. electricity generation mix. Approximately 28 per cent of U.S. nuclear output was provided by just three states: Illinois, Pennsylvania, and South Carolina. The average capacity factor for all nuclear units nationwide increased from 88.1 per cent in 2000 to over 90 per cent in 2002, an all-time record high utilization rate. Following the September 11, 2001 terrorist attacks on the United States, security at nuclear power plants around the United States was increased dramatically.

Nuclear power in the United States grew rapidly after 1973, when only 83 billion kWh of nuclear power was produced. As of 2002, nuclear power had grown nine-fold, with 104 licensed nuclear power units generating 780 billion kWh of electricity. This rapid growth in nuclear power generation, however, obscures serious underlying problems in the U.S. nuclear industry. After 1974, many planned units were canceled, and since 1977, there have been no orders for any new nuclear units, and none are currently planned. The 1979 Three Mile Island accident greatly increased concerns about the safety of nuclear power plants in the United States. The regulatory reaction to those concerns contributed to the decline in the number of planned nuclear units, with Watts Bar I (1996) the last plant completed. In late March 2000, the Nuclear Regulatory Commission (NRC), in a positive signal to the U.S. nuclear power industry, granted the first-ever renewal of a nuclear power plant's operating license. The 20-year extension (until 2034 and 2036 for two reactors) went to the 1,700-MW Calvert Cliffs plant in Maryland.

On July 9, 2002, the U.S. Congress voted to formally approve Yucca Mountain, located 100 miles north of Las Vegas, as the nation's permanent nuclear waste depository. Studies on Yucca Mountain as a possible nuclear power plant waste site have been going on for over two decades, with concerns centering on the dangers of transporting nuclear materials to the site via rail or highway. Nuclear utilities have complained that they are running out of nuclear waste storage capacity at their nuclear plants, with many being forced to resort to "dry cask" storage of spent fuel assemblies after water-storage pools reached capacity.

On November 7, 2002, South Carolina's Governor-elect, Mark Sanford, announced that he "would be inclined" to drop a legal suit against the Energy Department regarding plutonium shipments to the Savannah River nuclear site. The plutonium would be shipped from other nuclear weapons sites across the United States.

Hydroelectricity/Other "Renewables"

The United States consumed 5.9 quadrillion Btu of renewable energy in 2002, about 6 per cent of total domestic gross energy demand, with the largest component used for electricity production. As of February 2003, eleven states had adopted renewables portfolio standards (RPS) aimed at increasing the share of renewable power in the energy mix. Several other states are considering adoption of an RPS, while others with RPS already in place are looking to accelerate renewables development even faster.

Overall, hydropower made up around 45 per cent of total U.S. renewable consumption in 2002, with biofuels (including wood and waste), solar, wind, and geothermal making up most of the remainder. In the summer of 2002, the U.S. Northeast experienced a serious drought, calling into question the adequacy of hydroelectricity supplies during the summer cooling season. By early November, however, the drought had eased significantly following heavy rains in much of the region. Overall, total hydro generation rose by around 23 per cent during 2002 compared to 2001, a bad drought year. For 2003, total hydropower generation is expected to rise by 12 per cent as heavy rains in much of the eastern part of the country have raised water levels.

Wind, solar, biomass, and geothermal power, although growing, still supply only a tiny fraction of U.S. energy needs. In January 2000, however, the U.S. Department of Energy's National Renewable Energy Laboratory (NREL) released a report which said that the domestic photovoltaic (PV) industry could provide up to 15 per cent of "new U.S. peak electricity capacity expected to be required in 2020." In 2001, shipments of solar PV cells and modules expanded by 11 per cent, to around 98,000 peak kilowatts, according to EIA's Renewable Energy Annual 2001. Wind, geothermal, and biomass energy sources also have significant potential in the United States.

In 2002, 410 MW of wind power was installed in the United States, representing strong (10 per cent) growth, but down from the record growth (an additional 1,694 MW) of 2001, according to the American Wind Energy Association (AWEA). The slowdown in growth during 2002 was driven in

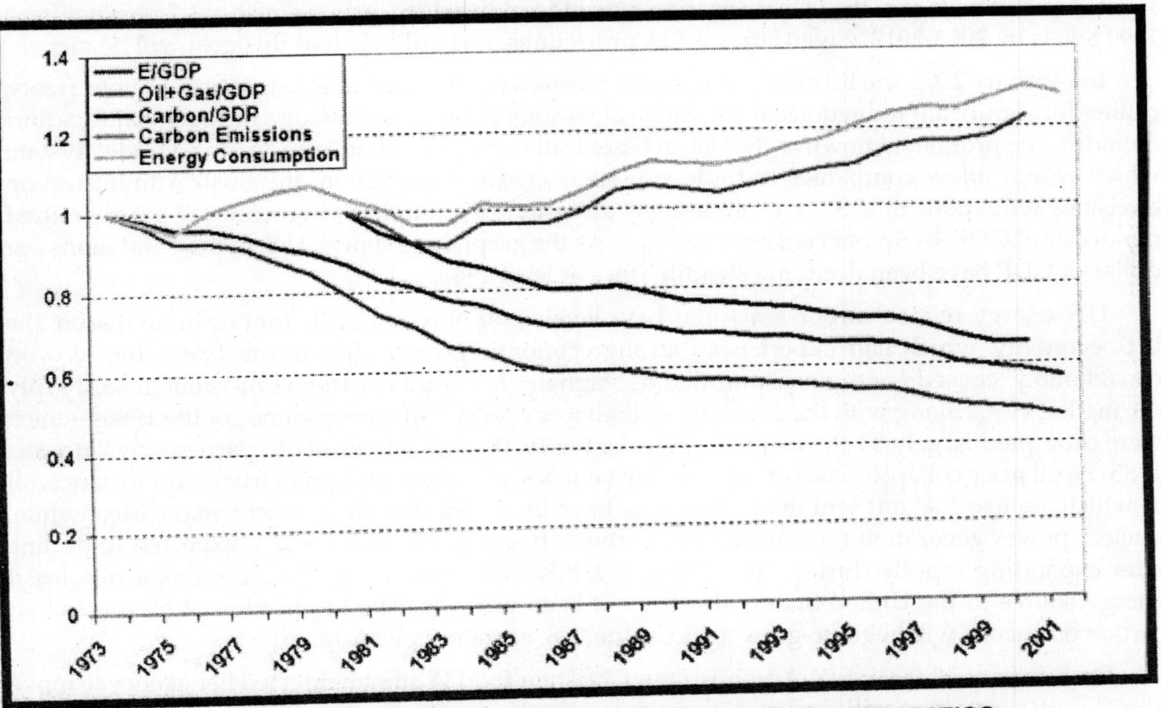

U.S. ENERGY CONSUMPTION, CARBON EMISSIONS, INTENSITY RATIOS
(Indexed: First year of Data = 1)

part by the uncertain status of a federal wind Production Tax Credit, or PTC, first established in 1992, and currently set to expire on December 31, 2003. The PTC, among other factors, has helped boost total U.S. installed wind generating capacity to 4,685 MW (as of December 31, 2002), up from 1,584 MW in 1992, with wind turbines now located in 26 states.

During 2002, wind power production increased 39 per cent from 2001, the fastest growing power source in the United States in percentage terms, but trailing far behind the growth in other major power sources in absolute terms. The first U.S. offshore windmill park, with a total capacity of 420 MW, reportedly is scheduled to be built off the Cape Cod coast, possibly by 2004. The project could power more than 200,000 homes in Cape Cod. Also, Iowa's largest utility (MidAmerican Energy) has announced plans for a 310-MW wind power facility, the country's largest to date. Both Cape Cod and Iowa are areas of the country considered to have large wind energy potentials. Iowa's governor, Tom Vilsack, has set a target for Iowa of reachying at least 1,000 MW in renewable power capacity by 2010.

Environment

The United States, with the world's largest economy, is also the world's largest single source of anthropogenic (human-caused) greenhouse gas emissions. Quantitatively, the most important anthropogenic greenhouse gas emission is carbon dioxide, which is released into the atmosphere when fossil fuels (*i.e.*, oil, coal, natural gas) are burned. Current projections indicate that U.S. emissions of carbon (mainly in the form of carbon dioxide) will reach 1,624 million metric tons in 2005, an increase of 287 million metric tons from the 1,337 million metric tons emitted in 1990, and around one-fourth of total world energy-related carbon emissions. At the December 1997 global warming summit in Kyoto, Japan, the U.S. delegation agreed to reduce U.S. carbon emissions 7 per cent from 1990 levels by 2008-2012. Given current EIA projections, it is unlikely that this goal will be met.

In February 2002, the Bush Administration released its proposed alternative to the Kyoto Treaty, calling for significant reductions in emissions of various pollutants (mercury, nitrogen oxide, sulfur dioxide). The program, known as the "Clear Skies Initiative," would utilize a "cap and trade" system which would allow companies to trade emissions credits. In addition, the Bush Administration envisions reductions in U.S. "greenhouse gas intensity"–the amount of greenhouse gases emitted per dollar of GDP–by 18 per cent over 10 years. As the graph here shows, U.S. carbon emissions per dollar of GDP have been declining steadily since at least 1980.

U.S. energy-related carbon emissions have leveled off in recent years for one main reason: the U.S. economy, which had experienced strong economic growth during the 1990s, has slowed considerably, caused energy consumption to stagnate. In contrast, carbon emissions rose sharply during the 1990s along with the economy, and also as energy "efficiency gains" of the 1980s, which were prompted largely by the oil price spikes of the 1970s, began to level off, particularly since the 1985/86 oil price collapse. Sales of sport-utility vehicles, minivans, and small trucks, for instance, all of which are less fuel efficient than small cars, have increased sharply in recent years. Meanwhile, nuclear power generation (which emits no carbon), has now stagnated and is expected to decline after expanding rapidly during the 1970s and 1980s. Hydroelectricity, the other major non-fossil energy source in the United States, also has not been growing. The implication of all this is that carbon emissions will begin to grow again as the U.S. economy picks up.

On February 28, 2001, EPA Administrator Christine Todd Whitman directed her agency to move ahead with a rule that will require U.S. refiners to reduce sulfur in diesel fuel from 500 parts per million currently, to 15 parts per million by 2006. On March 13, 2001, President Bush declared that his administration would not seek to regulate power plants' emissions of carbon dioxide, citing an EIA study that regulating these emissions could result in higher electricity prices. On March 27, the Bush administration declared that the United States had "no interest" in implementing or ratifying the Kyoto treaty, saying it would be too harmful to the U.S. economy, and that it would pursue other ways of addressing the climate change issue. On April 10, the EPA asked the U.S. Court of Appeals in Washington, DC to uphold a Clinton administration plan to regulate mercury pollution from coal-fired power plants, beginning in 2004. On April 12, the White House affirmed Clinton administration-approved energy efficiency standards for washing machines and water heaters. Under these standards, clothes washers would become 22 per cent more efficient by 2004 and 35 per cent more by 2007. The next day (April 13), the Department of Energy announced that it would require air conditioners to be 20 per cent more energy efficient by 2006. The Clinton administration had mandated a 30 per cent energy efficiency increase for air conditioners. In June 2001, President Bush announced that the federal government would lead an effort to reduce the use of energy by machines not in use (known as standby power, or "vampire," devices). In July 2001, the Interior Department announced that it would greatly reduce the scope of proposed oil leases in the Gulf of Mexico, and also would keep oil rigs at least 100 miles from the state's beaches. In January 2002, Energy Secretary Spencer Abraham announced an initiative, known as "Freedom CAR," to help automakers produce fuel-cell-powered electric vehicles. And in January 2002, President Bush proposed a new hydrogen fuel cell vehicle initiative. In February 2003, the Administration announced that U.S. businesses had pledged "ambitious commitments" to reduce greenhouse gas emissions over the next decade.

Country Overview

Independence: July 4, 1776

Population (July 2002E): 280.6 million

Location/Size: North America, between Canada and Mexico/9,629,091 sq. km (3,717,792 sq. miles)., the third largest country in the world, behind Russia and Canada

Major Cities: Washington, DC (capital), New York, Los Angeles, Chicago, Houston, Miami, Philadelphia, etc.

Languages: English, Spanish (spoken by a sizable minority)

Ethnic Groups (2000): White (77.1 per cent), Black (12.9 per cent), Asian (4.2 per cent), Native American (1.5 per cent), other (4 per cent). Note: Hispanics, who can be of any race, made up 11.8 per cent of the U.S. population as of 8/1/2000.

Religions (1997): Protestant (58 per cent), Roman Catholic (26 per cent), Jewish (2 per cent), other (6 per cent), none (8 per cent)

Defense (8/98): Army, 479,400; Navy, 380,600; Air Force, 370,300; Marine Corps, 171,300 (the United States also has nearly 1.35 million reservists)

Economic Overview

Currency: Dollar ($)

Exchange Rates, per Dollar (5/15/2003): British Pound (0.61831); Canadian Dollar (1.3743); Euro (0.8704);Japanese Yen (116.35)

Gross Domestic Product (GDP) (2002E): $10.4 trillion

Real GDP Growth Rate: (2001E): 0.3 per cent *(2002E):* 2.4 per cent *(2003F):* 2.6 per cent

Inflation Rate (consumer price index) (2001E): 2.8 per cent *(2002E):* 1.6 per cent *(2003F):* 2.2 per cent

Unemployment Rate (2001E): 4.8 per cent *(2002E):* 5.8 per cent *(2003F):* 5.9 per cent

Current Account Balance (2001E): -$393 billion *(2002E):* -$503 billion *(2003F):* -$546 billion

Merchandise Exports (2001E): $719 billion *(2002F):* $683 billion *(2003F):* $736 billion

Merchandise Imports (2001E): $1,146 billion *(2002F):* $1,167 billion *(2003F):* $1,257 billion

Merchandise Trade Balance (2001E): -$427 billion *(2002E):* -$484 billion *(2003F):* -$521 billion

Major Exports: Capital goods, automobiles, industrial supplies and raw materials, consumer goods, agricultural products

Major Imports: Crude oil and refined petroleum products, machinery, automobiles, consumer goods, industrial raw materials, food and beverages

Major Trading Partners: Canada, Japan, European Union, Mexico

Federal Budget Balance (2001E): $127 billion *(2002E):* -$158 billion *(2003F):* -$300 billion+

Energy Overview

Proven Oil Reserves (1/1/03E): 22.4 billion barrels

Oil Production (2002E): 8.1 million barrels per day (bbl/d), of which 5.8 million bbl/d was crude oil (NOTE: Including "refinery gain," US oil production in 2002 is estimated at 9.1 million bbl/d)

Oil Consumption (2002E): 19.7 million bbl/d

Net Oil Imports (2002E): 10.4 million bbl/d

Gross Oil Imports (2002E): 11.4 million bbl/d (of which, 9.05 million bbl/d was crude oil and 2.31 million bbl/d were petroleum products)

Crude Oil Imports from the Persian Gulf (2002E): 2.25 million bbl/d (around 20 per cent of total U.S. oil imports)

Top Sources of U.S. Crude Oil Imports (2002E): Saudi Arabia (1.52 million bbl/d); Mexico (1.49 million bbl/d); Canada (1.43 million bbl/d); Venezuela (1.20 million bbl/d)

Value of Gross Oil Imports (2002E): $102.8 billion (compared to $102.7 billion during 2001)

Crude Oil Refining Capacity (1/1/03E): 16.6 million bbl/d (133 refineries)

Total Oil Stocks (1Q03E): 1.47 billion barrels (including about 599 million barrels in the U.S. Strategic Petroleum Reserve)

Oil Wells Drilled (2002E): 4,964 (down from 8,060 during 2002)

Operating Oil and Natural Gas Rotary Rigs in Operation (2/03E): 907 (750 for natural gas and 153 for oil)

Natural Gas Reserves (1/1/03E): 183 trillion cubic feet (Tcf)

Dry Natural Gas Production (2001E): 19.4 Tcf *(2002E):* 19.0 Tcf *(2003E):* 19.3 Tcf

Natural Gas Consumption (2001E): 22.3 Tcf *(2002E):* 23.2 Tcf *(2003F):* 23.3 Tcf

Net Natural Gas Imports (2001E): 3.6 Tcf (over 90 per cent from Canada) *(2002E):* 3.5 Tcf *(2003F):* 3.8 Tcf

Natural Gas Wells Drilled (2002E): 15,947 (down from 22,083 in 2001)

Recoverable Coal Reserves (12/31/98): 275.1 billion short tons (54 per cent lignite and subbituminous; 46 per cent anthracite and bituminous)

Coal Production (2001E): 1,128 million short tons (Mmst) *(2002E):* 1,094 Mmst *(2003F):* 1,089 Mmst

Coal Consumption (2001E): 1,060 Mmst *(2002E):* 1,065 Mmst *(2003F):* 1,081 Mmst

Gross Coal Exports (2001E): 49 Mmst *(2002E):* 40 Mmst *(2003F):* 40 Mmst

Primary and Secondary Coal Stocks (closing; 2002E): 181 Mmst (down from 182 Mmst in 2001)

Electric Installed Capacity (2001E): 813 gigawatts (74 per cent thermal-fired, 12 per cent nuclear; 12 per cent hydroelectric, and 2 per cent "renewables")

Total Electricity Generation (2001E): 3,737 bkwh *(2002E):* 3,836 bkwh *(2003F):* 3,880 bkwh

Environmental Overview

Total Energy Consumption (2001E): 96.2 quadrillion Btu *(2002E):* 97.9 quadrillion Btu (25 per cent of world total energy consumption)

Energy-Related Carbon Emissions (2001E): 1,565 million metric tons of carbon (about 24 per cent of world total carbon emissions)

Per Capita Energy Consumption (2001E): 341.8 million Btu

Per Capita Carbon Emissions (2001E): 5.5 metric tons of carbon

Energy Intensity (2002E): 10,370 Btu/$1996

Carbon Intensity (2001E): 0.17 metric tons of carbon/thousand $1995

Sectoral Share of Energy Consumption (2002E): Industrial (34 per cent), Transportation (27 per cent), Residential (21 per cent), Commercial (18 per cent)

Fuel Share of Energy Consumption (2002E): Oil (39 per cent), Natural Gas (24 per cent), Coal (22 per cent), Nuclear (8 per cent), Hydroelectricity (3 per cent), Other "renewables" (3 per cent)

Fuel Share of Carbon Emissions (2001E): Oil (44 per cent), Coal (36 per cent), Natural Gas (20 per cent)

Renewable Energy Consumption (2002E): 5.9 quadrillion Btu (about 45 per cent of which was conventional hydroelectric power)

Number of People per Motor Vehicle (2000E): 1.3

Status in Climate Change Negotiations: Annex I country under the United Nations Framework Convention on Climate Change (ratified October 15th, 1992). Under the negotiated Kyoto Protocol (signed on November 12th, 1998–not ratified), the United States agreed to reduce greenhouse gases 7 per cent below 1990 levels by the 2008-2012 commitment period.

Major Environmental Issues: Air pollution resulting in acid rain in both the US and Canada; the US is the largest single emitter of carbon dioxide from the burning of fossil fuels; water pollution from runoff of pesticides and fertilizers; very limited natural fresh water resources in much of the western part of the country require careful management; desertification.

Major International Environmental Agreements: A party to Conventions on Air Pollution, Air Pollution-Nitrogen Oxides, Antarctic-Environmental Protocol, Antarctic Treaty, Climate Change, Endangered Species, Environmental Modification, Marine Dumping, Marine Life Conservation, Nuclear Test Ban, Ozone Layer Protection, Ship Pollution, Tropical Timber 83, Tropical Timber 94, Wetlands and Whaling. Has signed, but not ratified, Air Pollution-Persistent Organic Pollutants, Air Pollution-Volatile Organic Compounds, Biodiversity, Desertification, Hazardous Wastes.

The total energy consumption statistic includes petroleum, dry natural gas, coal, net hydro, nuclear, geothermal, solar, wind, wood and waste electric power. The renewable energy consumption statistic is based on International Energy Agency (IEA) data and includes hydropower, solar, wind, tide, geothermal, solid biomass and animal products, biomass gas and liquids, industrial and municipal wastes.

Energy Industries

Major U.S. Oil Companies (2002): ExxonMobil, ChevronTexaco, ConocoPhillips, Marathon, Amerada Hess, Anadarko, Unocal

Major U.S. Coal Companies (2001): Peabody Coal Sales.; Arch Coal; Kennecott Energy Co.; Consol Energy; RAG American Coal Holding; Horizon Natural Resources; A.T. Massey; Vulcan Partners; North American Coal; TXU

Oil Pipelines (2001E): Around 2 million miles *Natural Gas Transmission Pipelines (2000E):* 250,000 miles

Major Ports: Baltimore, Chicago, Hampton Roads, Houston, Los Angeles, New Orleans, New York, Philadelphia

90

Uruguay

After a prolonged economic downturn largely caused by the financial crisis in neighboring Argentina, Uruguay is expected to expand economically in 2004. Natural gas via new pipelines from Argentina is likely to play an increasing role in Uruguay's energy sector.

General Background

A small country of about 3.4 million people, Uruguay is sandwiched between two much larger neighbors: Brazil and Argentina. Despite its size and the absence of large quantities of natural resources, Uruguay's social conditions have traditionally ranked high in the region. For example, among Latin American nations, Uruguay had the second highest Human Development Index (HDI) score in 2002. Uruguay also has a more equitable income distribution than most countries in the region.

Argentina, Brazil, Paraguay, and Uruguay form the MERCOSUR common market (Chile and Bolivia are associate members). Uruguay's economy is highly connected to those of its MERCOSUR partners. In 2001, almost 50 per cent of Uruguay's exports flowed to them. Given this degree of economic exposure, it is not surprising that Uruguay has been hurt by its neighbors' financial instability. After averaging over 4 per cent real annual GDP growth between 1992 and 1998, Uruguay is now in its fifth year of economic contraction. In 2002, its real GDP growth rate was estimated at -10.9 per cent. Since the onset of this severe recession, Uruguay has lost its investment grade, which has made it dependent on external aid, and the withdrawal of foreign investors has produced liquidity problems in four major banks. The devaluation of both the Brazilian and Argentinean currencies has led to the abandonment of Uruguay's exchange-rate anchor and the adoption of a free-floating currency regime in June 2002. This has resulted in dramatically higher inflation in what was traditionally a stable economy.

The worst of Uruguay's crisis now appears to be over with real GDP expected to shrink by only 1 per cent in 2003, and forecasted to expand 3.2 per cent in 2004. Integral to this turnaround has been the International Monetary Fund's (IMF) decision to provide $204 million to the government as part of a $2.9 billion bailout. Uruguay also has benefited from a large-scale debt swap that will allow the government to postpone payment of its foreign obligations. The country's long-term economic prospects are still far from assured, however. The disbursement of future IMF funds, for instance, is conditional on widespread reform of Uruguay's tax structure and financial system. These may be difficult to accomplish with the end of President Jorge Batlle's (elected March 2000) term approaching in 2004 and the electorate divided about the desirability of these reforms.

Despite government intervention, the country's financial system remains fragile. Moreover, the public sector debt to GDP ratio continues to worsen, in large part because the public sector remains inefficient and unnecessarily large despite earlier cuts. The government's efforts at additional trimming have met with increasing resistance. The need to cut spending therefore has meant that major infrastructure projects have been put on hold. One of the most significant of these is the construction of a bridge connecting the Uruguayan city of Colonia with Argentina's capital, Buenos Aires. The project has received approval by both countries' legislatures, but has yet to be put out to tender because of their respective economic problems.

In recent years, the trend towards privatization in Uruguay has slowed. According to a law passed in January 2002, the Administracion Nacional de Combustibles, Alcohol y Portland (ANCAP), which controls the oil sector, was to have been opened to investment by private companies and eventually to have lost its fuel import monopoly. The ANCAP labor union, however, collected 650,000 signatures in order to force a plebiscite on whether the moves should go forward. The vote is scheduled for December 7, 2003.

Oil

Uruguay has no known oil resources, and must import the 42,000 barrels per day (bbl/d) it consumes. ANCAP owns Uruguay's only refinery, the 37,000-bbl/d La Teja refinery in Montevideo. In 2001, ANCAP initiated plans to modernize La Teja, but the project is linked to its search for a major foreign partner. Some of the firms that have expressed interest are Brazil's Petrobras, Enap of Chile, Argentina's Perez Companc, the Spanish firms Repsol YPF and Cepsa, ChevronTexaco, and Royal Dutch/Shell. Whether or not such a partnership will be possible will be decided in the aforementioned December plebiscite, and the modernization plans appear to have been put on hold.

In addition to its domestic downstream monopoly in Uruguay, ANCAP is active abroad. In April 2003, ANCAP's board of directors voted to take up its option to buy the shares it does not already own in Sol Petroleo, an Argentine fuel distributor. ANCAP has also extended its contract to manage Petrolera del Conosur, an Argentine downstream unit, through mid-February 2004.

ANCAP did not anticipate that international oil prices would remain at such high levels through 2003, and consequently was forced to announce an increase of about 32 per cent for fuel oil prices in July. Gasoline prices remain unchanged.

Natural Gas

Natural gas is likely to play a growing role in Uruguay's energy sector. The government has said that it would like natural gas to make up 30 per cent of the energy matrix. The region's recent economic deterioration is hindering this goal, however, because public spending has had to be cut and the instability of the different currencies makes prices hard to agree on. Despite such setbacks, projects already completed are having a major impact. The most recent development was the successful November 2002 opening of the \$160-\$170 million first phase of the Gasoducto Cruz del Sur (GCDS, Southern Cross pipeline). The GCDS is the second gas pipeline from Argentina. It stretches 120 miles from Argentina's western Neuquen basin and southern Austral basin fields to Montevideo, the capital of Uruguay. The initial capacity was 2-2.5 million cubic meters per day (71-88 million cubic feet per day–Mmcf/d), and it will eventually be capable of carrying twice that. Partners in the concession are British Gas (40 per cent), Pan American Energy (30 per cent), ANCAP (20 per cent), and Wintershall (10 per cent). They hope eventually to extend the GCDS to Porto Alegre, 540 miles north of Montevideo in southern Brazil.

In Uruguay, the GCDS is intended to provide energy to residential, industrial and commercial consumers as well as to power plants. Some of these uses will have to wait for more prosperous times, however. In April 2003, Uruguay's state-owned National Electric Power Generation and Transmission Administration (UTE) announced that it was going to delay taking bids to build a new 300-400 megawatt (MW) thermoelectric plant in Montevideo. It did this despite having agreed in 2000 to purchase natural gas from the GCDS for 15 years starting in December 2004. The delay in the start of construction means that it will not be able to make use of the gas at that time. Provided that UTE has 15 per cent of the financing required to start the plant by September 2004, the GCDS' owners have agreed to push back the start of the contract until December 2005. In July 2003, UTE reported that it was in advanced talks with engineering firms about the contract for the plant.

Despite their agreement to delay the gas contract's start-date, UTE and the pipeline continue to argue over the currency in which the contracts will be paid. Following the Argentinean peso's devaluation, Argentina's government converted export contracts to U.S. dollars, making the deal three times more expensive to the Uruguayans.

The GCDS currently supplies both of Uruguay's gas distributing firms, Conecta and Gaseba. A subsidiary of Gaz de France, Gaseba has had responsibility for power distribution in Montevideo since 1995. Since December 1999, the remainder of the country has been supplied by the Conecta consortium, which initially included Union Fenosa, Sempra Energy, and ANCAP, which has a 20-year contract to distribute gas in Uruguay. The consortium's investment in the project over a 5-year period was expected to be in the \$150-\$200 million range, with plans to distribute gas to 770,000 potential customers outside of Montevideo. At this time, only two firms, Fenosa and Ancap, remain in the consortium, and in September 2003, Fenosa announced that it would like to sell its 55 per cent stake. Conecta's prices have been lower than those in Montevideo. The arrival of the GCDS gas, however, has allowed Gaseba to cut its prices by about one third as it switches from more costly associated and manufactured gases.

The first natural gas pipeline that connected Uruguay and Argentina was the much smaller Paraná-Paysandú line. Completed in 1998, it was constructed by UTE and ANCAP and runs the 12 miles from Entre Rios, Argentina to Paysandu in western Uruguay. Its initial operating capacity was 4.875 million cubic feet per day (Mmcf/d).

Electricity

UTE is responsible for the country's generation, transmission and distribution of electricity. In so doing, it serves some 900,000 customers, who represent about 90 per cent of Uruguay's population. Since 1997, it has been overseen by the Unidad Reguladora de la Energia Eléctrica (UREE). Another 1997 law permitted independent producers to generate power, introducing competition to the sector. UTE has the option of taking 40 per cent stakes in any new power plants built by the private developers. The law also promised the creation of a wholesale power market to be monitored by the Electricity Market Administration (ADME). Despite the appointment of a chairman of ADME in March 2003, the market is still not functioning. Another aspect of the 1997 law which has not yet to be upheld is the division of UTE into separate companies for generation, transmission and distribution.

UTE is responsible for generating just over half of Uruguay's current capacity of 2.2 gigawatts (GW), around 70 per cent of which comes from four dams producing hydroelectric power. UTE derives its electricity from the Gabriel Terra, Baygorria, and Constitucion Dams, which are all on the Rio Negro. Their respective generating capacities are: 148 MW, 108 MW, and 333 MW. The rest of UTE's generation capability comes from thermal generating plants. The remainder of Uruguay's total installed capacity comes mostly from the Salto Grande Dam, which is shared with Argentina. This dam lies on the Rio Uruguay, and is Uruguay's single largest electricity generator. Construction on it began in 1973 and was not completed until 1984. Both countries receive 945 megawatts (MW) from their involvement. In September 2003, UTE announced that it was inviting bids for the supply and installation of a wind-power facility. The project is expected to increase capacity by 5-30 MW. The deadline for companies to express their interest is 31 October. There is a legal prohibition on the use of nuclear energy.

Despite the recession, Uruguay's demand for electricity continues to grow. UTE believes that 900 MW of thermal generation will be needed by 2010, but as noted above is having a difficult time initiating actual construction on additional power plants.

Under normal conditions, Uruguay's hydroelectric power should be sufficient to cover 80 per cent of the country's electrical demand. Hydroelectric power is particularly vulnerable to fluctuations in weather patterns, however, and a prolonged drought in 1999 and 2000 caused Uruguay to become an electricity importer in those years. At other times, hydroelectric production has exceeded demand, allowing Uruguay to export power to its MERCOSUR partners. To foster such exchanges, Uruguay has begun integrating parts of its transmission network with neighboring countries. It is promoting a series of transmission and generation projects intended to make the country a key link in the growing trade in electricity between Brazil and Argentina. Interconnection with Argentina's electric system–with a capacity to handle up to 2,000 MW–has been in effect since 1974. In January 1997, the presidents of Uruguay and Brazil signed an agreement which would allow the connection of the two countries' power grids. The Inter-American Development Bank agreed to support this project.

According to the Regional Electrical Integration Commission (CIER), in 2000 there were 2,131 miles of 100-160 kV high-tension lines and 480 miles of greater than 450 kV high tension lines. CIER's data also indicates that Uruguay has one of the highest rates of electrification (95 per cent) in Latin America.

Country Overview

Independence: August 25, 1828 (from Brazil)

Population (July 2003E): 3,413,329

Location/Size: Southern South America, bordering the South Atlantic Ocean, between Argentina and Brazil / 68,038.9 sq. miles

Capital: Montevideo

Language: Spanish (official), Portunol, or Brazilero (Portuguese-Spanish mix on the Brazilian frontier)

Ethnic Groups: White 88 per cent; Mestizo 8 per cent; Black 4 per cent; Amerindian are practically nonexistent

Religion: Roman Catholic 66 per cent, Protestant 2 per cent, Jewish 1 per cent, nonprofessing or other 31 per cent

Economic Overview

Currency: Uruguayan peso

Market Exchange Rate (September 26, 2003): US$1=28.05 Uruguayan pesos;

Gross Domestic Product (GDP, 2002E): $12.3 billion *(2003E):* $11.2 billion *(2004F):* $12.0 billion

GDP per Capita (2002E): $3,669 *(2003E):* $3,313 *(2004F):* $3,534

Real GDP Growth Rate (2002E): -10.7 per cent *(2003E):* -1.1 per cent *(2004F):* 3.2 per cent

Inflation Rate (2002E): 14.0 per cent *(2003E):* 19.5 per cent *(2004F):* 8.9 per cent

Unemployment Rate (2002): 19.4 per cent

Major Trading Partners: Brazil, Argentina, European Union, United States

Merchandise Exports (2002E): $1.9 billion *(2003E):* $2.1 billion

Major Export Products: meat, rice, leather products, wool, vehicles, dairy products

Merchandise Imports (2002E): $2.0 billion *2003E):* $1.9 billion

Major Import Products: machinery, chemicals, road vehicles, crude petroleum

Merchandise Trade Balance (2002E): -$103 million *(2003E):* $190 billion

Current Account Balance (2002E): $166 million *(2003E):* $183 million

Total Foreign Debt (2002E): $11.8 billion

Energy Overview

Oil Reserves and Production: None

Oil Consumption (2002E): 42,000 barrels per day (bbl/d), all of which is imported

Crude Oil Refining Capacity (1/1/03E): 37,000 bbl/d

Natural Gas Consumption (2001E): 1.4 billion cubic feet–all imported

Coal Production: None

Coal Consumption (2001E): 1.6 thousand short tons

Electricity Generation Capacity (2001E): 2.2 gigawatts (70 per cent hydroelectric, 30 per cent thermal)

Electricity Production (2001E): 8.0 billion kilowatt-hours (kWh)

Electricity Consumption (2001E): 6.15 billion kWh

Environmental Overview

Total Energy Consumption (2001E): 0.16 quadrillion Btu* (0.04 per cent of world total energy consumption)

Energy-Related Carbon Emissions (2001E): 1.69 million metric tons of carbon (0.03 per cent of world total carbon emissions)

Per Capita Energy Consumption (2001E): 46.8 million Btu (vs. US value of 341.8 million Btu)

Per Capita Carbon Emissions (2001E): 0.5 metric tons of carbon (vs. US value of 5.5 metric tons of carbon)

Energy Intensity (2001E–PPP): 6,178 Btu/$1995 (vs. US value of 10,805 Btu/$1995)

Carbon Intensity (2001E–PPP): 0.065 metric tons of carbon/thousand $1995 (vs. US value of 0.17 metric tons/thousand $1995)

Fuel Share of Energy Consumption (2001E): Oil (55 per cent), Hydroelectricity (44 per cent), Natural Gas (1 per cent)

Fuel Share of Carbon Emissions (2001E): Oil (99 per cent), Natural Gas (1 per cent)

Status in Climate Change Negotiations: Non-Annex I country under the United Nations Framework Convention on Climate Change (ratified August 18th, 1994). Ratified the Kyoto Protocol on February 5th, 2001.

Major Environmental Issues: water pollution from meat packing/tannery industry; inadequate solid/hazardous waste disposal.

Major International Environmental Agreements: Party to: Antarctic-Environmental Protocol, Antarctic-Marine Living Resources, Antarctic Treaty, Biodiversity, Climate Change, Climate Change-Kyoto Protocol, Desertification, Endangered Species, Environmental Modification, Hazardous Wastes, Law of the Sea, Ozone Layer Protection, Ship Pollution, Tropical Timber 94, Wetlands. Has signed, but not ratified, Marine Dumping, Marine Life Conservation, and the Nuclear Test Ban.

Energy Industries

Organization: Oil, natural gas, and refining–Administracion Nacional de Combustibles, Alcoholy Portland (ANCAP), the state oil and gas company; *Electric power*–Administracion Nacional de Usinas y Transmisiones Electricas (UTE), the state electricity company.

Port: Montevideo

Major Oil/Gas Producing Fields (2003): none

Refinery (Capacity–1/1/03E): ANCAP-La Teja, Montevideo (37,000 bbl/d)

91

Venezuela

Venezuela is important to world energy markets because it holds proven oil reserves of 77.8 billion barrels, including billions of barrels of extra-heavy oil and bitumen. Venezuela consistently ranks as one the top suppliers of U.S. oil imports and is among the top ten crude oil producers in the world.

General Background

After a period of modest economic growth in 2000 and 2001, the Venezuelan economy entered into recession in 2002. A loss of business confidence and the devaluation of the Venezuelan Bolívar started the country's economic downturn. Political conflict, particularly the nationwide strikes last December and January, further compounded the dire situation of the country's economy. As a result, Venezuela's real gross domestic product (GDP) in 2002 fell by an estimated 8.9 per cent.

On December 2, 2002, opponents of President Chávez organized a nationwide strike to call for an early referendum on the President's rule. In mid-December, the strikers shut down a large portion of the country's oil industry, drastically reducing the production of Venezuelan oil and its delivery to internal and external markets. President Chávez declared the strikers' demands unconstitutional and enlisted the help of the military to maintain production.

Although the general work stoppage ended on February 3, 2003 in non-oil sectors, there has been no resolution of the strike in Venezuela's oil sector, now in its sixth month. As of May 2003, Venezuelan crude oil production is widely believed–by striking workers and independent analysts–to be around 2.6 million barrels per day. In contrast, state oil company Petroleos de Venezuela S.A. (PdVSA) estimates current production at over 3 million barrels per day, close to pre-strike output levels. Venezuela's oil production has increased in recent months despite the dismissal of more than one-third of PdVSA's work force. U.S. crude oil imports from Venezuela have also continued climbing toward a more normal level.

The petroleum industry is the mainstay for Venezuela's economy, accounting for more than three-quarters of total Venezuelan export revenues, about half of total government revenues, and about one-

third of GDP. In February 2003, PdVSA had to drawdown $1.1 billion from its share in Venezuela's Macroeconomic Stabilization Fund (FIEM) to pay for gasoline imports, as well as to pay foreign companies that were operating in its fields. The FIEM fell to $3.3 billion in early October 2002, down $6.2 billion from the start of the year. The Venezuelan government created the fund to provide a cushion during hard economic times.

Since January 2003, President Chávez has expanded his administration's control over the economy by banning foreign-currency trading and imposing price controls on a number of products. Chávez has also been pressuring the Venezuela's central bank to reduce interest rates, and has indicated that he would like to restructure external debt under more favorable terms.

Venezuela's opposition movement continues to push for a constitutional referendum on Chávez's presidency. The country's constitution allows Venezuelans to petition for a referendum on whether a president should step down, after the midpoint of the president's six-year term. However after months of talks among Venezuelan government, the opposition, and diplomatic representation led by the "Group of Friends," which includes Brazil, Chile, Mexico, Portugal, Spain and the United States, the Organization of American States (OAS) and the Carter Center, relevant parties have been unable to reach an agreement to stage a referendum this year. The midpoint of Chavez's presidency is in August 2003.

Oil

Venezuela is home to the Western Hemisphere's largest conventional proven oil reserves at 77.8 billion barrels, as of January 2003. Substantial extra-heavy oil and bitumen deposits are not included

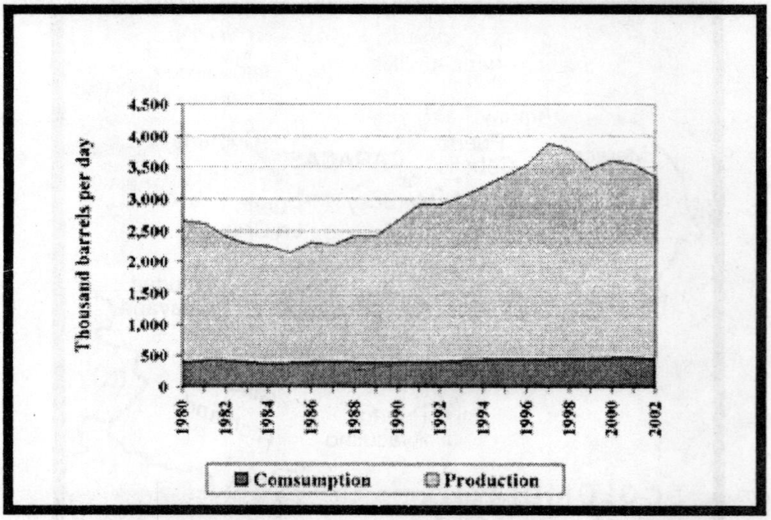

VENEZUELAN OIL PRODUCTION AND CONSUMPTION, 1980-2002

in this total. In 2002, Venezuela produced an estimated 2.9 million barrels per day (bbl/d), down almost 200,000 bbl/d from 2001 annual production figures. Of this 2.9 million bbl/d, about 453,000 bbl/d were consumed domestically, while the remaining 2.46 million bbl/d were exported. About 1.4 million bbl/d (62 per cent of total exports) were shipped directly to the United States in 2002 (note: this does not count crude oil sent to the Caribbean, refined there, and then re-exported to the United States).

Over the past several years, Venezuela has ranked consistently as one of the four top sources of U.S. oil imports (along with Canada, Mexico, and Saudi Arabia). Venezuelan exports to the United States peaked in 1997 at about 1.8 million bbl/d. While total U.S. petroleum imports have risen by about 1.2 million bbl/d since 1997, imports from Venezuela have decreased by about 390,000 bbl/d. In 1997, Venezuelan oil accounted for more than 19 per cent of total U.S. oil imports, compared to just 13 per cent in 2002.

Besides being a major supplier to the United States, Venezuela also provides significant quantities of oil to its neighbors. Under the auspices of the San Jose Accord, Venezuela and Mexico provide eleven Central American and Caribbean nations with a total of 160,000 bbl/d of crude oil and products under preferential terms. The San Jose Accord was originally implemented in 1980 and is renewed annually. In addition, Venezuela supplies Cuba with 53,000 bbl/d of oil on favorable financing terms under an agreement signed between President Hugo Chávez and Cuban President Fidel Castro in 2000. Trade with Cuba has at times been contentious, and oil shipments to Cuba were temporarily halted after the attempted coup in April 2002 that briefly ousted President Chavez. Shipments to Cuba resumed in September 2002.

Venezuela also supplies major Caribbean refineries with significant quantities of crude oil, the two largest being the 525,000-bbl/d Hovensa refinery on St. Croix (owned by PdVSA and Amerada Hess Corporation), and the 320,000-bbl/d Isla Refinery on Curacao (operated by PdVSA). Most of the

Venezuelan crude processed at these refineries is then marketed for export, with an estimated 200,000-300,000 bbl/d going to the United States.

PdVSA

Venezuela nationalized its oil industry in 1975-1976, creating Petroleos de Venezuela S.A. (PdVSA), the country's state-run oil and gas concern. PdVSA is one of the world's largest oil companies, as well as Venezuela's largest business and employer. PdVSA works with foreign investors in Venezuela under the country's hydrocarbons law of November 2001, which stipulates that PdVSA hold a 51 per cent stake in any new exploration and production agreement. The privatization of PdVSA is banned under Venezuela's 1999 constitution.

Political Role

PdVSA also plays an important role in Venezuelan politics. Along with being Venezuela's largest employer and responsible for approximately one third of the country's GDP, PdVSA is run by presidential appointees. Since taking office in 1998, Chávez has appointed five different directors of PdVSA. A previous attempt by Chávez to place political allies in top PdVSA positions resulted in a coup that briefly forced him out of office in April 2002. Management and laborers at PdVSA played a critical role in the brief overthrow of President Chávez. Ali Rodriguez is the current president of PdVSA.

Recent Strike

On December 2, 2002, PdVSA employees, along with business leaders, labor unions and opposition parties, launched a strike to back up demands that President Chávez resign or hold early elections. The strike crippled Venezuela's oil industry. On December 5, PdVSA declared force majeure on the export of petroleum products and on December 9, the Venezuelan National Guard took over gasoline distribution throughout the country. In addition, seven of PdVSA's eight highest executives, except Ali Rodriguez, submitted their resignations to President Chávez.

The strike paralyzed PdVSA's operations, with many wells being unmanned, refineries operating at partial capacity, and oil tankers waiting at sea to berth. Crude oil production dropped from 2.9 million bbl/d in November 2002 to about 600,000 bbl/d in January 2003. The loss of Venezuelan crude compelled OPEC to increase production in order to offset losses to the world oil market. The strike also forced the government to import oil products for domestic use, as well as to cut its budget due to the loss of the monthly oil revenue that it normally received from PdVSA. International companies, which have operations in Venezuela, also lost millions of dollars due to the political unrest. President Chávez reacted to the PdVSA strikes by dismissing nearly 17,000 employees.

In January 2003, the Venezuelan government decided to restructure PdVSA, splitting the company into two regional operating units: one responsible for all activities in the eastern Venezuela and the other, for activities in the western part of the country. The goal of the reorganization, according to PdVSA, was to decentralize the company and make it more efficient.

Resumption of Operations

As of May 2003, the Venezuelan government has made progress in returning oil production to pre-strike levels. Government and opposition sources continue to cite widely varying figures for the country's current oil production. Many observers put current output at 2.6 million bbl/d while PdVSA claims that production levels are over 3.0 million bbl/d. More than one-third of the company's employees have been terminated since the beginning of the strike, and President Hugo Chávez has said that they would not be rehired. The 1.3 million bbl/d Venezuela refining network is nearing full capacity after several months of partial operation.

In sum, the current status of PdVSA and Venezuela's oil industry remains unclear. According to many observers, it will most likely take PdVSA some time to restore fully its production activities on account of damage to reservoirs and equipment, lack of funds for sufficient investment particularly from foreign investors and the hiring of less experienced and qualified staff.

Venezuela and OPEC

Venezuela, a founding member of the Organization of Petroleum Exporting Countries (OPEC), regularly exceeded its OPEC-agreed oil production targets prior to President Chávez's December 1998 election. In a major policy reversal for Venezuela, the Chávez administration negotiated with Saudi Arabia and non-OPEC member Mexico to rein in production. These negotiations precipitated the 1998-1999 OPEC production cuts that saw oil prices rebound in 1999, after hitting historic lows.

Since his election, Chávez has maintained a policy of strict adherence to OPEC quotas. This requires PdVSA to shut in production, fill storage facilities, cut production at existing fields, and reducing investment and total production capacity. However, according to the government, PdVSA has been producing since March 2003 above its OPEC quota to make up for some of the production lost during the strike. The government indicated in late May 2003 that it would be willing to reduce its oil production at OPEC's upcoming meeting on June 11, 2003 in Doha, Qatar.

In the past, PdVSA has adjusted its own production to ensure that Venezuela as a whole meets its OPEC production targets. Thus, during periods of OPEC production cuts, private companies operating in joint ventures with PdVSA were able to maintain steady output. However, that policy has changed, as some private producers have been asked to hold in production. U.S. refiner Lyondell is suing PdVSA, charging that the company violated a crude oil supply agreement. Lyondell is seeking $90 million for a 30-million-barrel deficit in PdVSA deliveries to the refinery between April 1998 and September 2000. PdVSA enacted the cut to aid in OPEC compliance

Exploration and Production

Venezuela has four major sedimentary basins: Maracaibo, Falcon, Apure and Oriental. These fields contain reserves of 77.7 billion barrels of conventional oil, most of which has an API gravity of less than 20 degrees, making Venezuela's conventional crude oil heavy by international standards. Due to the maturity of many of these basins and their declining productivity, PdVSA plans to spend $45 billion to increase production at the country's existing oil wells, as well as to develop new non-conventional extra heavy crude oil and natural gas resources.

Venezuela would like to raise oil production to 5.1 million bbl/d by 2008. In order to achieve this goal, the government plans to launch a new licensing round in order to achieve this goal and will most likely need to attract significant amounts of foreign investment, particularly in the wake of lost oil revenue from the strikes.

Recent Developments

PdVSA announced in early May 2003 that it made two significant oil discoveries, which could raise reserves by 1–2.4 billion barrels. PdVSA's exploration division made both discoveries in the Chaguaramal and Furrial fields in the eastern state of Monagas. ConocoPhillips, along with PdVSA, AGIP, and OPIC Karimun Corporation, received approval in April 2003 to develop the Corocoro field in the Gulf of Paria in western Venezuela. The group plans to invest $480 million to produce eventually 55,000 bbl/d.

Extra Heavy Crude Oil

Venezuela contains billions of barrels in extra-heavy crude oil and bitumen deposits, most of which are situated in the Orinoco Belt, located in Central Venezuela (estimates range from 100–270 billion barrels of recoverable reserves). There are four congressionally approved joint ventures between PdVSA and foreign partners to develop extra-heavy crude oil. Thel four projects which are at different stages of development aim to convert the extra heavy crude from approximately 9° API crude to lighter, sweeter synthetic crude, known as syncrude. These projects normally produce about 450,000 bbl/d of synthetic crude oil (this is expected to increase to 700,000 bbl/d by 2005), much of which is destined for the U.S. Gulf Coast. Syncrude is considered by the International Energy Agency (IEA) as"non-conventional crude oil."

Conoco's Petrozuata produces extra-heavy crude oil from the Zuata region of the Orinoco Belt for transport to the port of Jose on Venezuela's northern coast. Conoco owns and operates two parallel 130-mile pipelines with a total capacity of 200,000 bbl/d to transport production from its wells and others in the region. Heavy crude oil is blended with a lighter crude oil for pipeline transportation to an upgrading facility. The upgrading facility processes the heavy oil into a higher value synthetic crude oil (with an API range between 19° and 25°), and associated byproducts: LPG; sulfur; petroleum coke and heavy gas oil. As production increases, pipeline capacity could be expanded to 500,000 bbl/d. Since 1997, Petrozuata has drilled more than 320 wells in 55,000 acres of the Zuata region, and production is currently 120,000 bbl/d.

ExxonMobil and PdVSA's joint venture at the Cerro Negro extra-heavy oil field started production in 2001. Extra-heavy crude oil from Cerro Negro is diluted with naptha and routed northward via pipeline to an upgrader complex at the port of Jose. The project's upgrader at the Jose complex processes 120,000 bbl/d of extra heavy crude oil into approximately 108,000 bb/d of syncrude and byproducts (sulfur and petroleum coke). Some of the syncrude is then exported to the partners' 180,000-bbl/d Chalmette refinery, located in Louisiana, near New Orleans, where the oil is refined and sold in

U.S. markets. Germany's Veba Oil and Gas was also a 16 per cent partner in the upstream component of the project. In May 2002, Petro-Canada acquired Veba. Petro-Canada, however, has not yet acquired ownership of Veba's Venezuelan assets. Exxon announced in May 2003 that it would like to increase production by 10 per cent to 20 per cent.

TotalFinaElf and *Statoil* are partners with PdVSA in the Sincor project, which began production in February 2002 and has been producing about 140,000-160,000 bbl/d of oil in recent months. Production is expected to plateau at 200,000 bbl/d, with about 35 years of operation. Sincor's extra-heavy crude is upgraded at a facility in the Jose complex, and then marketed for export, similar to the Petrozuata and Cerro Negro projects. Sincor's syncrude output comes in two grades, Zuata Sweet and Zuata Medium.

ConocoPhillip's and *ChevronTexaco's* joint venture with PdVSA, the Hamaca project, came onstream in November 2001 and is currently producing about 30,000 bbl/d of extra-heavy crude, most of which is diluted and shipped to refineries in the United States. Peak production of about 190,000 bbl/d is expected after an upgrading facility at the Jose complex is completed in early 2004. The crude will be upgraded to about 26° API, and the field is expected to pump for about 34 years.

Orimulsion

Orimulsion is a branded product that is used as a boiler fuel, similar to #6 fuel oil. It is an emulsion of approximately 70 per cent natural bitumen, 30 per cent water, and less than 1 per cent surfactants (emulsifiers). Bitumen is considered a non-oil hydrocarbon and is not counted towards Venezuela's OPEC crude oil production quota. Burning Orimulsion in conventional power plants produces emissions of carbon dioxide, sulfur dioxide, and nitrogen oxide roughly similar to emissions from fuel oil.

Bitor, a PdVSA subsidiary, manages the processing, shipping and marketing of Orimulsion. Bitor now operates one Orimulsion plant in Cerro Negro, with a capacity of 5.2 million metric tons per year, and hopes to produce 20 million metric tons per year by 2006. According to Bitor, more than 1.2 trillion barrels of bitumen exist in the Orinoco Belt. Economically recoverable reserves are now estimated at about 267 billion barrels. Canada, China, Denmark, Guatemala, Italy, Japan, South Korea and Lithuania either consume or are considering consuming Orimulsion.

Refining

PdVSA operates one of the Western Hemisphere's largest refining systems and is one of the world's largest oil refiners. Domestic refinery capacity stands at about 1.3 million bbl/d, with significant additional holdings in Curaçao, the United States (in Lake Charles, Lemont, Corpus Christi, Paulsboro, Savannah, and Lyondell), and Europe (in Germany, Sweden, Belgium, and the United Kingdom). About one-third of Venezuela's refined product exports are sold to the United States, where they are distributed mainly by Tulsa-based Citgo, PdVSA's U.S. refining and marketing subsidiary, and one of the largest U.S. gasoline retailers.

Natural Gas

Venezuela has proven natural gas reserves of about 148 trillion cubic feet (Tcf), the second largest in the Western Hemisphere (behind the United States) and the eighth largest in the world. The country produced about 1.1 Tcf in 2001, all of which was consumed domestically. About 60 per cent of the country's natural gas production is consumed by the oil industry, which either re-injects the gas into oil fields or flares it. About 10 per cent of Venezuela's natural gas is used for power generation; 6

per cent in petrochemical production; and the rest is used mainly by industrial or commercial customers in large cities. The Chávez administration has plans to increase both natural gas production and consumption.

Sector Organization: PdVSA and Private Companies

The natural gas unit of PdVSA has traditionally monopolized Venezuelan natural gas production, allowing for only a few joint ventures. However, in August 1999, legislation opened up the country's natural gas sector to foreign investment in exploration and production, distribution, transmission, or gasification (although no company would be allowed to explore, produce, and transport in the same region). Venezuela held its first non-associated natural gas exploration licensing round in 2001.

Exploration and Production

In February 2003, PdVSA selected ChevronTexaco and Statoil to operate two of its five offshore blocks located in the Deltana Platform. The Deltana Platform, which borders Trinidad's maritime border, is estimated to contain up to 40 Tcf of natural gas. ChevronTexaco will operate Block 2, the Loran field. Statoil will operate Block 4, Cocuina field. PdVSA expects to reveal soon a new bidding round for the remaining three blocks–1, 3, and 5. The company also has plans to open other bidding rounds for certain onshore tracts and for acreage in Lake Maracaibo.

Recent Developments

PdVSA announced in early May 2003 that its exploration division discovered a natural gas reserve of 2.5 to 3 Tcf at Urica, on the border of the eastern states of Monagas and Anzoategui. PdVSA plans to develop the well itself.

LNG

In December 2002, PdVSA, Royal Dutch/Shell, and Mitsubishi signed a preliminary development agreement to begin an initial feasibility study for the Mariscal Sucre Liquefied Natural Gas (LNG) project that would be fed by natural gas reserves off the Paria peninsula, as well as the Deltana Platform. However, recent political unrest has hindered progress on the project. Venezuela has also been in negotiations with Trinidad to develop and possibly combine the two countries' gas reserves in the region, a process called unitization. Unitization is an agreement focusing on setting acceptable parameters for sharing the resources to optimize production while avoiding damaging the reserve. These negotiations, however, have bogged down.

Pipelines

In April 2003, Colombia and Venezuela agreed to build a $120 million pipeline, allowing Colombia to export natural gas from the Guajira basin to the Maracaibo region of Venezuela. The $120 million natural gas pipeline would carry between 150 and 200 million cubic feet per day beginning in 2005. PdVSA has indicated that Venezuela would eventually use the pipeline to export natural gas to Colombia. Existing Venezuelan natural gas infrastructure consists of 3,000 miles of domestic pipeline.

Coal

Venezuela has recoverable coal reserves of approximately 528 million short tons (Mmst), most of which is bituminous. Venezuela is the second largest producer of coal in Latin America, after Colombia. Coal production in 2001 amounted to 8.9 Mmst, almost all of which was exported to other

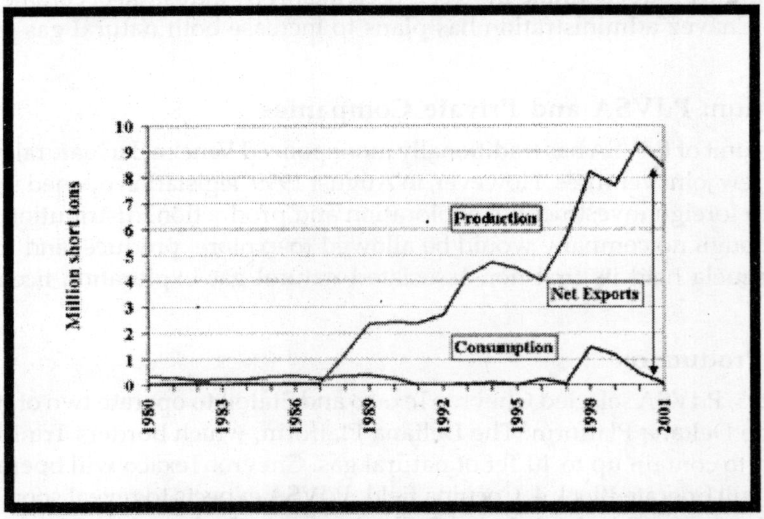

VENEZUELAN COAL PRODUCTION AND CONSUMPTION, 1980-2001

countries in the region, the eastern United States, and Europe. Domestic coal consumption in 2001 was about 74,000 short tons.

The Guasaré Basin, near the Colombian border, is the major coal producing region in Venezuela. Coal production has been limited during the last several years by infrastructure and transportation constraints. The government announced in April 1999 intentions to increase production of high-quality coal to 21 million tons per year by 2008. Venezuela's coal sector is dominated by Carbozulia, which is owned by PdVSA.

Electricity

Venezuela has about 21 gigawatts (GW) of electric generating capacity. Venezuela has one of the highest electrification rates in Latin America at over 90 per cent. Venezuelans also are the highest per capita users of electricity in Latin America. The country generated 87.6 billion kilowatthours of electricity generation in 2001, 68 per cent of which was hydropower, while the remainder was oil- and, to a smaller extent, natural gas-fired. Venezuela is home to the world's second largest operational hydroelectric dam (after Itaipu in Paraguay/Brazil), the 10,000-megawatt (MW) Raul Leoni dam on the Caroní River. Venezuela's grid is joined with Colombia, allowing trade between the two countries. In August 2001, an electricity interconnection between Venezuela and Brazil was inaugurated at Santa Elena de Uairen, Venezuela. The transmission line links Santa Elena de Uairen to Boa Vista in Brazil.

Sector Organization

The Venezuelan electricity sector is a mixture of state-owned utilities, comprising the majority of the sector, and some private companies. EDELCA is the largest generation company in Venezuela, generating over 70 per cent of the country's electricity. EDELCA operates the Guri plant (Raul Leoni dam) on the Caroní river, with an installed capacity of 10,000 MW and the 3,080-MW Macagua

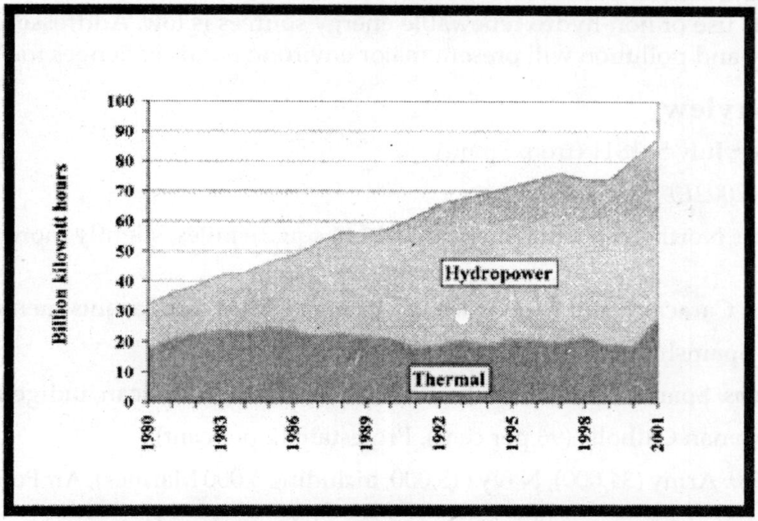

VENEZUELAN POWER GENERATION, 1980-2001

plant. EDELCA is also developing two new facilities on the Caroní River; Caruachi and Tocoma. The 2,160-MW Caruachi plant, located between Guri and Macagua, began partial operation in 2003. The Caruachi dam will not be completed until 2010. The 2,160-MW Tocoma hydroelectric dam will be EDELCA's fourth dam on the Caroní River and is scheduled to be completed by 2010. Venezuela's second largest state generating company is Cadafe. The country's largest private sector generating company is ELECAR.

Electric sector privatization was underway when the current administration came into power. As a final preparatory step before sales commence, the government had planned to "unbundle" electricity companies by September 2001, separating firms into generation, transportation, distribution, and marketing units. This unbundling now has been delayed indefinitely.

Generation Disruptions

After a decade of under-investment in Venezuela's electricity sector, the country is now facing a potential electricity crisis. The immediate reasons for the power shortfall include reduced hydropower capacity and electricity theft. Low rainfall levels have decreased hydropower capacity (as in neighboring Brazil), with diminished levels in reservoirs. The water level at the Guri dam recently rose a few centimeters for the first time since October 2002. Low water levels reportedly resulted in six significant power failures in 2002, and the frequency of outages has increased so far in 2003. Fossil fuel-fired generation is not expected to be able to make up the difference. Rampant electricity theft also is straining the infrastructure, with illegal hookups accounting for an estimated quarter of Venezuela's total consumption. The government has introduced a campaign to urge citizens and companies to reduce electricity consumption.

Environment

Venezuela's environmental problems include pollution and deforestation. Pollution from energy production and consumption is high relative to Venezuela's neighbors, as production is the mainstay of its economy and consumption is heavily subsidized. Therefore, it emits more carbon than many of its neighbors. Its use of non-hydro renewable energy sources is low. Addressing the high levels of energy intensity and pollution will present major environmental challenges for Venezuela.

Country Overview

Independence: July 5, 1811 (from Spain)

Population (2002E): 22.4 million

Location/Size: Northern South America/352,144 square miles, slightly more than twice the size of California

Major Cities: Caracas (capital), Maracaibo, Valencia, Maracay, Barquisimento

Languages: Spanish (official), Indian dialects in the interior

Ethnic Groups: Spanish, Italian, Portuguese, Arab, German, African, indigenous people

Religions: Roman Catholic (96 per cent), Protestant (2 per cent)

Defense (8/98): Army (34,000), Navy (15,000, including 5,000 Marines), Air Force (7,000), National Guard (23,000)

Economic Overview

Currency: Bolívar

Exchange Rate (5/15/03): US$1 = 1,599.5 Bolívars

Gross Domestic Product (2002E): $103 billion

Real GDP Growth Rate (2002E): -8.9 per cent *(2003F):* -10.0 per cent

Inflation Rate, per cent change in consumer prices (2002E): 22.4 per cent *(2003F):* 37.0 per cent

Unemployment Rate (2002E): 15.5 per cent

Merchandise Exports (2002E): $26.2 billion

Merchandise Imports (2002E): $12.3 billion

Current Account Balance (2002E): $7.6 billion

Major Trading Partners: United States, Colombia, Germany, Japan, Canada, and Italy

Major Export Products: Petroleum and derivatives (80 per cent), aluminum (4 per cent)

Major Import Products: Capital goods (20 per cent), consumer goods (20 per cent), and raw materials (60 per cent)

Foreign Debt (2002E): $38 billion

Energy Overview

Proven Oil Reserves (1/1/03E): 77.8 billion barrels (not including extra-heavy oil and bitumen)

Oil Production (2002E): 2.9 million barrels per day (bbl/d), of which 2.56 million bbl/d was crude

OPEC Crude Oil Production Quota (effective 6/1/03): 2.923 million bbl/d

Oil Consumption (2002E): 453,000 bbl/d

Net Oil Exports (2002E): 2.46 million bbl/d

Crude Oil Refining Capacity (1/1/03): 1.28 million bbl/d in Venezuela, with almost 2 million bbl/d of capacity in the Caribbean, the United States and Europe

Major Crude Oil Customers: United States, Canada, Germany, Spain

Oil Exports to the United States (2002E): 1.4 million bbl/d (not counting around 200,000-300,000 bbl/d of Venezuelan crude refined in the Caribbean and then sent to the United States)

Natural Gas Reserves (1/1/03E): 148 trillion cubic feet (Tcf)

Natural Gas Production/Consumption (2001E): 1.1 Tcf

Coal Reserves (2001E): 528 million short tons (Mmst)

Coal Production (2001E): 8.4 Mmst

Coal Consumption (2001E): 0.07 Mmst

Electric Generation Capacity (2001E): 21 gigawatts

Electricity Production (2001E): 87.6 billion kilowatthours (68 per cent hydroelectric and 32 per cent thermal)

Environmental Overview

Total Energy Consumption (2001E): 2.95 quadrillion Btu* (0.7 per cent of world total energy consumption)

Energy-Related Carbon Emissions (2001E): 38.6 million metric tons of carbon (0.6 per cent of world total carbon emissions)

Per Capita Energy Consumption (2001E): 119.9 million Btu (vs U.S. value of 341.8 million Btu)

Per Capita Carbon Emissions (2001E): 1.6 metric tons of carbon (vs U.S. value of 5.5 metric tons of carbon)

Energy Intensity (2001E): 35,984 Btu/$1995 (vs U.S. value of 10,736 Btu/$1995)

Carbon Intensity (2001E): 0.47 metric tons of carbon/thousand $1995 (vs U.S. value of 0.17 metric tons/thousand $1995)

Fuel Share of Energy Consumption (2001E): Natural Gas (45 per cent), Oil (34 per cent), Coal (0.07 per cent)

Fuel Share of Carbon Emissions (2001E): Natural Gas (55 per cent), Oil (45 per cent), Coal (0.13 per cent)

Status in Climate Change Negotiations: Non-Annex I country under the United Nations Framework Convention on Climate Change (ratified December 28th, 1994). Not a signatory to the Kyoto Protocol.

Major Environmental Issues: sewage pollution of Lago de Valencia; oil and urban pollution of Lago de Maracaíbo; deforestation; soil degradation; urban and industrial pollution, especially along the Caribbean coast.

Major International Environmental Agreements: A party to Conventions on Biodiversity, Climate Change, Desertification, Endangered Species, Hazardous Wastes, Marine Life Conservation, Nuclear Test Ban, Ozone Layer Protection, Ship Pollution, Tropical Timber 83, Tropical Timber 94, Wetlands and Whaling. Has signed, but not ratified, Marine Dumping.

The total energy consumption statistic includes petroleum, dry natural gas, coal, net hydro, nuclear, geothermal, solar, wind, wood and waste electric power. The renewable energy consumption statistic is based on International Energy Agency (IEA) data and includes hydropower, solar, wind, tide, geothermal, solid biomass and animal products, biomass gas and liquids, industrial and municipal wastes.

Oil and Gas Industries

Organization, Oil and Natural Gas: Petroleos de Venezuela, S.A. (PdVSA, state-held), with some foreign companies involved in joint ventures; *Coal:* Carbozulia, owned by PdVSA, with some foreign companies involved in joint ventures; *Electricity:* Several state-held and private utilities

Major Foreign Oil Company Involvement: BP, Chevron, CNPC (China), Conoco, ExxonMobil, Pennzoil, Phillips, Repsol-YPF, Shell, Statoil, Texaco, TotalFina, Union Texas, and Petro-Canada

Major Domestic Refineries (crude capacity-bbl/d) (1/1/03), PdVSA: El Palito, Puerto Cabello (126,900), Puerto de la Cruz (195,000), San Roque, Anzoategui (5,200); *Paraguana Refining Center:* Cardon/ Judibana, Falcon (940,000), Maracaíbo, Zulia (15,000)

92

Vietnam

Vietnam is important to world energy markets because of its potential to become a regional oil and natural gas supplier. Ongoing exploration has led to several oil and gas discoveries in recent years.

General Background

Vietnam's economy has maintained its relatively high levels of growth despite the global economic slowdows during 2001 and 2002, largely due to the fact that it is less integrated into the world economy and less vulnerable to declines in demand for its exports. Growth in the country's real gross domestic product (GDP) was 6.7 per cent in 2002, and is projected at 6.9 per cent for 2003. Vietnam has had Normal Trade Relations status with the United States since late 2001, although it will be several more years before it can join the World Trade Organization (WTO).

A significant development in the country's economic reforms was the first auction of a state-owned enterprise, which occurred in February 2003. Hai Phong Agricultural Mechanical Engineering Company was sold to the Viet Tin Construction Joint Stock and Trading Company for $300,000. The auction was financed by the Australian government with technical assistance from the World Bank in an attempt to speed up the privatization of inefficient state enterprises.

Much of Vietnam's large population relies heavily on non-commercial biomass energy sources such as wood, dung, and rice husks. Vietnam's per capita commercial energy consumption is among the lowest in Asia. While consumption is expected to remain low relative to other countries, Vietnamese energy consumption, especially natural gas, is predicted to rise substantially in coming years. Vietnam is involved in an ongoing territorial dispute with other regional states over the potentially hydrocarbon rich Spratly Islands. Vietnam, China, the Philippines, Brunei, Taiwan, and Malaysia all make claims to the Spratlys.

Oil

Vietnam has 600 million barrels of proven oil reserves, and further discoveries are likely. Crude oil production averaged 339,000 barrels per day (bbl/d) in 2002, a 5 per cent decline from 357,000 bbl/

d in 2001. The country has six operating oil fields, of which Back Ho (White Tiger), Rang Dong, Hang Ngoc, and Dai Hung (Big Bear) are the largest. Most oil exploration and production activities occur offshore in the Cuu Long and Nam Con Son Basin. Vietnam currently has no operating oil refineries–therefore almost all of its oil production is exported. Export markets include Japan (the largest importer of Vietnamese oil), Singapore, the United States, and South Korea. Vietnam had net exports of an estimated 153,000 bbl/d of oil in 2002, which accounted for 22.2 per cent of the nation's total merchandise exports in 2002.

The Vietnamese government controls both the upstream and downstream oil and natural gas industries. For upstream activities, Vietnam Oil and Gas Corporation (PetroVietnam), a government-owned company, is the only firm authorized to conduct petroleum operations. Any petroleum exploration and production activities by foreign investors must be conducted in cooperation with PetroVietnam. For downstream activities, several government-owned companies, such as Petrolimex and Petec under the Ministry of Trade, PetroVietnam Trading Company (Petechim) under PetroVietnam, SaigonPetro under Ho Chi Minh City People's Committee, and Vinapco under Vietnam Airlines, have been licensed to import petroleum products. Petrolimex is the largest petroleum products importer.

Vietnam has been considering an international public stock offering of a minority stake in PetroVietnam, similar to public stock offerings carried out by major Chinese oil and gas firms in recent years. In January 2003, Vietnam's Ministry of Finance announced that it would issue regulations by mid-year that will allow domestic firms to offer bonds to

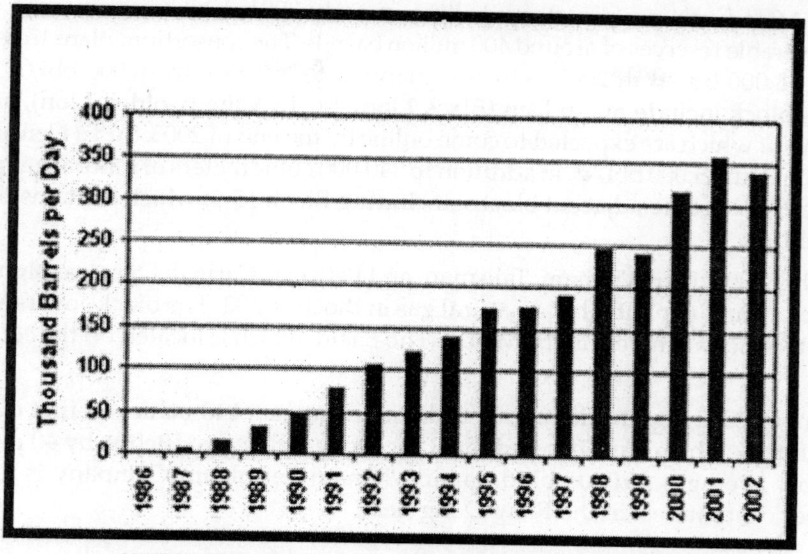

VIETNAM'S CRUDE OIL PRODUCTION, 1986-2002

the public. PetroVietnam is expected to be one of the first firms to take advantage of the new regulations because it is involved in a number of large projects (including the nation's first oil refinery) for which it needs to raise capital.

The government has issued around 44 investment licenses for oil and gas exploration since the industry was opened to foreign partners in 1998. About 30 companies, including American, European, Korean, and Japanese firms, now operate in offshore Vietnam. However, several foreign companies have withdrawn from their contracts. Reasons include problems with the Vietnamese regulatory framework and disappointment at recovering smaller quantities of oil and gas than anticipated. The most active foreign company in Vietnam is Zarubezhneft, a Russian oil company whose joint venture with PetroVietnam, called Vietsovpetro, is one of the largest actors in the sector. In January 2003, PetroVietnam and Petronas Carigali Overseas Company of Malaysia signed a contract to explore and exploit oil and gas in lot 01-02/97, located on Vietnam's continental shelf. First drilling is expected to be carried out in 2004.

Some of the new entrants to Vietnam's upstream have made substantial finds. During 2002, large oil and gas deposits were discovered in the Ca Ngu Vang (Golden Tuna) and Voi Trang (White Elephant) fields. The Ca Ngu Vang well, drilled by SOCO Vietnam, was the first well to be drilled on Block 9-2. Designed as a basement exploration well, and drilled to a depth of about 4,500 meters, the test well showed indications that oil deposits could be located at the 15 meter level as well. Oil from the test well flowed at a rate of approximately 2,500 barrels per day (bbl/d), and SOCO's estimates state that there could be as many as 250 million barrels of oil in that area. The Voi Trang well, drilled by the Hoang Long Joint Operating Company, was the second well drilled on Block 16-1. It tested successfully at a rate of 3,500 bbl/d and produced high quality, 42° API crude oil.

A major oil find in Block 15-1 of the Cuu Long Basin was announced in October 2000 by a consortium including Conoco, the Korean National Oil Company (KNOC), SK Corporation of South Korea, and the French firm Geopetrol. The first exploratory well drilled by the consortium tested at

12,600 bbl/d. After further exploratory drilling in early 2001, KNOC announced that Block 15-1 contained recoverable reserves of around 400 million barrels. The consortium plans to begin commercial production of 38,000 bbl/d in 2003, which is planned to increase to 70,000 bbl/d by 2005. Fields located on this block include Su Tu Den (Black Lion), Su Tu Vang (Golden Lion), and Su Tu Trang (White Lion), all of which are expected to come online by the end of 2003. Su Tu Den alone is expected to produce more than 25,000 bbl/d, in addition to 340,000 cubic meters of natural gas per day. Conoco also holds interests in some adjacent blocks, including Block 15-2, which contains the 40,000 bbl/d Rang Dong field.

In December 2002, PetroVietnam, Talisman, and Petronas Carigali Overseas signed a $20 million contract to explore and exploit oil and natural gas in Block 46/02. The block covers an area of 12,000 square kilometers (square miles) in the Ma Lai-Tho Basin, which is located off the country's southern coast.

In November 2002, Japan Vietnam Petroleum Company, Ltd achieved first oil from two new wellhead platforms, which have boosted the Rang Dong field's production by 40 per cent. Gross oil production now averages 65,000 bbl/d. Japan Vietnam Petroleum Company is the operator and holds a 46.5 per cent interest in the Rang Dong field.

Refining

Currently, Vietnam does not have a major refinery, but it is in the process of building its first. The $1.3 billion Dung Quat Refinery, which is located in Quang Ngai province, will have a capacity of about 140,000 bbl/d. The project began as a joint venture between PetroVietnam and the Russian firm Zarubezhneft. Initial construction has begun, and the facility was scheduled to be completed by late 2002, however, delays have pushed the completion date back to 2004, and it may not be fully operational until 2005. In December 2002, Russian state oil company Zarubezhneft announced that it had withdrawn from the oil refinery construction project, but will still perform some construction work as a subcontractor. Vietnamese firm PetroVietnam has reimbursed Zarubezhneft $230 million it had invested in the venture, and some of that money will be re-invested in Vietnam's oil and gas industry.

A second refinery project is under consideration. Mitsubishi and JGC Corporation signed a memorandum of understanding with PetroVietnam in October 2001 covering a feasibility study for the project. A collaboration agreement for the feasibility study was signed by Mitsubishi and ABB Lummus Global (an American technical consulting firm), in January 2003. The location being considered is Ngai Son, in Thanh Hoa province.

Natural Gas

As with oil, Vietnam's natural gas consumption is rising, with further increases expected as additional fields come onstream. The Cuu Long basin is the largest Vietnamese natural gas production area, and is a source of associated natural gas from oil production. An existing 100-kilometer (62-mile) pipeline from the Bach Ho field is operating near peak capacity. The Ruby and Rang Dong oil fields, both of which have considerable amounts of associated natural gas, are near the pipeline. However, the Bach Ho pipeline has insufficient capacity to carry natural gas from these fields, so the natural gas is flared.

In December 2002, a South Korean consortium headed by Korea National Oil Corporation (KNOC) signed an agreement to install facilities to be used to pump and supply up to 130 million cubic feet per day of natural gas to Vietnam. The natural gas, located in the Rong Doi and Rong Doi Tay fields on Block 11-2 of the Nam Con Son Basin, will be bought by PetroVietnam for 23 years. Purchasing will begin in 2005, when facilities are expected to be completed. PetroVietnam is in turn expected to sell the natural gas to Electricity of Vietnam (EVN).

In September 2002, it was announced that a new 5,000 km (3,100 mile) natural gas pipeline, which will be Asia's largest, will pass through Vietnam. The pipeline, which will be built through the Asia-Pacific Economic Cooperation (APEC) forum's Partnership for Equitable Growth (PEG), will link an Indonesian natural gas field with Vietnam, Malaysia, Thailand, and China. Construction costs are estimated at $8 billion, with 30 per cent of funding expected to come from the five nations' oil and gas industries. The rest is expected to be provided by international organizations and other countries.

In 1992, BP, Conoco, and ONGC (India) formed a joint venture with PetroVietnam to develop natural gas resources in the Nam Con Son basin, in the Lan Tay and Lan Do fields. Conoco purchased its stake from Norway's Statoil, which sold off its Vietnamese assets in October 2001. The fields contain an estimated 2 Tcf of natural gas. A 370-kilometer (230-mile) pipeline, which cost an estimated $565 million to construct, was completed in June 2002. It now connects the fields to the Vietnamese mainland at Vung Tau, and natural gas delivery began in November 2002. At 247 billion cubic feet (Bcf) per year, pipeline capacity currently exceeds production from the fields, although the extra capacity will eventually transport gas from the nearby Hai Tach, Moc Tinh, and Rong Doi fields, as well as from probable future gas finds in the area. The pipeline comes ashore near the Dinh Co Gas Terminal, and an extension takes some of the gas inland about 20 miles to the Phu My power plant.

The Malay basin is another potentially hydrocarbon-rich area. TotalFinaElf and Unocal have both found natural gas in exploratory drilling in the area. The region is near the Mekong Delta, which is sparsely populated. Although companies are considering constructing a natural gas pipeline to the Delta, demand in the region might not be able to support the construction. A feasibility study for a pipeline to bring natural gas ashore from the PM3 Block, shared by Vietnam and Malaysia, was commisioned in July 2001. In September 2001, Vietnam's prime minister approved study, paving the way to build the $230 million project in southernmost Ca Mau province. The pipeline will stretch for 289 kilometres (about 180 miles) under the sea from the PM3 Block to the mouth of the Doc River in Tran Van Thoi district, then 43 kilometres (27 miles) over land to the Ca Mau industrial zone. It is designed to have an annual capacity of approximately 70 Bcf and will be put into operation during the first quarter of 2005. Gas from the pipeline will be used to supply a 720-MW power plant and a nitrogenous fertiliser plant.

Liquefied Petroleum Gas

Vietnam is a growing consumer and exporter of liquefied petroleum gas (LPG), with almost 70 per cent of urban households using LPG for cooking and other purposes. Japan is the major consumer of Vietnamese LPG exports, receiving the country's first export shipment in May of 1999. According to UK government estimates, domestic consumption is increasing rapidly, and hit a new high of 400,000 tons in 2001. Estimates for 2002 put the country's consumption level at 480,000 tons. There is only one local producer of LPG in Vietnam, called the Dinh Co Plant, which is able to meet about 70 per cent of national demand.

Vietnam's LPG sector has been open to foreign companies since 1998. Saigon Petro, Elf Gas, Petrolimex, and Mobil Unique (a consortium of Mobil, Mitsui, and Unique Gas & Petrochemical), and PTT of Thailand are the major companies involved in the sector. Unique Gas & Petrochemical Plc (UGP), which is involved in the Thai LPG market, could enter the Vietnamese market in the near future. The country's first liquefaction plant at Dinh Co receives gas from the Bach Ho field. It is 100 per cent foreign-owned and is now operating near capacity.

Coal

Vietnam contains coal reserves estimated at 165 million short tons (Mmst), the majority of which is anthracite. Production has increased dramatically since the mid-1990s, peaking at 12.6 Mmst in 1997. In 2001, Vietnam produced just over 11 Mmst, a very slight increase from 10.98 Mmst the previous year. These increases have resulted in an increase in exports (primarily to Japan) and an increase in coal stockpiles. However, export markets have been shrinking during the past few years, which has exacerbated Vietnam's oversupply problem. The state coal company, Vinacoal, cut back production in 1999 in response to the oversupply, but raised it again in 2000 and 2001 as domestic demand increased.

The Vietnamese government is promoting the construction of coal-fired power plants. Vinacoal plans to build as many as seven new power plants over the next decade, with a total capacity of 2,170 megawatts (MW). One 100-MW plant at Na Duong is under construction and scheduled to come online in 2004, and six more are under consideration. Vietnam's coal consumption is expected to increase as it becomes a larger electricity producer, with coal-fired power plants eventually accounting for 25 per cent of the country's total electricity production. The Vietnamese government estimates that it will need to use 10.2 million tons of coal per year by 2010 to meet increasing domestic demand, which it projects will be as high as 20,000 MW by then. Vietnam previously had focused much more on hydropower, and the shift to coal marks an important change in Vietnam's energy sector.

In December 2002, Vinacoal and the Polish Mine Machinery Import-Export Company (Kopex) signed a contract that will provide Vietnam with $90 million in credit from Poland, to be used in the construction of three coal mines in northern Vietnam. The new coal pits would be in the form of underground pits, whereas Vietnam's current coal mines are open-cast. The money for construction will be distributed in early 2004.

Electricity

Although per capita consumption of electricity in Vietnam is among the lowest in Southeast Asia, demand has been on the rise for the past several years, straining the country's generating capacity. As an emerging market, Vietnam has experienced rapid commercial growth, mass migration to major cities, and rising living standards, all of which have contributed to the country's growing demand for electricity. In 2000, Vietnam had a total electric generating capacity of 5 gigawatts (GW). The same year, hydropower accounted for roughly 56.4 per cent of electricity generation, while thermal power accounted for about 43.6 per cent.

The state power company, Electricity of Vietnam (EVN), is working on a plan to develop a national electricity grid by 2020, patching together serveral regional grids. By 2005, EVN aims to build hydropower plants in the central and central highland regions. Three hydroelectric dams, with capacities of between 285 megawatts (MW) and 370 MW, are planned, and construction of the first at Dai Ninh began in 2001. EVN also plans to increase Vietnam's natural gas consumption, using gas from offshore fields to fuel new power plants. Two small gas-fired plants are currently in operation. Construction of Vietnam's first nuclear power plant is included in the plan, to be completed by 2020.

In January 2003, EVN announced plans to spend $1.13 billion on the construction of eight power plants by the end of the year. The total capacity of the plants would be 1,800 MW, with 500-kilovolt (KV) transmission lines. The plants to be built include 70-MW hydropower plant in Quang Tri, a 600-MW thermal power plant in O Mon, a 600-MW thermal plant in Hai Phong, a 110-MW hydropower plant in Plei Krong, a 280-MW hydropower plant in Buon Kuop-Chupong Kong, a 170-MW hydropower plant in A Vuong, a 150-MW steam turbine hydropower plant in Phu My, and a 300-MW hydropower plant in Dai Ninh.

Foreign companies are becoming involved in the growing Vietnamese power market. EVN and a consortium including Tokyo Electric Power (TEPCO), Sumitomo, and Electricite de France (EdF) plan to construct a 715-MW plant ("Phu My 2-2") in the Mekong Delta. Construction of the facility began in January 2003, and it is planned to begin commercial operation in 2004. The plant is intended to be fueled by gas from Nam Con Son Basin, which is planned to be tapped by the BP, Conoco, and ONGC consortium. BP also received approval for the 700-MW Phu My 3 project, located in the same area, in April 2001. Siemens will be undertaking the construction work for that facility.

Both of these projects are to be built as Build-Operate-Transfer (BOT) projects, where the foreign investors operate the plants for a set period of time. These projects were held up, though, by disagreements over pricing. The foreign investors considered the prices offered by EVN to be below the levels which would give them a sufficient rate of return on capital. The Vietnamese government eventually stepped in to resolve the issue, and the price was set at 4.09 cents per kilowatt hour (kWh).

Foreign governments and NGOs are also providing assistance so that Vietnam can expand its electricity generating capacity. In December 2002, the Dutch government approved a non-refundable aid package to help Vietnam optimize its electricity transmission system and to reduce electricity losses to 10 per cent. The funds will be used towards the purchase of equipment that can identify and overcome breakdowns in electricity transmission.

A North-South power cable transmits electricity from Vietnam's largest generator, the Hoa Binh hydropower plant in the north, to large population centers in the south, linking the country into one electricity grid. The cable has helped to alleviate an electricity shortage in Ho Chi Minh City. Plans are underway to build an underground electric cable, an above-ground cable, and a transformer station in Tao Dan (in the Ho Chi Minh City area). Construction began in mid-2000 and is to finish in 2003. The $56 million project is being funded by the World Bank. The World Bank also is funding a $201 million project to extend distribution infrastructure into rural areas currently without electricity. The government currently is considering building more cables, including a 500-KV power line from Pleiku to Danang city. If built, the power line would be 300 km (188 miles) long and cost an estimated $130 million. EVN submitted a project feasibility study to the government in January 2003, which is under review.

Renewable energy consumption is growing in Vietnam, and the country has great solar power potential. About 50 per cent of Vietnamese in rural areas do not have access to electricity. A solar power cooperation program installed a "Vietnamese-French Friendship Solar Station" in Ho Chi Minh City and aims to bring electricity to the southern provinces of Gia Lai, Quang Nam, and Binh Phuoc. The program plans to include other kinds of renewable energy, such as small hydroelectric plants and wind power. The program is operated by Solarlab and is funded by the French Ministry of Foreign Affairs, Electricite de France, and the European Union.

Country Overview

Independence: September 2, 1945 (from France)

Population (2002E): 81.1 million

Location/Size: Southeast Asia / 127,545 square miles, slightly larger than New Mexico

Major Cities: Hanoi (capital), Ho Chi Minh City, Haiphong, Da Nang, Can Tho, Nha Trang, Hue, Nam Dinh, Vung Tau

Languages: Vietnamese, French, Chinese, English, Khmer, tribal languages

Ethnic Groups: Vietnamese (85-90 per cent), Chinese, Muong, Thai, Meo, Khmer, Man, Cham

Religion: Buddhist, Confucian, Taoist, Roman Catholic, indigenous beliefs, Islamic, Protestant

Defense (8/98): Army, 412,000; Navy, 42,000; Air Force, 15,000; Paramilitary, between 4 million and 5 million

Economic Overview

Currency: Dong

Official Exchange Rate (3/14/03): US$1 = 15,428 dong

Nominal Gross Domestic Product at Market Exchange Rate (2002E): $35 billion

Real GDP Growth Rate (2002E): 6.7 per cent *(2003E)*: 6.9 per cent

Inflation Rate (2002E): 3.8 per cent *(2003E)*: 4.1 per cent

Merchandise Exports (2002E): $15.9 billion

Merchandise Imports (2002E): $16.5 billion

Merchandise Trade Balance (2002E): -$900 million

Major Export Products: Crude oil, rice, marine products, garments and textiles, coffee, coal, rubber, nuts

Major Import Products: Petroleum products, tractors, tires, steel products, foodstuffs, cotton, textiles, sugar

External Debt (2001E): $12.9 billion

Energy Overview

Proven Oil Reserves (1/1/03E): 600 million barrels

Oil Production (2002E): 339,000 barrels per day (bbl/d), all of which is crude oil

Oil Consumption (2002E): 186,000 bbl/d

Net Oil Exports (2002E): 153,000 bbl/d

Natural Gas Reserves (1/1/03E): 6.8 trillion cubic feet (Tcf)

Natural Gas Production (2001E): 45.9 billion cubic feet (Bcf)

Natural Gas Consumption (2001E): 45.9 Bcf

Coal Reserves (1/1/01E): 165 million short tons (Mmst)

Coal Production (2001E): 11.0 Mmst

Coal Consumption (2001E): 7.3 Mmst

Net Coal Exports (2001E): 3.7 Mmst

Electric Generation Capacity (1/1/01E): 5 gigawatts (GW)

Electricity Generation (2001E): 29.8 billion kilowatthours (43.6 per cent thermal, 56.4 per cent hydroelectric)

Environmental Overview

Total Energy Consumption (2001E): 0.76 quadrillion Btu* (0.19 per cent of world total energy consumption)

Energy-Related Carbon Emissions (2001E): 12.6 million metric tons of carbon (0.2 per cent of world total carbon emissions)

Per Capita Energy Consumption (2001E): 9.6 million Btu (vs. U.S. value of 341.8 million Btu)

Per Capita Carbon Emissions (2001E): 0.16 metric tons of carbon (vs. U.S. value of 5.5 metric tons of carbon)

Energy Intensity (2001E): 24,525 Btu/$1995 (vs U.S. value of 10,736 Btu/$1995)**

Carbon Intensity (2001E): 0.4 metric tons of carbon/thousand $1995 (vs U.S. value of 0.17 metric tons/thousand $1995)

Fuel Share of Energy Consumption (2001E): Oil (50 per cent), Coal (19.7 per cent), Natural Gas (6.6 per cent)

Fuel Share of Carbon Emissions (2001E): Oil (59.2 per cent), Coal (33.7 per cent), Natural Gas (6.8 per cent)

Status in Climate Change Negotiations: Non-Annex I country under the United Nations Framework Convention on Climate Change (ratified November 16th, 1994). Signatory to the Kyoto Protocol (signed December 3rd, 1998–not yet ratified).

Major Environmental Issues: logging and slash-and-burn agricultural practices contribute to deforestation and soil degradation; water pollution and overfishing threaten marine life populations; groundwater contamination limits potable water supply; growing urban industrialization and population migration are rapidly degrading environment in Hanoi and Ho Chi Minh City.

Major International Environmental Agreements: A party to Conventions on Biodiversity, Climate Change, Desertification, Endangered Species, Environmental Modification, Hazardous Wastes, Law of the Sea, Ozone Layer Protection, Ship Pollution and Wetlands. Has signed, but not ratified, the Nuclear Test Ban.

The total energy consumption statistic includes petroleum, dry natural gas, coal, net hydro, nuclear, geothermal, solar, wind, wood and waste electric power. The renewable energy consumption statistic is based on International Energy Agency (IEA) data and includes hydropower, solar, wind, tide, geothermal, solid biomass and animal products, biomass gas and liquids, industrial and municipal wastes.

Oil and Gas Industries

Organization: State-owned PetroVietnam reports directly to the Cabinet. The company was reorganized in 1990 and now oversees the activities of eight subsidiaries that control functions such as administration, exploration and production, marketing (PetroVietnam Processing and Distribution Company (PVPDC)), training, gas production and distribution (Vietgas), petrochemicals, and information collection.

Major Foreign Oil Company Involvement: BP Amoco, ARCO, BHP, CanOxy, Conoco, Enterprise, Fina, Idemitsu, IPL, Japan National Oil, Mitsubishi, Mobil, OMV, Occidental, Pedco, PetroCanada, Petronas Carigali, Statoil, Sumitomo, TotalFina

93

Yemen

Yemen is important to world energy markets because of its oil and natural gas resources as well as its strategic location at the Bab el-Mandab strait linking the Red Sea and the Gulf of Aden, one of the world's most active shipping lanes.

General Background

Yemen's government continues to implement an economic reform program which it agreed to in 1995 as a condition for lending from the International Monetary Fund (IMF). The program includes banking reform, privatization of state-run industries, major infrastructure investment, and reduction or elimination of government subsidies, including wheat, flour, gasoline, and electricity. Yemen's economy has benefited from relatively high oil prices over the last year, which increased its hard currency receipts and remittances from Yemeni workers in the wealthier Persian Gulf countries. Oil income makes up about 70 per cent of total Yemeni government revenues. Yemen's real gross domestic product grew 3.4 per cent in 2002, and it is projected to grow 4.1 per cent in 2003.

Under terms of its agreements with the IMF and World Bank, Yemen is required to initiate a privatization program. The government has said that it will encourage private investment in the agriculture, fisheries, and oil, and sell off its stake in companies throughout the Yemeni economy. Some companies will be offered for tender or auction, while others will be sold by private subscription. The government also stresses that it is seeking both foreign and local investors. State-owned businesses cited as candidates for privatization include farm and agricultural cooperatives, construction companies, power stations, public housing facilities, refineries, the state's petroleum retail network, shipping companies, and the state telecommunications company. Progress toward privatization, however, has been slow. The Yemeni government decided in June 2001 not to fully privatize the Aden Refinery, but rather to divest only part of its ownership stake. Price controls also have been reintroduced for some commodities, mainly foodstuffs, within the last year.

.Security remains a major concern of foreign firms doing business in Yemen, particularly after the French-flagged tanker Limburg was attacked off the coast of Yemen in October 2002. The Canadian oil company Nexen, which operates the ash-Shihr oil export terminal, agreed in Januray 2003 to provide assistance to the Yemeni government in improving security. Kidnappings of foreigners, including oil workers, have been a problem. Besides kidnappings, there have been periodic attacks on an oil pipeline in the Marib region of eastern Yemen, which is operated by U.S.-based Hunt Oil.

Yemen and Saudi Arabia have resolved their longstanding border dispute. In a treaty signed in June 2000, the two governments agreed on the delineation of sections of their common border which had been in dispute since the 1930s. The deal is expected to open up opportunities for increased Saudi trade and investment in Yemen, and has made possible the award of oil and gas exploration rights for areas in Yemen adjacent to previously disputed areas of the border.

Oil

Yemen's oil output, which averaged 443,288 barrels per day (bbl/d) in 2002, provides the country's main source of hard currency revenue. The country contains proven oil reserves of 4 billion barrels. Reserves are concentrated in five areas: Marib-Jawf Block 18 (490 million barrels) in the north, East Shabwa Block 10A (180 million barrels) and Masila Block 14 (500+ million barrels) in the south, and the Jannah Block 5 (345 million barrels) and Iyad Block 4 (135 million barrels) in central Yemen. Nexen, which owns 52 per cent of the Masila block, produces around 230,000 bbl/d at Masila. Hunt oil has output in Yemen of around 165,000 bbl/d–140,000 bbl/d from Marib-Jawf and

25,000 bbl/d from Jannah. TotalFinaElf produces around 20,000 bbl/d from its East Shabwa concession. The Yemeni government has said the country's overall oil production capacity should reach 500,000 bbl/d by the end of 2003.

Oil exploration activity in Yemen has accelerated since 1997, after a downturn during the mid-1990s due to Yemen's brief civil war, though current security concerns may again slow exploration activity somewhat. In April 1998, Nexen entered an agreement with Kerr-McGee Yemen, acquiring interests in two exploration blocks near the Nexen-operated Masila Block. Each company obtained a 47.5 per cent interest in Block 50 and a 43.75 per cent interest in Block 51, with the Yemen Ministry of Oil holding the balance for both blocks. Block 51 is adjacent to the Masila Block and consists of two million acres. Block 50 covers more than 8 million acres and is located northwest of the Masila Block.

Norwegian independent DNO began production in November 2000 at the Tasour field in Block 32, which currently is producing about 17,500 bbl/d. The company plans to increase output from Tasour to 20,000 bbl/d by the end of 2003. Dove Energy of Britain currently produces around 25,000 bbl/d at the Sharyoof field in Hadramawt province. Russia's Rosneft and Avirex of the United Arab Emirates were granted drilling rights in the eastern al-Mahrah province in February 2002. The British independent Capital Oil and Gas was awarded exploration rights for an area in Al-Mahra province in January 2003.

Yemen's border demarcation treaty with Saudi Arabia has opened up new areas to exploration. Four new blocks have been demarcated along the border, and several companies have expressed interest in exploring them. Nexen signed a memorandum of understanding with Yemen in January 2001 covering Block 59, located adjacent to the Saudi border. Nexen holds a 60 per cent interest, with the other 40 per cent held by Occidental Petroleum (of which Nexen is no longer a subsidiary). Austria's OMV, along with Cepsa of Spain and PanCanadian, concluded an exploration and production contract with the Yemeni government for Block 60 in December 2001. The area is near the Saudi border, and was previously unexplored due to the border dispute.

Refining

Yemen currently has a crude refining capacity of 130,000 bbl/d from two aging refineries. The refinery in Aden, operated by Aden Refinery Company (ARC), has a capacity of 120,000 bbl/d, while capacity at the Marib refinery, operated by Yemen Hunt Oil Company, is 10,000 bbl/d. The Aden refinery, which had a design capacity of 170,000 bbl/d, sustained significant damage during the country's 1994 civil war, but was later partially rebuilt. The Yemeni government has backed away from plans to privatize the Aden refinery, but may offer a partial stake to private investors at some point.

Yemen signed an agreement in December 2002 with the Hadramawt Refinery Company, backed by Saudi investors, for a $450 million facility with a capacity of 50,000 bbl/d to be completed at Mukalla within two years.

Natural Gas

With natural gas reserves of 16.9 trillion cubic feet (Tcf), Yemen has considerable potential as a natural gas producer and exporter. The bulk of Yemen's gas reserves are concentrated in the Marib-Jawf fields, operated by the Yemen Exploration and Production Company (YEPC). There is currently no production of natural gas in Yemen, and the gas which Hunt Oil extracts as a by-product of oil production at these fields is reinjected.

From the mid-1990s until 2002, the primary interest in natural gas development in Yemen was focused on the export of liquefied natural gas (LNG). TotalFinaElf, along with several other major multinational companies, had established the Yemen Liquefied Natural Gas Company ("Yemen LNG"). The proposed project suffered a major blow in June 2002, however, when ExxonMobil and Hunt Oil announced that they were leaving the consortium. The Yemeni government extended Yemen LNG's approval for an additional four years shortly thereafter, but with no progress has been reported since then.

In the last year, attention has been focused on possible development of a modest amount of natural gas production for domestic consumption. Discussions have reportedly been held with potential investors about a pipeline from Marib-Jawf to Aden, with the output to be used for electric power generation and a petrochemical plant, but no contracts have yet been signed.

Electricity

Yemen generated 3.0 billion kilowatthours of electricity in 2001 on installed capacity of 810 megawatts (MW), all oil-fired. Yemen's two largest power plants are the 165-MW power station at Ras Kanatib, near Hodeidah, and the 160-MW Mukha station. Yemen's generating capacity is inadequate for the country's needs, and a rolling blackout schedule is maintained in many cities. Government plans for the power sector call for privatizing all generators having a capacity of less than 5 MW, and also for selling generators of 5 MW-20 MW through public offerings.

Yemen has started on some modest projects to expand and improve its power sector. On July 8, 1997, Yemen's Public Electricity Corporation (PEC) completed work on linking the southern and northern electricity grids. The grid linkage was funded by the Kuwait-based Arab Fund for Economic & Social Development, which provided $54 million of the $64 million required for the project. Yemen undertook a World Bank-funded upgrade of the Dhaban power plant to 50-MW total capacity.

The Mukalla power project involved the construction of a 40-MW diesel-fired plant, six substations, and the laying of 62 miles of transmission lines. Wartsila NSD Corporation, the Dutch branch of Finland's Wartsila Diesel, started work on the plant in mid-1997 and construction was completed in 1998. Wartsila also completed the Aden power project, which involves building a 30-MW plant to serve the city's port. The new plant is part of the redevelopment of Aden, which was heavily damaged in the 1994 civil war. The company won the construction contract in late 1997 from Yemen Investment and Development International, the consortium leading the redevelopment of Aden port. Adair International was awarded a contract for a 21-MW plant near the port in January 2001, but construction has yet not begun.

Yemen continues to face serious power shortages, and has announced plans to reform the country's power sector, and to double power generating capacity. Currently, fewer than one-third of households in Yemen have access to electricity. Yemen also has said that it plans to invite offers on the country's first private power station. Current plans call for the construction of a 400-MW privately-owned gas-fired plant near the Marib oil and gas field, but the project has yet to make any tangible progress.

Country Overview

Independence: May 22, 1990 (reunification)

Population (2002E): 18.7 million

Location/Size: Southwest corner of the Arabian Peninsula/527,790 sq. kilometers (203,730 sq. miles); approximately the size of Wyoming and Colorado

Major Cities: Sanaa (capital), Aden, Al Hudaydah, Taizz

Language: Arabic

Defense (8/98): Army (37,000), Navy (1,500), Air Force (3,500), Paramilitary Forces (50,000)

Economic Overview

Currency: Yemeni Rial (YRI) US$1 = YR178 (average for 2002)

Gross Domestic Product (GDP) (2002E)–market exchange rate: $8.7 billion

Real GDP Growth Rate (2002E): 3.4 per cent *(2003F):* 4.5 per cent

Consumer Price Inflation (2002E): 10.0 per cent *(2003F):* 9.4 per cent

Major Trading Partners: China, Japan, Saudi Arabia, Singapore, South Korea, United Arab Emirates, United States

Major Export Products: Crude oil, cotton, coffee, hides, vegetables, dried and salted fish

Major Import Products: textiles and other manufactured consumer goods, petroleum products, sugar, grain, flour, other foodstuffs, cement, machinery, chemicals

Energy Overview

Proven Oil Reserves (1/1/03): 4 billion barrels

Oil Production (2002E): 443,288 barrels per day (bbl/d)

Oil Consumption (2002E): 74,000 bbl/d

Net Oil Exports (2002E): 369,288 bbl/d

Crude Oil Refining Capacity (1/1/03): 130,000 bbl/d

Natural Gas Reserves (1/1/03): 16.9 trillion cubic feet (Tcf)

Natural Gas Production/Consumption: None.

Electric Generating Capacity (1/1/01): 810 megawatts (100 per cent oil-fired)

Electricity Generation (2001E): 3.0 billion kilowatthours

Environmental Overview

Total Energy Consumption (2001E): 0.15 quadrillion Btu* (0.04 per cent of world total energy consumption)

Energy-Related Carbon Emissions (2001E): 2.6 million metric tons of carbon (0.04 per cent of world carbon emissions)

Per Capita Energy Consumption (2001E): 7.6 million Btu (vs. U.S. value of 341.8 million Btu)

Per Capita Carbon Emissions (2001E): 0.1 metric tons of carbon (vs. U.S. value of 5.5 metric tons of carbon)

Energy Intensity (2001E): 9,293 Btu/$1995 (vs U.S. value of 10,736 Btu/$1995)

Carbon Intensity (2001E): 0.17 metric tons of carbon/thousand $1995 (vs U.S. value of 0.17 metric tons/thousand $1995)

Fuel Share of Energy Consumption (2001E): Oil (100.0 per cent)

Fuel Share of Carbon Emissions (2001E): Oil (100.0 per cent)

Status in Climate Change Negotiations: Non-Annex I country under the United Nations Framework Convention on Climate Change (ratified February 21st, 1996). Not a signatory to the Kyoto Protocol.

Major Environmental Issues: Very limited natural fresh water resources; inadequate supplies of potable water; overgrazing; soil erosion; desertification

Major International Environmental Agreements: A party to Conventions on Biodiversity, Climate Change, Desertification, Environmental Modification, Hazardous Wastes, Law of the Sea, Nuclear Test Ban and Ozone Layer Protection

The total energy consumption statistic includes petroleum, dry natural gas, coal, net hydro, nuclear, geothermal, solar, wind, and wood and waste electric power. The renewable energy consumption statistic is based on International Energy Agency (IEA) data and includes hydropower, solar, wind, tide, geothermal, solid biomass and animal products, biomass gas and liquids, industrial and municipal wastes.

Oil and Gas Industries

Organizations: Yemen Petroleum Company (YPC)–production and refining; General Corporation for Oil and Mineral Resources (GCOMR)–investment and holding company; Yemen Refining Company (YRC)–refining; General Department of Crude Oil Marketing (GDCOM)–handles government shares of exports; Yemen Exploration and Production Company (YEPC)–contracts

Major Oil Fields: Alif, Asaad Al-Kamil, Camaal, Azal

Foreign Company Involvement: Capital Oil and Gas; Cepsa, Dove Energy, DNO, Nexen, Hunt Oil, PanCanadian, Rosneft, TotalFinaElf, TransGlobe Energy

Major Refineries (Capacity): Aden (120,000 bbl/d), Marib (10,000 bbl/d)

Major Ports: Aden, Hisn an Nushaymah, Al Khalf, Mocha, Nishtun, Ra's Isa, Ra's Kathib, Salif

Major Pipelines: Marib-Ra's Isa Pipeline (pipeline between the Marib fields and the deep sea port of Ra's Isa on the Red Sea), Shabwa-Rudhum Pipeline (pipeline linking the Shabwa fields to the Rudhum terminal on the Gulf of Aden at Hisn an Nushaymah)

94

Monthly Energy Chronology 1996

The following information contains a listing of the major energy events that occurred in 1996.

January 1996

January 1

Russia temporarily stops oil flows through the 1.2-million b/d capacity Druzhba pipeline after Ukraine imposes a 13-per cent increase in transit tariffs from $4.53 to $5.20 per ton per kilometer. Talks between the two countries are suspended shortly thereafter, and oil shipments to Ukraine's 78,000-barrel per day (b/d) Drogobych refinery are cut. In 1995, Russia shipped about 600,000 b/d through the Druzhba pipeline to Eastern Europe.

January 3

Under a new hydrocarbon law which brings the country inline with European Union regulations, Greece opens bidding for six onshore and offshore exploration concessions. Onshore concessions are located in northwest Peloponnese, Aetolo-Archarnania, and Ioannina. Offshore blocks are near Katakolo in the Ionian Sea's northwest Peloponnese, the Patras Gulf, and near Paxi Island. Greece produces about 10,000 b/d from its offshore Prinos field near Thasos Island in the northern Aegean Sea. However, it has not been able to develop other Aegean Sea oil and gas fields because of territorial disputes with Turkey.

January 9

In Cairo, Israeli Energy Minister Gonen Segev meets with Egyptian Oil Minister Hamdy al-Banbi to discuss possible construction of a $500-million pipeline to carry 250 million cubic feet per day (Mmcf/d) of Egyptian natural gas to Israel, a similar amount to Jordan, and 50 Mmcf/d to the Palestinian-controlled Gaza Strip. In 1995, Egypt augmented its 22-trillion cubic feet (Tcf) of natural gas reserves with new discoveries in the Mediterranean Sea. Amoco and Agip hold the concessions which would supply gas for the Israeli pipeline.

January 13

According the Iran's official Islamic Republic News Agency (IRNA), Iran and Nigeria have agreed to cooperate in oil exploration, production, and marketing. The deal comes after oil officials from the two countries met in Lagos. Both Iran and Nigeria have various forms of economic sanctions imposed on them by members of the international community.

January 14

Qatar's state news agency announces that Qatari and Dubai officials are discussing the possible construction of a pipeline to carry Qatari natural gas to Dubai's power plants, most of which are gas-fired. In the past, Qatari Energy Minister Sheikh Abdullah bin Hamad al-Attiya has advocated the creation of a GCC gas grid, which would allow regional countries to raise oil exports by substituting domestic consumption of oil with natural gas.

January 15

The British Government approves the $2.47-billion Eastern Trough Area Project development. The Eastern Trough is located east of the Shetland Islands and contains an estimated 400 million barrels of oil, 35 million barrels of natural gas liquids, and 1.1 Tcf of natural gas. British Petroleum, Royal Dutch/Shell, Exxon, Agip, BHP, Murphy Oil, and Mitsubishi Oil will be partners in the development. The area is located near the 1-billion barrel Foinaven and Schiehallion heavy oil fields currently under development.

January 16

In Peru, state-owned Petroperu workers stage a 24-hour strike to protest the government's privatization plans. While union representatives declare that Petroperu "is totally paralyzed," operations and oil production appear to remain unaffected. Privatization could result in loss of about 1,500 of the company's current 5,500 jobs. Peru produces about 130,000 b/d, most of which is consumed domestically.

January 16

South Africa's Central Energy Fund reports that an environmental study designed to assess the impact of storing Iranian oil in South Africa would be delayed by six months until September 1996. In July 1995, South Africa agreed to allow Iran to store up to 15 million barrels at the 45-million barrel Saldanha Bay storage facility near Cape Town. The deal is opposed by the United States and comes after the United States imposed a unilateral embargo on Iranian oil exports in April 1995.

January 17

Iraq agrees to talks concerning a U.N. plan to allow for the Iraqi sale of $1 billion of oil for 90 days for a 180-day trial period. Under U.N. Resolution 986, proceeds from the sale would be used for humanitarian purposes. In the past, Iraq has opposed clauses 6 and 8b contained in Resolution 986. Clause 6 stipulates that oil exports under this plan must pass through the 1.6-million b/d Iraq-Turkey pipeline, which currently is unusable because of sludge build-ups and pumping station damage. By most estimates, the line would take a minimum of three months to repair. Clause 8b states that part of the proceeds from the sales would be disbursed under U.N. supervision to Kurdish provinces in northern Iraq. Negotiations between Iraq and the United Nations are scheduled to begin February 6, 1996.

January 17

Romanian officials announce that the country's new petroleum law will take effect in March 1996 and will provide a framework for the country's upcoming second exploration licensing round. Romania currently produces 135,000 b/d of oil, with reserves of about 1 billion barrels. Amoco, Shell, and Enterprise Oil actively are exploring in the country.

January 22

Venezuela starts a bidding round which offers 10 concessions with a resource potential of between 7 and 23 billion barrels of oil and up to 10 Tcf of natural gas reserves. Five blocks are located near the Furrial field in the Orinoco Delta, two in the northern Andes near Lake Maracaibo, two in the eastern Anzoategui and Monagas states, and one in the Barinas region near the Colombian border. The government anticipates that production in these blocks May be able to increase the country's light and medium crude oil production by as much as 1 million b/d by 2003 and bring in tax revenues of between $15 to $20 billion. Over 75 companies, including British Petroleum, Exxon, and Shell, are involved in the auction. Awarded acreage will allow companies to develop the concessions on a profit sharing basis.

January 23

Bahrain agrees to accept International Court of Justice (ICJ) jurisdiction in a longstanding dispute over the potentially oil-rich Hawar Islands and the Fahst al-Dibal and Jarada sandbars. In May 1993, Bahrain passed a law which decreed its sovereignty over the islands and reefs, some of which are only a mile from Qatar's coast. After Qatar filed a case with the ICJ in February 1994, Bahrain refused to acknowledge the Court's jurisdiction. In February 1995, both sides then requested Saudi mediation outside of the ICJ. Bahrain now states that it will accept the ICJ's jurisdiction if a related dispute over northwestern Qatar's Zubara region is included in the case. The ancestors of Bahrain's ruling family are buried in Zubara.

January 26

Colombian rebels dynamite the country's 230,000-b/d Cano Limon-Covenas pipeline. The pipeline is bombed simultaneously in two locations 200 miles apart, resulting in an estimated spill of 10,800 barrels. The line is the primary transport route for crude oil production from the country's prolific Llanos basin fields to the Covenas export terminal on the Caribbean coast. So far, the line has been bombed eight times in 1996, 46 times in 1995, and around 400 times since its inauguration in 1986. In late 1995, the Colombian government released a report stating that oil companies have spent $500 million since 1991 on security measures to protect oil infrastructure from guerilla attacks.

January 29

In Nigeria, Shell workers stage strike after the company announces plans to cut 10 per cent of its 5,000-person workforce. Initial reports state that oil production remains unaffected. The strike also comes a day after Shell admits to importing weapons for Nigeria's police force. The company says that it provided handguns to police in order to provide adequate protection for its oil facilities. In other developments during the month, the son of Nigerian ruler General Sani Abacha is killed in a plane crash on January 17. The crash is followed several days later by bomb blasts at the Kano airport where Ibrahim Abacha's plane attempted to land.

January 30

In Ecuador, employees of state-owned Petroecuador stage a 24-hour solidarity strike to protest the government's intended privatization of the electricity sector. Electricity workers took over the country's seven hydroelectric power plants on January 17. Operations subsequently were cut to a minimum, including those at the Paute plant, which supplies an estimated 75 per cent of the country's electricity. President Sixto Duran-Ballen recently declared a state of emergency and sent troops to surround the power stations.

January 30

Vice Admiral Scott Redd, commander of the U.S. Fifth Fleet based in the Persian Gulf, states that Iran test-fired a new anti-ship missile near the Strait of Hormuz on January 6. The missile reportedly has a range of 60 miles and is viewed as a threat to regional security by U.S. naval forces operating in the area. Oil tankers carry about 15 million b/d through the Strait.

January 30

A Dubai-based oil tanker sinks approximately 1.5 miles off Iran's Lavan Island in the Persian Gulf. The 803-ton ship was en route from Dubai to the Iranian port at Bushehr when it suffered an unspecified "accident" and began taking on water.

January 31

In Mexico's southeastern Tabasco region, about 1,000 protestors demand compensation for pollution caused by state-owned Pemex's oil operations. Demonstrators take control of 18 oil drilling rigs and prevent drilling and maintenance activity at 21 other rigs by blocking regional highways. Mexico has an estimated 4,700 oil wells nationwide, and the country's oil production remains unaffected.

January 31

In Yemen, France-based Total and U.S.-based Hunt Oil sign a memorandum of participation regarding a recent $3-billion liquefied natural gas (LNG) export deal recently ratified by the Yemeni parliament. In September 1995, Total and Yemen's government agreed to build a 5-million ton per year LNG plant. However, natural gas to supply the LNG plant would come from the Marib concession already awarded to a Hunt Oil-led consortium, which includes Exxon. After months of talks between the various companies, Total and the government agree to give the Hunt consortium a 38 per cent stake in the project.

February 1996

February 5

Turkmenistan cuts natural gas shipments to Armenia by 20 per cent, to 140 million cubic feet per day. The move comes after Armenia's arrears reach $50 million. Since the 1988 outbreak of hostilities between Armenia and Azerbaijan, Armenia's gas supply has been disrupted frequently. The country currently receives its gas supplies through a Georgian pipeline and from Iran.

February 5

Clashes between Nigeria and Cameroon over the disputed Bakassi Peninsula result in two deaths. The Bakassi Peninsula is a 600-square mile region bordering the two countries which contains oil and natural gas reserves. Nigeria claims that an 1858 treaty places the Peninsula in its territory. Cameroon contends that Nigeria offered it the region after Cameroon refused to help Biafran rebels during Nigeria's 1967 civil war. More recently, Nigeria has accused Cameroon of siphoning off 130,000 barrels per day (b/d) of oil from the producing fields in the region. The International Court of Justice in the Hague is currently reviewing the case.

February 9

Ecuador's state-owned Petroecuador announces that it will offer nine exploration blocks in the country's eastern Amazon basin region called the Oriente. Despite transportation problems and environmental concerns, the Oriente region has been a focus of Ecuador's last two bidding rounds for exploration acreage. Petroecuador believes the region could hold between 100 million to 1 billion barrels of additional oil reserves, which could yield new production of at least 100,000 b/d between 2002 and 2004, depending on development timetables.

February 10

Mexico's state-owned Petroleos Mexicanos (Pemex) announces that protests led by the leftist Democratic Revolutionary Party in the Tabasco region have shut-in 28,000 b/d of oil production, resulting in cumulative losses of $8.5 million during the past 13 days. Demonstrations have occurred at 61 sites in the Tabasco region, which accounts for about one-quarter of Mexico's oil production. About 1,000 protestors are demanding compensation for pollution caused by state-owned Pemex's oil operations in the region. In 1995, Mexico produced roughly 2.6 million b/d of oil.

February 11

In the South China Sea, Philippine navy patrol boats forcibly seize a pirate ship which violates Philippine territorial waters. The vessel is piloted by a "Chinese-looking" crew and is stopped in the same region as two ships were three weeks earlier. Those ships were of a similar class as those used by the Chinese navy. The Philippines and China currently are involved in a territorial dispute over the South China Sea's Spratly Islands. In 1995, the Spratly's were estimated by a Russian geological team to contain up to 1 billion barrels of hydrocarbon reserves.

February 12

Turkey and Turkmenistan sign a protocol to build a pipeline which could carry up to 350 billion cubic feet per day of Turkmen natural gas to European markets. Possible routes are either through Georgia or Iran. However, financing for the $8-billion project still remains an issue.

February 14

At the 62,000-b/d Talara refinery in northern Peru, over 2,000 employees of state-owned Petroperu end a 13-day strike protesting the company's planned privatization. As preparation for Petroperu's privatization, about 1,500 of the company's 5,500 employees recently were laid-off, with about 80 per cent of those workers employed at the Talara refinery. Striking employees return to work after a judge issues an order rescinding the lay-off notices.

February 15

In Wales, a Liberian-flagged oil tanker carrying 900,000 barrels of light crude oil runs aground. Force 10 gale storms lead to evacuation of the tanker's 35-member crew. The tanker had been transporting crude from the North Sea to Texaco's refinery at Milford Haven Harbour on the Welsh coast. During the following week of salvage operations, the tanker spills an estimated 450,000 barrels of oil.

February 15

Yemen accepts a France-brokered proposal to end the conflict between Yemen and Eritrea over Greater Hanish Island in the Red Sea. Press reports state that the French proposal would allow both countries to station troops on the disputed island. In December 1995, Yemen and Eritrea clashed for several days over Greater Hanish Island, which is strategically-located in the Red Sea's Bab al-Mandab strait.

February 16

In Algeria, U.S.-based ARCO signs a $1.5-billion production sharing contract to conduct enhanced oil recovery work on the Rhourde el-Baguel oil field. Rhourde el-Baguel is the Algeria's second largest oil field, containing 2.5 billion in remaining reserves. ARCO plans to drill additional wells and introduce gas injection at the field in the hopes of raising the 25,000 b/d of current output to 125,000 b/d within 10 years. The deal is the latest in a series of large commitments by foreign oil companies to invest in Algeria's oil sector.

February 19

Mexican Energy Minister Jesus Reyes Heroles states that Pemex will not be privatized and that such action is still against the country's constitution. A week earlier at the Tula Hidalgo oil refinery, Mexican President Ernesto Zedillo had reiterated his support for selling part of the state-owned oil company. Due to changes in Mexican law in 1995, some of Pemex's operations, including petrochemicals, could be privatized.

February 19

In New York, the United Nations and Iraq end almost two weeks of negotiations over Iraq's possible sale of $1 billion of oil for 90 days for a 180-day trial period. Under U.N. Resolution 986, proceeds from the sale would be used for humanitarian purposes. Concurrent with the talks, Iraqi and Turkish officials meet to determine the status of the 1.6-million b/d Iraq-Turkey pipeline, through which a majority of the oil exports would flow. The U.N.-Iraq talks, which end without a decision, are scheduled to restart in March 1996.

February 22

In Qatar, authorities arrest 60 army and police officers after a failed coup attempt to restore the deposed emir, Sheik Khalifa bin Hamad al-Thani. The Sheik is the father of the current emir, Sheik Hamad bin Khalifa al-Thani, who overthrew him while he was vacationing in Switzerland in June 1995. Policies regarding Qatar's oil sector remained unchanged after the son's coup.

February 22

In the Caucasus, suspected Chechen rebels blow up a natural gas pipeline near the town of Shelkovskaya, cutting off Russia gas shipments to the region. Simultaneously, a separate attack on the pipeline occurs in neighboring Dagestan. The pipeline carries gas from Stavropol to Makhachkala. The attacks are followed shortly thereafter by the region's heaviest fighting since Russian forces invaded Chechnya in December 1994. Historically, Chechnya has been Azerbaijan's primary route for its oil exports to Russia

February 23

Morocco begins privatizating its 129,000-b/d SociJtJ Marocaine de l'Industrie du Raffinage (Samir) oil refinery by listing the refinery on the country's stock exchange. A projected 25 per cent of the Samir refinery's shares will be sold in what is expected to be the country's largest privatization to date. The Samir refinery was built in 1962 and satisfies about 75 per cent of Morocco's domestic demand for refined products. In another development, Mobil freezes investments at the 225 service stations it operates in Morocco. The company makes the move after stating that the government's control over margins has adversely affected its profitability. Mobil had planned to double its current investment in Morocco to $5 million.

February 26

In Congo, the Angolan government and the Front for the Liberation of the Enclave of Cabinda–Renovado (FLEC-R) agree to extend their present truce by another 120 days. Cabinda is the source of over 50 per cent of Angola's oil production, which reached almost 700,000 b/d at the end of 1995. The enclave is separated from the Angolan mainland by a 16-mile wide coastal strip of Zairian territory. Various FLEC factions have fought for Cabindan autonomy since Angola's 1975 independence from Portugal. Since most of Cabinda's oil fields are located offshore, however, oil production has remained relatively unscathed.

February 27

Russian First Deputy Prime Minister Vladmir Kadannikov states that the Russian duty on oil exports will be reduced shortly from ECU 20 per ton to ECU 14 per ton, before its elimination in July 1996. However, the government will raise excise taxes on oil producers and institute a new pipeline levy to compensate for any loss in revenue to the government.

February 27

In Nigeria, an explosion occurs on a pipeline carrying refined products to a depot about 40 miles from Lagos in the southwestern area of the country. Local repair workers notice that digging was evident around the pipeline, leading officials to suspect sabotage.

February 27

The Azerbaijan International Operating Company (AIOC) approves a $231-million export pipeline for part of the initial oil production from the consortium's new Caspian Sea oil fields. The AIOC is a consortium of international oil companies which is developing the Azeri, Chirag, and Guneshli oil fields off Azerbaijan in the Caspian Sea. In late 1995, the consortium also approved the use of existing Russian pipelines to export part of the 70,000 b/d of new oil production. Even though Turkey and Russia have lobbied aggressively for pipeline routes through their respective countries, no decision has been made yet on the export route for later oil production, which could reach 700,000 b/d by 2005.

February 28

Oil industry sources report that Kuwait National Petroleum Corporation (KNPC) plans to temporarily shut down two of the country's three refineries for maintenance activities. The 415,000-b/d Mina al-Ahmadi refinery will be closed in mid-March for 40 days. Subsequently, the 145,000 b/d Shuaiba refinery will be shut for a similar time period beginning in mid-May 1996. The refinery closures are expected to result in a temporary shortage of refined products, especially from the Min al-Ahmadi refinery. Kuwait plans to restore the Shuaiba refinery's pre-Gulf War operating capacity of 195,000 b/d by 1997.

February 29

Canada-based MacDonald Oil Exploration announces that it will sign an exploration agreement covering a 6,000-square mile area on Cuba's southeast coast. The deal comes amidst increased Canadian concern over possible U.S. sanctions against non-U.S. companies which conduct business in Cuba. In response to Cuba's downing of two U.S. civilian aircraft, the U.S. government is considering punitive measures which might allow ligation by U.S. citizens and corporations against foreign companies operating in Cuba as well as denying U.S. entry visas to foreign executives.

March 1996

March 3

In Peru, negotiations between state-owned Petroperu and a Shell/Mobil consortium are extended until April 26. Since July 1995, the companies have been negotiating development terms for the Amazon Basin's Camisea gas field, which is believed to hold an estimated 11-trillion cubic feet (Tcf) of non-associated gas and 700 million barrels of condensate. The original deadline for the conclusion of negotiations was December 1995, but was later extended until March 1996. The Camisea development could involve connecting the field to the Bolivia-Brazil gas export pipeline now under construction. However, present talks are rumored to have stalled over potential tax breaks for Shell. Chevron and Elf Aquitaine also are interested in developing the Camisea field.

March 6

Afghan President Burhanuddin Rabbani visits Turkmenistan to discuss two proposed pipelines which could export up to 1 million barrels per day (b/d) of crude oil and 2 billion cubic feet per day (Bcf/d) of natural gas from fields in southern Turkmenistan to Pakistan. The project is being evaluated by a venture formed by Unocal (40 per cent) and Saudi Arabia's Delta Oil (60 per cent). The pipeline would be supplied with gas from the giant Dauletabad field, which holds a probable 45 Tcf of Turkmenistan's estimated 113 Tcf of reserves. The proposed pipelines are anticipated to run through territory controlled by several of Afghanistan's 13 rebel factions, which have been fighting since the Soviet withdrawal in 1989.

March 7

In offshore Nigeria, Shell makes a potentially large discovery with its Bongo-1 well in over 3,000 feet of water. The company estimates oil reserves at up to 500-million barrels. However, extensive appraisal drilling will be necessary before any accurate reserve estimates are compiled. Shell accounts for about half of Nigeria's oil production, which averaged roughly 1.9 million b/d in 1995.

March 7

The Caspian Pipeline Consortium agrees to a reorganization reducing Oman Oil Company's (OOC) stake in the venture from 25 per cent to 10 per cent. The CPC consortium is dominated by the Russian (25 per cent) and Kazak (25 per cent) governments and plans to begin work shortly on a $1.5 billion pipeline to export crude oil from Chevron's Tengiz field to the Black Sea. The project has been stalled because of financing difficulties. In 1995, another venture comprising Chevron, Mobil, British Gas, and Agip developed an alternative pipeline export scheme. Tengiz output currently is limited to under 80,000 b/d because of insufficient space on Russia's Transneft pipeline network.

March 18

Iran rejects a recent Gulf Cooperation Council (GCC) proposal stating that Iran's dispute with the United Arab Emirates (UAE) over the Abu Musa and Tunb Islands should be settled by the International

Court of Justice. Iran forcibly seized the two islands from the UAE in the early 1970s, although joint sovereignty was maintained until 1994. Shortly after Iran's statement, the United States announces that in April 1996, it will conduct its first military maneuvers with the UAE.

March 18

In Taiwan, Conoco and state-owned Chinese Petroleum Corporation (CPC) sign a joint venture deal in which Conoco will invest at least $65 million to drill 13 wells over the next five and a half years. Conoco has a 75 per cent stake in the venture, but CPC has the future right to increase its stake to 50 per cent. CPC will purchase all oil and gas produced from the venture over the next 20 years. The agreement comes amidst Taiwan's presidential elections and Chinese military maneuvers near offshore acreage in which development is planned.

March 18

In New York, the United Nations and Iraq end a second round of negotiations over Iraq's possible sale of $1 billion of oil for 90 days for a 180-day trial period. Under U.N. Resolution 986, proceeds from the sale would be used for humanitarian purposes. While both sides have reached agreement on several key issues, the details concerning distribution of aid to Kurds in northern Iraq remains unresolved. Due in part to market uncertainty surrounding the talks, oil prices rise to their highest levels since the 1991 Gulf War. The U.N.-Iraq talks are scheduled to restart on April 8.

March 19

After internal disputes over the extent of Israeli government involvement, Prime Minister Shimon Peres temporarily freezes the pipeline project to export natural gas from Egypt to Israel. At present, the Israeli government is divided over whether to take up to a 30 per cent stake in the project. The proposed $500-million pipeline will carry 250 million cubic feet per day (Mmcf/d) of Egyptian natural gas to Israel, a similar amount to Jordan, and 50 Mmcf/d to the Palestinian-controlled Gaza Strip. Amoco and Agip hold the concessions which would supply gas for the Egyptian-Israeli pipeline.

March 20

The state-owned oil company of Azerbaijan (SOCAR) cedes a quarter of its 40 per cent stake in the offshore Shakh Deniz field development to Elf Aquitaine. Iran's NIOC as well as Russia's Yukos, Sidanko, and Rosneft also are seeking stakes in the $4 billion project. Shakh Deniz contains possible reserves of 700 million barrels of oil and up to 14 Tcf of natural gas. In addition to SOCAR and Elf, current partners in the project are BP, Statoil, and Turkey's TPAO.

March 20

Turkmen President Sapamurat Niyazov and Georgian President Eduard Shevardnadze meet in Tbilisi to discuss energy cooperation between the two countries. The meeting focuses on allowing a proposed Turkmen gas export pipeline to cross Georgian territory. The line would carry up to 1.5 Mmcf/d of natural gas to Turkey and Western Europe by 2020. Also at the meeting, President Niyazov agrees to write-off about one-fifth of Georgia's current $480 million debt for Turkmen gas it already has received.

March 21

Greek Foreign Minister Theodoros Pangalos states in an Izvestia newspaper article that Russia, Greece, and Bulgaria are considering the construction of a $2.5-billion oil pipeline which would bypass Turkey's Bosporus Strait. The proposed pipeline would run from the Bulgarian city of Burgas to Alexandroupolis in Greece. If built, the Bulgarian-Greek pipeline could compete against a proposed plan to export new Caspian Sea oil production through a proposed Turkish pipeline.

March 22

Nigeria implements two new procedures to curtail corruption with respect to oil export revenues. The first measure will have "internationally recognized" agents inspect all crude oil exports. The second measure involves announcing oil export earnings on a monthly, rather than yearly, basis. A government-sponsored study released in 1995 stated that the government could not account for about $12 billion worth of oil export revenues between 1988 and 1994. Both of the new changes will be implemented in April 1996.

March 22

Kenyan Minister for Finance Musalia Mudavadi states to World Bank representatives that his government has "initiated the process" of privatizing the country's oil industry. Possible privatizations would include Kenya Petroleum Refineries Limited and Kenya Pipeline Corporation. The former owns a 50 per cent stake in the 90,000-b/d Mombasa refinery. The latter company has so far invested an estimated $400 million in the construction of a liquefied petroleum gas import pipeline which would have an extension to Kampala, Uganda.

March 24

Kuwaiti Oil Minister Abdul Mohsen al-Mudej states that his government will sell off 30 Kuwait National Petroleum Corporation (KNPC) owned gas stations. The move is part of the Kuwaiti government's efforts to encourage private involvement in the country's oil sector. KNPC currently owns about 90 retail gas stations in Kuwait.

March 25

In Bolivia, employees of state-owned Yacimentos Petroliferos Fiscales Bolivianos (YPFB) begin a strike in protest of the government's planned privatization of the country's "strategic industries." In anticipation of the strike, army troops and police are stationed at refineries, pipelines, and gas distribution facilities in order to ensure that oil and gas output remained unaffected. No initial disruptions in production are reported. Bolivian President Gonzalo Sanchez de Lozada currently is meeting with opposition leaders to reach an agreement on the country's new hydrocarbon law. YPFB workers are demanding a one year severance pay for every year of employment with the company.

March 26

Jordanian Energy and Mineral Resources Minister Hashem Dabbas states that Jordan currently is negotiating with three oil companies to build a $2-billion, 250,000-b/d refinery at Aqaba on the Red Sea. Under optimal conditions, the refinery could be operational by 2000. Jordan currently has a 100,000-b/d refinery at Zarqa, which runs at around 60 per cent of capacity. The country imports its 75,000 b/d crude oil requirements from Iraq under a U.N.-approved plan.

March 26

Arco and Lukoil, one of Russia's largest independent oil companies, form a joint venture to develop oil and gas reserves in Russia and other former Soviet republics. Total investment could reach $3 billion over a three year period. Arco will contribute almost all of the projected capital needs, even though it has a 54 per cent stake in the venture. In September 1995, Arco purchased $250 million worth of convertible bonds in Lukoil, giving it a 6 per cent stake in the Russian company.

March 29

The Pakistani government officially agrees to allow a proposed Iranian pipeline to run over Pakistani territory on its way to India. The $3-billion pipeline will carry an estimated 1.6 billion cubic feet (Bcf/d) from Iran's South Pars gas field to markets in Karachi. Negotiations concerning a possible extension to western India recently were stalled over whether the pipeline route would run through offshore or onshore Pakistani territory. Recent discussions between the Indian and Pakistani governments also focused on the location of compressor stations.

March 28

In the Philippines, President Ramos signs legislation ending 25 years of government control over petroleum product prices. The new deregulation law implements a 4 per cent tariff on refined product sales. This tariff subsequently will be eliminated by 2004. The law also encourages new investment in the Philippine refining sector, which is dominated by three large refining companies.

April 1996

April 1

In Australia, a federal court upholds a 1994 decision regarding oil exploration permits in the Timor Gap between Australia and Indonesia. Prior to the signing of the Timor Gap Zone of Cooperation Treaty in 1989, Australia-based WMC held a permit for exploration in what was then Australian territorial waters. The court's rejection of the recent government appeal now clears the way for three other permit holders to file monetary claims for compensation from the government. WMC's claim is estimated at around $120 million.

April 3

U.S.-based Pennzoil sells half of its roughly 10 per cent stake in a large offshore Caspian Sea oil field development project. The $8-billion development of the offshore Azeri, Chirag, and Guneshli fields is being undertaken by a consortium of Western companies led by Amoco and British Petroleum. Pennzoil's partial stake is bought by Japan-based Itochu Exploration, which will now be required to invest approximately $400 million in future field development costs. The three Caspian Sea fields contain estimated reserves of up to 5 billion barrels of oil and 2.5 trillion cubic feet (Tcf) of natural gas. Pennzoil also retains a 30 per cent interest in the nearby 1-billion barrel Karabakh oil field development.

April 4

Russia-based Lukoil purchases a 10 per cent stake in the Caspian Sea's Shakh Deniz oil field development. Lukoil's share leaves Azerbaijan's state-owned oil company, SOCAR, with a 20 per cent stake in the project. The field is estimated to contain up to 700 million barrels of oil and 14 Tcf of natural gas. The development is being undertaken by British Petroleum/Statoil (45 per cent), SOCAR (20 per cent), Turkey's TPAO (15 per cent), and Elf Aquitaine (10 per cent).

April 4

U.S. based Enron Corporation files a lawsuit in Texas against the owners of the U.K.'s Central Area Transmission System (CATS). Enron, which signed a "take-or-pay" contract for 240 million cubic feet per day (Mmcf/d) of natural gas from the offshore J-Block area fields, has refused to accept or pay for gas shipments since September 1995. Subsequently, Phillips Petroleum, an owner of CATS, has been reinjecting the gas into the J-Block fields in order to boost condensate production. Enron argues that the CATS pipeline does not meet the standards necessary to fulfill the contract and that reinjection could jeopardize reservoir integrity and future gas shipments. Due to recent gas market deregulation in the United Kingdom, gas prices have fallen below those to which Enron agreed to pay when it signed its take-or-pay contract with CATS.

April 4

The Ecuadorean government awards exploration rights to U.S.-based ARCO and two Argentinean companies. The award gives ARCO right to the block 24 concession in Ecuador's Amazon basin. Although Ecuador has raised its oil production above 390,000 barrels per day (b/d), output is capped because of limited pipeline access from the Amazonian Oriente region to the Esmeraldes export terminal on the Pacific Ocean. This month, ARCO agreed to construct an 80,000-b/d pipeline to connect Oriente fields to the 350,000-b/d Trans-Ecuadorean pipeline at Baeza.

April 6

According to press reports in Yemen, government sources state that the country's oil production will reach 400,000 b/d by 1997. The projected 40,000 b/d increase over current levels will come from a new discovery in Hadramawt in southern Yemen as well as from new production at the East Shabwah field near Sanaa. The Hadramawt area's Jannah block is expected to yield about 30,000 b/d from the four area fields.

April 9

In the South China Sea, Indonesia sends eight warships and a helicopter to Natuna Island after the captain of a private boat is shot and killed by unidentified gunmen. Indonesia's Pertamina and Exxon currently are undertaking a $35-billion gas field development in the area, which could include construction of a liquefied natural gas (LNG) plant on Natuna Island. The offshore Natuna gas field is estimated to have roughly 45 Tcf of recoverable reserves.

April 12

In Vietnam, U.S.-based Conoco is awarded exploration rights to offshore blocks 133 and 134, which are located approximately 200 miles off the country's southeastern coast. The two tracts overlap with Chinese territorial claims in the South China Sea and acreage which previously was awarded by China to U.S.-based Crestone Energy in May 1992. Crestone has conducted limited exploration in its Wan'an Bei 21 concession. After the Vietnamese award, a Chinese foreign ministry spokesman states that "China has indisputable sovereignty in this area of the South China Sea."

April 15

The Vietnamese government approves the creation of a 100,000-acre industrial park at Dung Quat on Vietnam's central coast. The facility will be the site of country's first refinery, even though Vietnam's crude oil reserves and primary markets for refined products are located in the south. After government insistence in late 1995 that the refinery be located at Dung Quat, Total pulled out of the $1-billion project and stated that a refinery at that location was not viable financially. Subsequently, Conoco, Malaysia's Petronas, and South Korea's Lucky Goldstar agreed to continue with the 125,000-b/d project.

April 16

The Colombian army is put on high alert after rebels from the leftist guerilla organization, Revolutionary Armed Forces of Colombia (FARC), attack a six-truck military convoy near the southern town of Puerres. The attack results in the deaths of 31 soldiers who were assigned to guard the Trans-Andean pipeline. The pipeline transports about 40,000-b/d of Ecuadorean crude oil to the Colombian terminal at Tumaco on the Pacific Ocean.

April 17

The Iranian government-run Islamic Republic News Agency (IRNA) states that Iran will soon receive its first shipment of five new 300,000-Dwt oil tankers from South Korea's Daewoo. The deliveries will boost state-owned National Iranian Tanker Company's (NITC) fleet capacity to 3.8 million tons. The tanker acquisitions have become especially important to Iran since the mid-1995 imposition of the unilateral U.S. embargo of Iranian oil exports. In the second half of 1995, FOB liftings of Iranian crude at domestic terminals fell dramatically, requiring NITC to increase its capabilities to transport Iranian crude oil to spot markets.

April 17

In Congo, Elf Aquitaine states that its offshore N'Kossa field development has fallen three months behind schedule. Production of 50,000 b/d from the 400 million barrel field will be pushed to July after a delay in the arrival of a floating production platform. Output is expected to reach 100,000 b/d within two years after field start-up. By 1998, Congo's oil production is expected to rise from 175,000 b/d at present to around 260,000 b/d. Additional production will come from Agip's offshore Kitina field. The N'Kossa field is being developed by Elf (51 per cent), Chevron (30 per cent), state-owned Hydro-Congo (15 per cent), and South Africa's Engen (4 per cent).

April 18

Mobil agrees to purchase a 25 per cent stake from the Kazak government in the giant Tengizchevroil consortium. The sale decreases the Kazak government's share in the venture to 25 per cent and leaves Chevron's 50 per cent stake unchanged. Tengizchevroil's development of the 6.6-billion barrel Tengiz oil field is expected to cost close to $20 billion. Chevron has invested about $700 million in the field so far, which now produces about 80,000 b/d. Output has been constrained up until now because of lack of space on Russia's Transneft pipeline network.

April 18

Press reports state that the Laotian government has signed a memorandum of understanding with an unnamed U.S. company regarding construction of an oil refinery in Borikhamxay province east of Vientiane. The proposed refinery would be supplied via a crude oil pipeline from Vietnam. As of early 1996, several Western oil companies, including Hunt Oil, were conducting exploration activities in Laos.

April 18

The United Arab Emirates (UAE) criticizes Iran for starting-up a power plant on Greater Tunb Island, which is located strategically in the Persian Gulf's Strait of Hormuz. Iran seized control of the Greater and Lesser Tunb islands in the early 1970s. Joint control over nearby Abu Musa was maintained until 1994, at which time Iran forcibly took the island. The UAE has recently called for International Court of Justice resolution of the dispute. However, Iran does not recognize the Court's authority.

April 19

Kenyan Minister of State Jackson Kalweo states that his country would like to have the International Court of Justice settle a territorial dispute with Sudan over a triangular region of desert and swamps around the town of Elimi. The 2,400-square mile Elimi region, which is recognized internationally as part of Sudan, is thought by Kenyan officials to hold an unspecified amount of oil.

April 20

At an oil refinery in the central Russian city of Ryazan, an underground oil tank explodes, killing two workers who were conducting maintenance work on the structure. The blast occurs after the workers penetrate the top of the 350,000-cubic foot tank and cause gas vapors to ignite.

April 24

In New York, the United Nations and Iraq end a third round of negotiations over Iraq's possible sale of $1 billion of oil for 90 days for a 180-day trial period. Under U.N. Resolution 986, proceeds from the sale would be used for humanitarian purposes. While both sides have reached agreement on most of the key issues, chief Iraqi negotiator Abdul Amir al-Anbari says that the United States and the United Kingdom have fundamentally altered the text of a proposed agreement which he had received from the United Nations early in the third round. Al-Anbari states that the changes have postponed any possible deal. The U.N.-Iraq talks are scheduled to restart on May 10.

April 25

Clashes between Nigeria and Cameroon over the disputed Bakassi Peninsula erupt for the second time since February 1996. Both countries claim the other side started the latest confrontation. No casualties are reported. The Bakassi Peninsula is a 600-square mile region bordering the two countries which contains oil and natural gas reserves. Nigeria claims that an 1858 treaty places the Peninsula in its territory. Cameroon contends that Nigeria offered it the region after Cameroon refused to help Biafran rebels during Nigeria's 1967 civil war. More recently, Nigeria has accused Cameroon of siphoning off 130,000 barrels per day (b/d) of oil from the producing fields in the region. The International Court of Justice in the Hague is currently reviewing the case.

April 27

The Caspian Pipeline Consortium (CPC) agrees to build a 900-mile pipeline which could carry oil production from the Tengiz oil field to export terminals on the Black Sea. In early 1996, CPC was reorganized, resulting in a decreased stake for Oman Oil Company and an option for Chevron to acquire a 15 per cent share in the pipeline. The consortium includes the governments of Kazakstan, Russia, and Oman as well as eight oil companies. The $1.2 billion pipeline deal will extend Russia's existing pipeline system from Tikhoretsk to the Novorossiisk terminal on the Black Sea. Second phase development will include a new line from Tengiz to Tikhoretsk.

April 29

In Congo, the government announces that bidding documents will be issued shortly for the privatization of Hydrocarbures de Congo (Hydro-Congo) and Congolaise de Rafinage (CORAF). Although Hydro-Congo was previously active in upstream exploration and production, its activities recently were limited by a lack of financial resources to the distribution of petroleum products. CORAF is the state-owned refining company. The privatizations are expected to spur foreign investment in the country's ailing downstream sector and allow for adequate processing of increased domestic oil output. In 1995, Congo produced about 175,000 b/d. Oil production is expected to rise to around 260,000 b/d by 1998 after Elf's N'Kossa and Agip's Kitina fields come online.

April 30

In the United States, President Clinton approves the sale of $227 million of crude oil from the Strategic Petroleum Reserve. At current oil prices, roughly 12 million barrels would be sold. The Clinton Administration hopes that the sale will lower gasoline prices in the United States, which are at their highest levels in five years.

May 1996

May 1

Bolivian President Gonzalo Sanchez de Lozada signs a new hydrocarbon law allowing for the privatization of state owned Yacimientos Petroliferos de Fomento Boliviano (YPFB). YPFB is one of six large state-owned companies which are undergoing privatization. Fifty-two foreign companies have pre-qualified to bid for YPFB's exploration, transportation, and product marketing operations. YPFB will retain control of three domestic oil refineries and all domestic pipelines. Bolivia currently is involved in a $1.5 billion export pipeline project to carry natural gas to Brazil. A number of foreign companies, including Enron and Tenneco, are building the privately-financed line.

May 3

Following French diplomatic initiatives, Yemen and Eritrea agree to form an arbitration panel to solve their territorial dispute over three islands located 100 miles north of the Red Sea's Bab al-Mandab strait. In December 1995, the two countries fought briefly over control of Greater Hanish Island, resulting in twelve deaths. Eritrea now controls Greater Hanish Island, while Yemeni forces hold Lesser Hanish and Zugar Islands. All of the islands lie astride major oil shipping routes as well as near offshore oil exploration acreage held by foreign companies such as Houston-based Anadarko Petroleum.

May 4

Eight hundred Norwegian oil workers stage a strike in sympathy of a similar month-long action by a construction workers' union demanding revised contract laws. The Federation of Trade Unions of Oil Workers' (OFS) strike cuts 1 million barrels per day (b/d) of oil production from the 190,000 b/d-Statfjord B, 320,000-b/d Statfjord C, 150,000-b/d Gullfaks A, 150,000-b/d Gullfaks C, Ula, Heimdal, and Oseberg C fields. The shut-in amounts to about one-third of Norway's oil output. Subsequently, Statoil, a major Norwegian oil company, declares force majeure on one cargo of crude oil deliveries. The strike ends after six-days due to lack of unanimity among OFS members.

May 10

In Vietnam, Royal Dutch/Shell relinquishes its Block 10 concession in the Nam Con Son basin off the country's southeastern coast. Shell's decision is made after drilling four unsuccessful exploration wells on the acreage, which was long-thought to hold promising potential. The move is made amidst the Vietnamese government's rejection of Australia-based BHP's efforts to renegotiate the terms for its Dai Hung oil field in Block 5-1. BHP sought a revised contract after appraisal drilling reduced Dai Hung's estimated reserves from 500 million to 200 million barrels. Due to unexpectedly complex geology, Dai Hung's production since its October 1994 start-up has fallen from 35,000 b/d to 12,000 b/d.

May 11

The Vietnamese government states its intention to build a second 120,000-b/d refinery in Thanh Hoa province, roughly 75 miles south of Hanoi. Vietnam is planning to build its first 120,000 b/d at

Dung Quat on the country's central coast. Foreign companies such as France-based Total have criticized the government's decision to build the two new refineries far away from the country's crude oil reserves and main product markets in the south. A consortium including Conoco is developing plans to build the first refinery, which could become operational by 2000. The government hopes to finish the second refinery by 2002.

May 11

Kazakstan and Iran sign an agreement in which Kazakstan will ship up to 80,000 b/d of crude oil from the Tengiz oil field in exchange for a comparable amount of Iranian crude oil supplied at the Kharg Island oil terminal in the Persian Gulf. It is unknown how U.S. sanctions against Iran will impact on the agreement, which involves U.S.-based Chevron. In March 1996, output at the Chevron-operated Tengiz field reached 100,000 b/d, up dramatically from the 50,000-60,000 b/d which the field was limited to prior to a recent export pipeline agreement with Russia. Production is slated to reach 130,000 b/d the end of 1996. Peak output from the 6-billion barrel field could reach more than 700,000 b/d by 2010.

May 14

The Peruvian government and a Shell/Mobil consortium sign a $2.8-billion deal to develop the Camisea wet natural gas field. The Camisea field contains estimated reserves of 11 trillion cubic feet (Tcf) of natural gas and 700 million barrels of condensate. Development will include construction of a 600-megawatt (MW) gas-fired power plant and 373 miles of oil and gas pipelines to link the Amazonian field to the coastal capital of Lima. Construction of other gas-fired power stations is possible, as is a link to the proposed gas export pipeline between Bolivia and Brazil. Concurrent with agreement, a car bomb explodes at a Shell office building in Lima. Maoist Sendero Luminoso rebels opposing foreign involvement in Peru's natural resource development are blamed for the attack.

May 14

Iran signs an agreement to acquire a 10 per cent stake in the Caspian Sea's Shakh Deniz oil field. However, the Iranian share will be taken by OEIC, a 60 per cent privately-owned engineering and construction company, rather than by state-owned National Iranian Oil Company. In April 1996, Russia-based Lukoil purchased a 10 per cent stake in Shakh Deniz, which contains up to 700 million barrels of oil and 14 Tcf of natural gas. The new development consortium now comprises British Petroleum/Statoil (45 per cent), SOCAR (10 per cent), Turkey's TPAO (15 per cent), and Elf Aquitaine (10 per cent).

May 15

In offshore Myanmar, U.S.-based ARCO is awarded the block M-7 concession, which lies adjacent to Total's and Unocal's $1.1-billion Yadana gas field development. The Yadana field contains estimated gas reserves of 5.7 Tcf. Total and Unocal currently are planning to construct a pipeline to supply 525 Mmcf/d of Yadana gas to Thailand, starting in 1998. In nearby acreage, a Texaco consortium is negotiating a similar gas sales contract with Thailand for the 1.5-Tcf Yetagun field.

May 16

Chevron and state-owned China National Offshore Oil Corporation sign a production sharing contract for promising exploration acreage in the South China Sea. Chevron's Contract Area 63/15 lies south of Hainan Island off the Vietnamese coast. The concession also is adjacent to Arco's large

Yacheng 13-1 gas field, which in January 1996, began supplying natural gas to Hongkong and mainland China via pipeline.

May 16

In Kazakstan, Mobil acquires a $1.1-billion, 25 per cent stake in the Tengiz field development from the Kazak government. Chevron (50 per cent), Mobil (25 per cent), and the Kazak government (25 per cent) are now the parties involved in the project. Chevron has spent almost $1 billion so far in field development work, which has suffered delays because of difficulties finding a solution to the problem of export routes. Chevron is expected to spend another $750 million for its share of a new export pipeline, which will run to Russia's Novorossiysk oil terminal on the Black Sea. Mobil's commitment will require an initial investment of $500 million, with the balance of $600 million due in installments by 2000.

May 20

In New York, the United Nations and Iraq agree to U.N. Resolution 986, which provides Iraq with the opportunity to sell $1 billion of oil for 90 days for a 180-day trial period. Under the resolution, proceeds from the sale would be used for humanitarian purposes. The agreement comes following months of heated negotiations. Iraqi oil exports are expected to begin by the Fall of 1996, after a pumping station on the Iraq-Turkey pipeline is repaired and U.N monitoring and aid distribution facilities are put in place. Shortly after the agreement, the White House announces its decision to allow U.S. oil companies to purchase Iraqi oil exports.

May 22

In Cabinda, Front for the Liberation of the Enclave of Cabinda–Armed Forces of Cabinda (FLEC-FAC) rebels fight with Angolan government troops only a week after FLEC-FAC signed a cease-fire agreement with the government. Since 1975, FLEC-FAC's 3,000-man army has fought the Angolan government for the 2,880-square mile Cabinda province, which is located on the coast between Zaire and Congo. Cabinda's offshore oil fields are run primarily by Chevron and account for over half of Angola's 700,000 b/d oil production. FLEC-FAC's struggle is separate from Jonas Savimbi's UNITA conflict. In 1995, UNITA and the government signed a peace accord and presently are demobilizing their armies.

May 22

In Equatorial Guinea, Planning Minister Antonio Fernando Nve Ngu states that oil production from Mobil's offshore Zafiro field will begin production by August 1996, only 18 months after the field's discovery. A floating production, storage, and offloading (FPSO) vessel will be used in a fast-track development program, which is slated to yield 40,000 b/d initially and 80,000 b/d 1998. The Zafiro field is a boon for the country's economy, which is one of the poorest in Africa.

May 23

State-owned Petroleum Authority of Thailand (PTT) signs agreements to increase its purchases of gas from foreign-operated Gulf of Thailand gas fields. In the two separate deals, PTT agrees to buy 330 Mmcf/d of gas from Unocal's 1-Tcf Pailin field by 2001 and 550 Mmcf/d of gas from Total's 3.3-Tcf Bongkot field by mid-1998. Thailand's domestic gas demand is estimated at 1 Bcf/d and is projected to double over the next five years. To meet its increased requirements, Thailand also is planning to import gas from Malaysia, Myanmar, and Indonesia.

May 28

Following a May 7 ambush of Nigerian troops by Cameroonian army forces, Nigeria and Cameroon agree to allow a U.N. mission to travel to the disputed Bakassi peninsula. Clashes between Nigeria and Cameroon have erupted several times on the peninsula since February 1996. The Bakassi Peninsula is a 600-square mile region bordering the two countries. It contains oil and natural gas reserves and is near Shell production facilities in southeastern Nigeria. Nigeria claims that an 1858 treaty places the Peninsula in its territory. Cameroon contends that Nigeria offered it the region after Cameroon refused to help Biafran rebels during Nigeria's 1967 civil war. More recently, Nigeria has accused Cameroon of siphoning off 130,000 barrels per day (b/d) of oil from the producing fields in the region. The International Court of Justice in the Hague also is reviewing the case.

May 29

In Vietnam, Unocal signs a contract for the Block B concession off the country's southwestern coast. The move follows Conoco's acquisition last month of offshore acreage claimed by both Vietnam and China. Unocal's acquisition is the first by a foreign company in Vietnam's territory in the Gulf of Thailand. The company believes that natural gas May be present in Block B. Any sizable gas discoveries by Unocal likely would be exported to Thailand via Unocal's other existing gas processing complexes in the Gulf of Thailand.

May 29

In Russia, Mobil and Texaco sign a production sharing agreement (PSA) for the Sakhalin III venture. Located offshore Sakhalin Island in Russia's Far East, Sakhalin III is one of three large oil and gas developments being planned by foreign oil companies. In the deal, both companies agree to spend $150 million in exploration activities over a six year period. Since last year, many foreign oil companies have been waiting for revisions to Russia's new PSA law, which is widely viewed as not offering adequate -legal protection for foreign investment in the country's oil and gas sectors.

June 1996

June 6

Representatives of the Organization of Petroleum Exporting Countries (OPEC) convene in Vienna. Overproduction by member countries, Gabon's departure from the Organization, and Iraq's anticipated oil exports under U.N. Resolution 986 are the major issues discussed at the session. Iraq's production quota is raised by 800,000 barrels per day (b/d) to 1.2 million b/d. Other member's production quotas remain unchanged, except for Gabon, which officially leaves OPEC.

June 7

In Vietnam, Australia-based Anzoil announces the discovery of a potential 900-billion cubic foot (Bcf) natural gas field located onshore about 60 miles southeast of the capital. The find is unique because very few oil and gas exploration efforts have surveyed Vietnam's onshore or northern areas. The only other known onshore oil or gas field in Vietnam is the small 50-Bcf Tien Hai gas field developed in the 1970s. It now supplies local industrial customers with about 10 Million cubic feet per day (Mmcf/d) of gas.

June 11

Exxon states that it will soon begin work on its $15-billion Sakhalin I oil and natural gas development in Russia's Far East. The Sakhalin I project will develop an estimated 5 billion barrels of oil and 15 trillion cubic feet (Tcf) of gas located in three offshore hydrocarbon fields. The $300 million appraisal program will include drilling one exploration well and conducting a 3-D seismic survey. The U.S. company says that it will start working despite ongoing differences with the Russian government over the country's new production sharing law, which is widely viewed as not offering adequate legal protection for foreign investment in the country's oil and gas sectors.

June 14

In Manila, the Philippine government announces that full deregulation of the downstream oil sector will occur by March 1997. This deregulation program call=s for a complete phase out of the Oil Price Buffer Fund, which has been used to protect consumers from world oil price fluctuations. Past increases in refined product prices, such as the 13 per cent hike in February 1996, have resulted in physical threats against government officials. Freeing of prices will likely spur investment in the country's downstream sector by foreign companies like British Petroleum and Mobil.

June 17

Petronas, Malaysia's state-owned oil and gas company, acquires a 30 per cent stake in Engen, South Africa's domestic refining and retailing company. The acquisition price is $436 million. The move will help to secure Petronas' downstream market access in southern Africa.

June 17

The Middle East Economic Digest reports that Iraq's State Oil Marketing Organization (SOMO) will soon sign a 3-month sales contracts with foreign oil companies. In another development, the United Nations selects France-based Banque Nationale de Paris (BNP) to hold the escrow account for Iraqi oil sales, the proceeds of which will be used by the U.N. for humanitarian purposes. Iraq could begin exporting oil under U.N. Resolution 986 as early as late September 1996.

June 17

In one the few oil field developments moving forward in Vietnam, Mitsubishi announces plans to bring its offshore Rang Dong field into production by May 1998. The 250-300 million barrel oil field is expected to come online with output of 45,000 b/d, with a later peak of 100,000 b/d. The field will use a floating, production, storage, and offloading (FPSO) vessel to decrease capital costs. The Rang Dong project comes amidst difficulties recently encountered by foreign companies with other oil field developments, such as Australia's BHP at Dai Hung, Mobil at Thanh Long, and Petronas at the Ruby field.

June 19

The U.S. House of Representatives passes a bill calling for the imposition of punitive sanctions on foreign companies which invest over $40 mil'ion in Iran's oil industry. The bill also penalizes companies and individuals who sell weapons or oil equipment to Libya. The U.S. Senate already has approved a similar measure. President Clinton is expected to support the bill after a committee works out differences between the House and Senate versions of the measure.

June 20

The Venezuelan Congress approves eight, multi-billion dollar, profit-sharing deals which allow foreign oil companies to explore and produce oil in Venezuela for the first time since the country's 1975 nationalization of the oil industry. The deals could boost Venezuela's current oil production by 500,000 b/d by 2005. Foreign oil companies such as Amoco and British Petroleum are expected to sign final deals with state-owned PdVSA within 10 days and May begin working on their new acreage by the third quarter of 1996. The eight blocks are estimated to hold between 7 to 11 billion barrels of light crude oil reserves.

June 20

In Ecuador, U.S.-based Arco announces that it will build a 105-mile secondary crude pipeline to connect its Villano field to the main 350,000-b/d Trans-Ecuadorean export pipeline. This move is significant because it follows the Ecuadorean government's recent abandonment of a 100,000-b/d expansion of the Trans-Ecuadorean pipeline. The proposed expansion would have boosted the country's ability to carry heavier crude oil from the Amazonian Oriente region, where the Villano field is situated. After coming online in 1999, Arco's new pipeline will transport up to 80,000 b/d of heavy 21° API crude oil.

June 21

The Brazilian government says it will create an independent organization to regulate foreign and domestic investment in the country's oil sector. In November 1995, a constitutional amendment was passed ending state-owned Petrobras' monopoly over the Brazil's oil industry. However, no legal framework has been developed to allow for foreign investment, especially in upstream oil exploration and production. Brazil currently produces 785,000 b/d of oil, enough to meet only about half of domestic demand. Foreign investment is viewed as a necessity to increase domestic production so as to meet as much of future consumption requirements, which are projected to rise to 2.3 million b/d by 2005.

June 21

In Lima, the Peruvian Defense Ministry states that between 8,000-10,000 Ecuadorean troops have moved into the northern Cordillera del Condor demilitarized zone. Border skirmishes have occurred in the region ever since a 1942 war between Peru and Ecuador, in which the latter country lost over half of its territory. The most recent conflict took place in early 1995, resulting the deaths and injury of over 1,000 soldiers from both sides. The Cordillera del Condor region lies near Peru's Block 1A-B, which accounts for about half of Peru's 125,000 b/d of oil output.

June 24

China National Petroleum Corporation (CNPC) states that oil production from the Tarim Basin, which is located in China's western Xinjiang Uygur Autonomous Region, will reach 64,000 b/d in 1996. This represents a 25 per cent rise in output as compared to 1995. By 2005, CNPC predicts that the Tarim Basin will become the country's second largest oil production region after Daqing.

June 24

Following a failure to obtain long-term gas purchase agreements with power producers, U.S.-based Tenneco suspends its plan to build a $689-million, 750-mile pipeline to carry Argentinean natural gas over the Andes to Chile. Canada's NovaGas currently is building a competing $350-million, 290-mile natural gas pipeline to link the two countries. However, NovaGas' pipeline has provoked criticism by environmental groups and local residents, who recently demonstrated along the main highway near Santiago that provides access to the pipeline route.

June 24

Qatar's government and Arco sign an agreement to bring the offshore al-Rayyan field online in September 1996. A test well drilled at al-Rayyan in September 1995 resulted in a flowrate of 10,000 b/d. Initial output of 35,000 b/d will come from temporary production facilities. Subsequently, a determination whether to install permanent platforms will made by mid-1997. Qatar currently produces about 500,000 b/d, more than 30 per cent above the country's OPEC quota of 378,000 b/d.

June 27

In Bolivia, Enron agrees to a revised contract for its role in the construction of a $1.5 billion pipeline to transport Bolivian natural gas to markets in southeastern Brazil. The revision substantially decreases Enron's profit margin and requires that the U.S. company have all financing completed and construction begun by June 15, 1997. The changes come after critics charged that the original 1994 contract was contrary to Bolivia's national interests.

June 27

Following arrests of pro-democracy activists by Burma's military government, the U.S. Senate Appropriations Committee approves mandatory unilateral sanctions against Burma. In September 1996, the Senate is expected to vote on the 1995 Free Burma Bill as an amendment to the U.S. State Department's foreign aid appropriations bill. If passed, proposed U.S. sanctions would bar any future U.S. investment in Burma until democracy is restored. This would likely affect U.S.-based Unocal, which is undertaking a $1.1-billion development of its 5.8-Tcf Yadana field located in Burma's Gulf of Martaban. The Yadana project also will entail constructing an export pipeline to ship up to 550-Mmcf/d of gas to Thailand, starting in 1998.

July 1996

July 1

The United States rejects an Iraqi plan for distributing food and medicine under United Nations Resolution 986, on the grounds that it would allow Saddam Hussein=s government to evade certain sanctions as well as to give it control over distribution of supplies to separatist Kurds in northern Iraq. Resolution 986 would allow Iraq to sell $1 billion worth of oil every 90 days for an initial period of 6 months.

July 1

Norway=s offshore oil unions reach a new wage agreement with their employers, averting a planned strike which could have resulted in a 330,000 barrels per day (b/d) cut in oil production. The strike would have initially affected operations at Statoil=s Statfjord A platform and then spread to the Gullfaks C platform later in the week.

July 1

Chevron formally takes control of Venezuela=s Boscan oilfield, which currently produces 80,000 b/d of heavy crude. Chevron plans to invest $400 million over the next 3 years in order to boost capacity by 50 per cent, to about 115,000 b/d.

July 2

Mobil announces that Norway=s government has approved a $7 billion project to develop 2.2 billion barrels of oil and natural gas in the Norwegian Sea off the country=s west coast. When complete, the AAasgard project will be one of the world=s most extensive subsea oil projects. Norway=s state-owned oil company, Statoil, holds a 60.5 per cent share and will be the operator of the project, which is expected to begin producing 200,000 b/d of oil by the end of 1998. Natural gas output from of 1 billion cubic feet per day is expected to begin in the year 2000

July 2

Shell Oil announces that it has finalized plans with 3 other oil companies on a $1.45 billion project to drill at record depths in the Gulf of Mexico. The project, known as AUrsa, is expected at its peak to produce 150,000 b/d from a depth of 4,000 feet.

July 5

The Financial Times reports that six international consortia have bid for oil exploration licenses in disputed waters surrounding the British-owned Falkland Islands. Britain and Argentina fought a short war over the islands in 1982, but in recent years have improved relations and attempted to cooperate in developing the potentially large oil reserves in the Falklands area.

July 7

OPEC issues a resolution announcing Gabon=s withdrawal from the organization, effective January 1, 1995. Gabon had an OPEC quota of 287,000 b/d.

July 11

China=s official Xinhua news agency announces that China is planning to open up additional oil and natural gas resources–in the western Tarim Basin and elsewhere–to foreign energy companies. Faced with rapidly rising energy demand, China is attempting to significantly expand its oil and gas exploration and production efforts. According to the report, China to date has signed 35 oil contracts worth $500 million with 32 foreign oil companies.

July 11

Shell Nigeria reopens four flow stations at the Nembe Creek crude oil pipeline in southeastern Nigeria after a shutdown of nearly a week. The pipeline and four pumping stations have a combined capacity of 125,000 b/d of Bonny Light crude oil.

July 13

Three explosions at a natural gas well in 65 miles southeast of Austin, Texas kill two workers and create a 40-foot fireball with 1,000 degree heat.

July 16

A $1 billion, class-action lawsuit is filed against Texaco on behalf of Ecuador=s Amazon Indians and settlers. The suit seeks to recover damages incurred from Texaco=s activities over 25 years in Ecuador=s rain forest, as well as compensation for the families living in the area, located in eastern Ecuador. Texaco pulled out of Ecuador in 1991.

July 17

The United States Senate passes legislation which would penalize foreign energy companies investing in or trading with Iran or Libya. The U.S. House of Representatives passed a similar bill in June, with differences in the two bills to be worked out in House-Senate conference.

July 17

Anadarko Petroleum announces that Algeria has issued a AProvisional Exploitation Authorization for initial crude oil production from the Hassi Berkine and Hassi Berkine South oilfields in the Sahara Desert in southeastern Algeria. Anadarko expects production from the fields to reach 200,000 b/d by 1999. The 42-degree API crude oil will be shipped by pipeline to Mediterranean ports.

July 18

The United Nations formally approves an Iraqi aid distribution plan, a major step forward in the direction of allowing Iraq to sell oil under Resolution 986.

July 19

Ecuador=s newly-appointed energy minister, Alfredo Adum, announces that Ecuador will need additional oil pipelines to increase the country=s crude oil delivery capacity. Adum will take over the energy ministry on August 10, when Ecuador=s president-elect, Abdala Bucaram, replaces Sixto Duran-Ballen.

July 26

Explosions fed by a propane leak at a major natural gas processing plant in the southern Mexican state of Chiapas kill six workers and injure 39 others. Estimates are that the accident could cost state oil company Petroleos Mexicanos (Pemex) more than $200 million. The Cactus facility processed about one-third of the propane and butane gas used for cooking and heating in Mexico.

July 29

In Colombia, around 5,000 rural residents block entrances to Ecopetrol facilities in the Orito region, where 112,000 b/d of oil are produced. The protestors are demanding government assistance in improving their living conditions.

July 30

Mobil announces that it has acquired a 26 per cent stake in the giant Natuna natural gas field, located 800 miles north of Jakarta, Indonesia, in the South China Sea. Mobil joins Exxon, which owns 50 per cent of the field, and Indonesia=s state oil company Pertamina, which has a 24 per cent stake. Natuna was discovered in 1973 and is estimated to contain at least 45 trillion cubic feet of recoverable natural gas.

July 30

Occidental Petroleum Corporation announces that it has sold all of its oil interests in Congo back to the Congolese government for $215 million. Occidental=s production from its holdings in Congo amounted to 9,000 b/d during 1995.

August 1996

August 1

Kuwait Oil Company (KOC) renews its technical service agreement with British Petroleum (BP). The new contract follows an original 1992 contract under which BP provided assessments of reservoir damage to Kuwait=s war-damaged oil fields. Under the new contract, BP will continue to provide technical support related to reservoir management and field development. Although Kuwait=s upstream is closed to private equity investment, two other major oil companies, Chevron and Shell, have similar technical service contracts.

August 1

State Oil Company of Azerbaijan (SOCAR) officials announce that a recently signed deal between Houston-based Brown & Root and Chevron=s Tengizchevroil venture will lead to exports of 20,000 barrels per day (b/d) of oil from the 5 billion barrel Tengiz oil field in Kazakstan to the Black Sea. Due to begin in October 1996, the exports will account for about 20 per cent of the field=s production. Under the plan, oil will be shipped by barge from Aktau in Kazakstan across the Caspian. It subsequently will be carried by rail through Azerbaijan to Georgia=s Black Sea export terminal at Poti. Tengiz crude currently is exported through Russian pipelines to the Black Sea terminal of Novorossiysk.

August 4

Near the southeastern town of Georgievka in Kazakstan, a section of gas pipeline explodes, killing two people and temporarily cutting gas supply to the nation=s capital of Almaty. An investigation has begun to determine the cause of the blast.

August 5

The Vietnamese government announces that it will raise import tariffs on a variety of refined petroleum products, beginning on August 15. The move is made after tax and tariff collection efforts fall below target levels. The price increase reverses recent attempts to lower imported product tariffs, some of which are as much as 60 per cent of landed costs.

August 6

President Clinton signs a new bill imposing sanctions on non-U.S. companies which invest over $40 million a year in the energy sectors of either Iran and Libya. Under the law, the President would be required to impose at least two of the following sanctions: import and export bans; lending embargoes from U.S. banks; a ban on U.S. procurement of goods and services from sanctioned companies,; and a denial of U.S export financing. The European Union has stated its opposition to the U.S. law and threatened retaliation.

August 8

In Norway, 280 members of the YS skilled workers= union threaten to strike unless an agreement on general working terms is reached with the Norwegian Oil Industry Association. The proposed strike would shut the Mongstad terminal, which handles crude oil from Norsk Hydro=s Troll and Statoil=s Heidrun fields as well as smaller deliveries from the Gullfaks and Statfjord production systems. The Mongstad terminal has a storage capacity of 8 million barrels. The strike also would close the 150,000 b/d Mongstad refinery, the country=s largest.

August 8

The United States withdraws its opposition to United Nations Resolution 986, which would allow Iraq to sell $1 billion worth of oil every 90 days for an initial period of 6 months. Under the plan, proceeds from the oil sales would be used for humanitarian purposes. In recent days, the United States had been the only nation on the 15-member U.N. Security Council to oppose the plan. It had done so ostensibly to ensure proper monitoring procedures were in place to allow for the equitable distribution of humanitarian supplies to civilians.

August 8

U.S.-based Unocal and the government of Turkmenistan sign a deal to build a $2-billion, 900-mile long pipeline to carry gas from Turkmenistan=s 45-trillion cubic foot Dauletabad gas field through Afghanistan to markets in Pakistan. The proposed pipeline will carry 1.9 billion cubic feet (Bcf/d) initially and 3.8 Bcf/d after final completion in 2002. Unocal and Saudi Arabia-based Delta Oil will hold an 85 per cent stake in the project, with the Gazprom (10 per cent) and Turkmenrusgaz (5 per cent) holding the remainder.

August 11

Australia-based BHP reportedly pulls out of negotiations with Iran, regarding construction of a $3-billion, 1.6-Bcf/d, 1,000-mile overland pipeline to carry gas from Iran=s South Pars gas field to Pakistan and possibly India. BHP=s withdrawal follows President Clinton=s passage of a new sanctions against Iran and Libya.

August 12

India and Oman agree to set up a joint working group to evaluate the technical aspects of a proposed $5-billion, 810-mile, 2-Bcf/d natural gas pipeline between the two countries. In 1995, Oman Oil Company (OOC) officials stated that such a line, which would run offshore at depths of up to 13,000 feet, was not technically feasible. OOC also expressed doubts that Oman had enough gas reserves to support both the pipeline and Oman=s proposed liquefied natural gas (LNG) project.

August 16

Azerbaijan Foreign Ministry officials reject a recent Russian proposal to resolve Caspian Sea territorial issues by using a 1941 treaty between the Soviet Union and Iran as a starting point. Both Azerbaijan and Kazakstan oppose Russia=s plan to use the 12-mile territorial limit normally applied to international waters. The five Caspian states plan to meet in Askhabad in November 1996 to discuss their respective differences.

August 16

U.S-based Mobil agrees to pay $18.8 million for Petrolube, a state-owned Peruvian lubricants company. The sale represents one of the first in the Peruvian government=s efforts to privatize its downstream oil sector.

August 19

South Africa=s Minister of Mineral and Energy Affairs Penuell Maduna states that his government is discussing new oil supply arrangements with Saudi Arabia and Iraq in an effort to decrease South Africa=s dependence on Iranian oil imports. The majority of South Africa=s 230,000 b/d of oil imports come from Iran. Minister Maduna also added that another deal in which South Africa would allow Iran to store 15 million barrels of oil at a 45-million barrel facility near Cape Town was still under consideration, but that a number of environmental issues were still outstanding.

August 19

Malaysia=s state-owned oil company, Petronas, agrees to take a 30 per cent stake in France-based Total=s $600-million project to develop Iran=s offshore Sirri A and E oil and gas fields. Total signed the development deal in July 1995, after the U.S. government barred U.S.-based Conoco from continuing advanced negotiations with Iran concerning the project. The recent farm-in by Petronas will require the Malaysian company to spend $180-million over a four-year period. It is unclear how the recently-passed U.S. sanctions on Iran will affect Petronas or Total.

August 20

A Dubai-based shipping company states that Iraq recently loaded a trial cargo of crude oil from the Mina al-Bakr terminal onto a 36,000-dwt tanker. Mina al-Bakr has four 400,000 b/d berths capable of handling very large crude carriers (VLCCs). In mid-1996, Iraqi officials stated that 1.2-million b/d capacity at the terminal had been restored. However, it is reported that mines and sunken ships in the terminal=s main shipping channel still present a danger to large tankers. Under U.N. Resolution 986, about half of Iraq=s limited oil exports would be shipped through Mina al-Bakr.

August 20

Iran and Pakistan agree to build a $1-billion, 120,000-b/d refinery in western Pakistan. When completed, the proposed refinery will be supplied with Iranian crude oil. Both countries will have a 50 per cent stake in the project, which still requires approval by Pakistan=s cabinet.

August 21

In Venezuela, a subsidiary of state-owned Petroleos de Venezuela (PdVSA), Corpoven, signs a memorandum of understanding (MOU) with U.S.-based ARCO. The MOU provides for a $3.5-billion joint venture to develop and upgrade roughly 200,000 b/d of crude oil from the country=s 270-billion Orinoco Heavy Oil Belt. The project will produce 9°API gravity crude oil in the Hamaca region and upgrade it to 25° API for export to U.S. refineries. The project will be implemented in three phases, the last of which will be completed in 2006. Another PdVSA subsidiary, Maraven, recently signed another, similar deal with Conoco.

August 22

Nigerian Oil Minister Dan Etete states that National Nigerian Petroleum Corporation=s (NNPC) joint venture partners, which include Shell, Mobil, and Chevron, do not show accountability and transparency in their operations. He states that the Nigerian government shortly will introduce a wide array of changes to day-to-day operations and accounting procedures which will provide Nigeria with a more equitable return from the country=s oil production. The move comes amidst delinquent payments by NNPC to its joint venture partners of almost $1 billion during the first half of 1996. Oil Minister Etete statements also occur after a World Bank review criticizes the Nigerian government for failing to account recently for $700 million in oil revenues.

August 23

Argentina=s YPF reaches a preliminary agreement with the Ecuadorean government that would convert a services contract signed in the 1980s to a participation contract which provides the government with a greater return on oil production from the company=s field in Block 16 of the Oriente region. Several days earlier, the Ecuadorean government sent troops in to take control of the fields, which are operated by U.S-based Maxus, a YPF subsidiary. Output was cut from 34,000 b/d to 18,000 b/d. Most recent oil deals in Ecuador are participation contracts, which were introduced in the early 1990s.

August 24

Oman begins mediation of a territorial dispute between Bahrain and Qatar involving the potentially oil-rich Hawar islands and reefs. Qatar originally filed the case with the International Court of Justice (ICJ) in the Hague in 1991. Bahrain then rejected the ICJ=s jurisdiction. Subsequent mediation attempts by Saudi Arabia also failed.

August 25

Equatorial Guinea begins commercial oil production with the start-up of its Zafiro field. With initial output of 7,000 b/d, the Mobil-operated Zafiro field is expected to reached a peak production of 40,000 b/d by the end of 1996. Mobil discovered the field in mid-1995 and brought the field on-stream through a fast-track plan which utilizes a floating production vessel.

August 27

In Congo, Exxon signs a production sharing contract (PSC) for the Mer Profund Nord concession, located 50-75 miles offshore in an area where water depths reach up to 6,000 feet. Exxon will begin seismic exploration in the 1-million acre block in early 1997. Exxon and Elf currently are involved in the development of the nearby Mer Profund Sud concession. Work on these concessions represents an important trend in deepwater exploration activity offshore West Africa as well as a significant part of Congo=s ongoing effort to boost its oil output above current levels with the addition of offshore fields.

August 28

In Colombia, the National Liberation Army (ELN), a rebel group which has fought the government for over 30 years, offers to suspend its attacks on the country=s oil infrastructure in exchange for a dialogue. The ELN historically has opposed foreign involvement in Colombia=s oil sector. In July 1996, it carried out a 23 bombings on oil pipelines in Colombia as a precursor to its peace initiative. State-owned Ecopetrol recently estimated that rebel bombings in 1995 caused $13 million worth of damage to Colombia=s oil pipelines.

August 30

Iran and Turkey sign deal to export $23 billion worth of natural gas through a proposed 800-mile long a friendship pipeline between the two countries. The deal comes amidst new U.S. sanctions on Iran.

August 30

The United States presents a conclusive evidence to the U.N. Sanctions Committee that Iran is a complicitous in the smuggling of petroleum products from Iraq through the Persian Gulf. According to U.S. allegations, Iran uses barges and small ships to carry oil products from southern Iraq into Iranian territorial waters. Shipping documents then are forged to show that the cargo is of Iranian origin.

September 1996

September 3

In Yemen, Oil Minister Mohammed Attar states that the country will raise its oil production from 350,000 b/d at present to 400,000 b/d by 2000. Initial boosts in output will come from the 27,000 b/d Jannah and 20,000 b/d East Shabwa fields, both of which are due to come online in 1997.

September 3

Officials from Nigeria and Equatorial Guinea meet to discuss territorial rights concerning Equatorial Guinea=s Zafiro field, which Mobil recently brought online in a fast-track development program. The field currently produces about 10,000 b/d and is expected to reach an output level of 40,000 b/d in 1997. Nigeria previously had awarded Zafiro development rights to Elf. Nigeria also is in dispute with Cameroon over the oil-rich Bakassi Peninsula. A United Nation=s team is expected to depart soon for the region in the hopes of mediating a settlement.

September 4

India=s Foreign Investment Promotion Board rejects four proposed oil and petrochemical projects, including Occidental=s plan to invest up to $2-billion in enhanced oil recovery (EOR) technology for the mature Bombay High field. About half of India=s oil production comes from the Bombay High region. Perceived endangerment to national security was one of the Board=s reasons for rejecting Occidental=s EOR proposal. September 4 Trade press reports that Azerbaijani and Iranian officials still disagree over a construction timetable of a $70-million pipeline to supply Iranian natural gas to Azerbaijan=s enclave of Nakhichevan, which is separated from the majority of the country by Armenia. The two countries signed a pipeline construction agreement in 1992. However, further progress has been stalled by Iran=s refusal to accept oil products, rather than cash, from Azerbaijan for 8.8 billion cubic feet of Iranian gas already delivered to Nakhichevan.

September 5

Bolivia and Brazil sign a $1.8-billion natural gas export pipeline construction contract. Brazil=s state-owned Petrobras will oversee construction of the proposed 1,900-mile pipeline. Under the gas supply contract, Petrobras will purchase 280 million cubic feet per day (Mmcf/d) for the first seven years of the 20 year sales period. Gas purchase volumes will double after the first seven years. After coming online in late 1998, the pipeline=s first phase will link Santa Cruz, Bolivia with Campinas, Brazil. A second phase will extend the Brazilian line from Campinas to Porto Alegre by 1999.

September 5

Following U.S. cruise missile strikes on military facilities in southern Iraq, crude oil prices rise as the market speculates when Iraq will begin exporting oil under U.N. Resolution 986. Benchmark Brent Blend for October rises above $22/barrel amidst the uncertainty. The U.S. attack follows an Iraqi-supported invasion of Kurdish safe haven areas in the country=s northern area. Subsequently, President Bill Clinton states that the U.N. oil-for-food sale should be postponed indefinitely.

September 6

In Kazakstan, British Gas confirms that it will reduce its 42.5 per cent stake in the giant Karachaganak gas field development project. Karachaganak, which is an extension of Russia's Orenburg field and located in the northwestern part of the country, contains about 46 trillion cubic feet (Tcf) of natural gas. In 1992, British Gas and a gip of Italy won a tender allowing them exclusive development rights to the field. Russia=s Gazprom later joined the project with a 15 per cent stake. It is rumored that Lukoil will acquire any interest relinquished by British Gas. The Karachaganak project will require a total investment of $10 billion over 40 years.

September 8

The Revolutionary Armed Forces of Colombia (FARC) launch a new series of attacks in the country, killing at least 23 people and blowing up a section of the 200,000-b/d Cano Limon-Covenas oil pipeline. The resulting pipeline rupture spills 8,000 barrels of oil. Since it came into operation in 1986, this strategic pipeline has been attacked over 400 times.

September 9

Malaysia=s state-owned Petronas says that it will proceed with its contract to develop Iran=s offshore Sirri A and E oil and gas fields. After pressure from the U.S. government, Conoco withdrew from the project in March 1995. Subsequently, Total was awarded the development. Recently, Petronas acquired a 30 per cent stake in the $600-million venture, which will develop the fields= estimated 600 million barrels of oil reserves. Peak production is expected to reach 120,000 b/d. In early 1996, Total was involved in talks with Dubai to ship 100 Mmcf/d of associated gas through a proposed 15-mile pipeline to link with the emirate's gas grid. Sirri gas then would be used for injection at Dubai's Fateh field.

September 12

In Burma, Unocal and Total, the partners undertaking the 5.7-Tcf Yadana gas field development in the Gulf of Martaban, release the proposal=s feasibility study. A $700-million domestic project will complement a controversial 525-Mmcf/d gas export pipeline which will link the offshore field to Thailand. The domestic project, called the AThree-in-One, comprises a 90-mile gas pipeline link between the field and Rangoon, a 200-megawatt gas-fired power station, and a 1,750-ton per day urea fertilizer plant. Controversy over the gas export pipeline has been fueled by alleged human rights violations by Burma=s military government. Currently, a unilateral sanctions bill against Burma is pending in the U.S. Congress.

September 16

The Norwegian Petroleum Directorate (NPD) states that it will aid Angola=s Ministry of Petroleum in formulating the country=s future oil development strategy. The NPD specifically will project future production levels and export revenues, advise on appropriate safety and environmental programs, and provide relevant legal assistance for Angola=s petroleum sector. The new agreement will last three years and follows previous cooperation activities under which Norway has advised Angola on topics such as deepwater oil production.

September 16

Representatives from Arco and Indonesia=s state-owned Pertamina travel to South Korea in the hopes of obtaining a long-term sales contract for a proposed two-train, 6-million ton per year liquefied natural gas (LNG) plant that would be located on Irian Jaya, in eastern Indonesia. The LNG plant would be fueled by Arco=s recently appraised Wiriagar discovery, which holds up to 12 Tcf of natural gas reserves. Although it contains a smaller reserve base, a Wiriagar LNG project potentially could be more economical than Indonesia=s current for the 45-Tcf (recoverable) Natuna LNG proposal, since the field holds high levels of high carbon dioxide.

September 19

U.S. officials tell the International Court of Justice in the Hague that the United States should not be held liable for damages resulting from U.S. naval attacks on three offshore Iranian oil platforms during the Iran-Iraq war in 1988. The United States launched the attacks after Iran fired a Silkworm missile at a U.S.-flagged Kuwaiti tanker in 1987. That incident, along with another in which a U.S. frigate struck a mine in the southern Persian Gulf, resulted in injuries to U.S. naval personnel. The U.S. contends that the three oil platforms were armed military bases and used to facilitate Iranian attacks on neutral shipping during the war. Iran counters that the oil platforms were of Avital commercial and economic importance and targeted to inflict Amaximum financial, commercial, and economic damage.

September 19

After extensive negotiating efforts by Unocal, Afghanistan=s main rebel factions sign an agreement to ensure the safety of a proposed 890-mile, 2-Bcf/d capacity pipeline which would carry natural gas from Turkmenistan=s 45-Tcf Dauletabad gas field to Pakistan. The proposed $2-billion pipeline tentatively would run from the Dauletabad field south to the Afghan border and through Herat, Qandahar, and Quetta, Pakistan. The line would then link with Pakistan=s gas grid at Sui. Gas shipments are projected to start at 700 Mmcf/d in 1999 and rise to 1.4 Bcf/d or higher by 2002. Following a memorandum of understanding signed in August 1996, a consortium of companies comprising Unocal and Saudi Arabia=s Delta Oil (85 per cent combined), Russia=s Gazprom (10 per cent), and state-owned Turkmenrusgaz (5 per cent) is planning to build the line, but political risk, security, and financing concerns May stall the project.

September 20

In Russia, Arco signs several agreements with Lukoil to jointly invest $5-billion in energy projects over the next 18 years. The deals also create a new company, LukArco, which will be held jointly by Lukoil (54 per cent) and Arco (46 per cent). In late 1995, Arco bought an 8 per cent stake in Lukoil. The recent moves will provide Arco with a prominent position in Russian upstream activities.

September 20

After pressure from 20 multinational oil companies belonging to the Colombian Petroleum Association (CPA), the Colombian government agrees to review the terms of the sliding scale production sharing contracts which most foreign oil companies signed between 1989 and 1994. The contracts provide state-owned Ecopetrol with as much as 70 per cent of the oil produced from large fields. The CPA states that the unfavorable terms have led to a decrease in exploration activated in Colombia. The government=s decision is made despite threats from the National Liberation Army (ELN), one of Colombia=s two insurgent factions, to assassinate Minister of Mines and Energy Rodrigo Villamizar if British Petroleum=s contract to develop the Volcanera, Pauto, and Florena fields was revised.

September 25

In Nigeria, Finance Minister Anthony Ani states that the government should divest its majority equity interest in the country=s six main joint ventures with foreign oil companies. Minister Ani also says that the government should PSCs, under which companies would pay taxes and royalties directly to the government. Since 1993, Nigeria=s upstream oil sector has faced a continuing budget crisis, and financial resources allocated to Nigerian National Petroleum Corporation (NNPC) have been below those necessary to bring about an expansion of oil production capacity.

September 25

In New York, visiting Colombian President Ernesto Samper states that a new 700 million barrel oil field has been discovered by state-owned Ecopetrol in central Colombia. Later, Ecopetrol President Luis Bernardo Florez says that while the field May contain as much as 600-700 million barrels of high-quality crude oil, further appraisal work is needed. If accurate, the find would raise Colombia=s proven oil reserves by 20 per cent.

October 1996

October 2

Yemen=s Oil Ministry states that it is examining plans to build a 60,000 barrel per day (b/d) refinery at Ras Katnib, northwest of Aden. The $250-million refinery project would be financed by private investors from Yemen and the Persian Gulf. It would be the first private investment in Yemen=s oil industry. A second phase would raise the refinery=s capacity to 100,000 b/d. Yemen now has two refineries with a combined capacity of 110,000 b/d. The country=s current crude oil production is around 350,000 b/d, of which 80,000 b/d is consumed domestically.

October 3

In Paris, Yemen and Eritrea sign an agreement to allow international arbitration of their territorial dispute over the Hunaish and Zuqar Islands. These islands are located in the Bab al-Mandab strait at the mouth of the Red Sea. The recent dispute flared in December 1995 after Eritrea forcibly took control of Greater Hunaish Island. In August 1996, Eritrea seized Lesser Hunaish Island, but withdrew after French mediation efforts. The Bab al-Mandab is a strategic shipping lane and also an area targeted for oil exploration by international companies.

October 4

U.S.-based Unocal and France=s Total are sued in a federal court in Los Angeles for their alleged complicity in human rights abuses by Burma=s military government. The plaintiffs state that the two companies had knowledge that the military was relocating villages, using forced labor, and committing a variety of abuses during efforts to clear a path for an onshore natural gas export pipeline. The companies maintain that no such actions have occurred along the pipeline route and that local workers are paid above average wages. The pipeline is part of a $1.2-billion project to carry 525 million cubic feet per day (Mmcf/d) of gas from the 5.7-trillion cubic foot (Tcf) Yadana field to power companies in Thailand. Shipments will begin in 1998, after the anticipated completion of the onshore pipeline.

October 4

Azerbaijan=s parliament ratifies the $4-billion Shakh Deniz field development. The project is being undertaken by a consortium comprising British Petroleum (25.5 per cent), Statoil (25.5 per cent), Elf (10 per cent), Lukoil (10 per cent), Iran=s OEIC (10 per cent), and TPAO (9 per cent). The offshore Shakh Deniz field contains an estimated 1 billion barrels of oil and up to 17.6 Tcf of natural gas.

October 4

In Irian Jaya, Indonesia, British Gas says that its Mogoi Deep-1 well on its Muturi concession resulted in flow rates of 40 Mmcf/d of natural gas. This find is the latest in a number of gas discoveries on the island in eastern Indonesia. U.S.-based Arco is planning to build a liquefied natural gas (LNG) plant on Irian Jaya. It would be supplied with reserves from Arco=s Wiriagar field, which May hold up to 12 Tcf of reserves.

October 8

Kuwait Oil Company (KOC) and Exxon sign a joint technical study agreement concerning possible development of the Karaa al-Maru oil field in northwestern Kuwait. The field contains estimated reserves of between 300-350 million barrels of light 49° API oil. This would be the first development of oil fields in this region of Kuwait. While Kuwait=s upstream sector remains closed to foreign investment, several other foreign oil companies have technical service agreements with the government. These companies include British Petroleum, Chevron, and Royal Dutch/Shell.

October 8

Qatari Foreign Minister Harned Ben Jasem al-Thani states that his country has frozen all business dealings with Israel because of lack of progress in the Middle East peace process. Earlier in the year, Enron had proposed constructing a terminal at Eilat on the Gulf of Aqaba to receive LNG from the company=s planned project in Qatar.

October 8

After more than a year of review, Oman cancels its plans to build a $10-billion, 1-billion cubic foot per day (Bcf/d) pipeline to supply India with natural gas. The pipeline faced technical and cost problems, since it would have been laid at underwater depths of up to 13,500 feet. Also, Oman is planning to build an LNG export facility, and questions were raised as to whether the country=s gas reserves could support both projects. Oman=s proven gas reserves are around 25 Tcf.

October 9

In Congo, France=s Elf Aquitaine states that its Moho Marine structure May contain larger reserves than the 440-million barrel N=Kossa field, the country=s largest. The Moho Marine is located offshore and is under development by a consortium comprising Elf (51 per cent), Chevron (30 per cent), Hydro-Congo (15 per cent), and Energy Africa (4 per cent). In 1996, the Elf-led consortium brought N=Kossa online with 50,000 b/d of output. This has boosted Congo=s total oil production to around 225,000 b/d.

October 9

In southern Algeria, the Armed Islamic Group (GIA) blocks a road near Laghouat inside the Hassi R=Mel security zone and kills 34 people after their vehicles are stopped at a false police checkpoint. In April 1995, the government created four Aexclusion zones to protect oil and gas facilities and personnel in the producing centers of El Oued, Laghouat, Illizi, and Ouargla. Within these zones, traffic and shipments are regulated by army and police units. The conflict in Algeria, which began in 1992, has claimed an estimated 60,000 lives.

October 9

In Venezuela, PdVSA President Luis Giusti announces that the state-owned oil company=s third round of marginal field bidding will involve 20 fields in Monagas, Anzoategui, and Western Venezuela. Giusti also states that this round will differ from the previous two in that it will involve already producing oil fields as well as those in need of secondary recovery technology. This third round marks continuing efforts by PdVSA to boost oil production above the current 3 million b/d.

October 10

Workers employed by Bolivia=s state-owned oil company, YPFB, stage a 24-hour strike to demand passage of a new land reform bill. The strike shuts down the country=s three refineries, which have a combined capacity of 45,000 b/d.

October 11

In Georgia, the 11-member Azerbaijan International Operating Company (AIOC), which includes British Petroleum, Amoco, and Exxon, sign a pipeline engineering contract with John Brown Engineers and Constructors, a British subsidiary of Norway=s Kvaerner. The $300-million pipeline project will involve renovating an existing line that links Baku to Supsa, on the Black Sea. The pipeline will have a new capacity of 105,000 b/d and will be completed in 1998. It will be used to carry a fraction of the anticipated oil exports from AIOC=s $8-billion, 4-billion barrel Azeri, Chirag, and Guneshli field development. AIOC still has not reached a decision on the route for the main 700,000 b/d export pipeline.

October 14

After months of protests by workers, Mexico=s state-owned oil company, Pemex, cancels further plans to privatize its 61 state-owned petrochemical processing plants. Alternatively, Pemex plans to raise capital by creating subsidiaries to manage the 61 plants. Subsequently, it will sell off 49 per cent of its ownership in the subsidiaries. A previous effort to sell the plants was canceled in 1992.

October 16

In Geneva, the U.N. Compensation Committee rejects an Iraqi request to use revenues from the proposed oil sales under U.N. Resolution 986 to fund a legal defense against claims resulting from the 1991 Gulf War. Resolution 986 would allow Iraq to sell up to $1 billion worth of oil every six months. The deal originally was delayed by Iraq=s incursion into the Kurdish safe haven zone in September 1996. More recently, the United States has requested time to review the pricing structure for the Iraqi oil sales, money from which must be used for humanitarian aid.

October 18

After four years of development, U.S.-based Occidental writesoff its $105-million investment in an exploration and production joint venture in the Komi Republic in Russia=s Arctic region. The move was made due to continued high taxes, tariffs, and transportation fees imposed by both Russian and local governments. In addition, the recent increase in world oil prices has resulted in decreased access to Russia=s export pipeline network by foreign ventures.

October 24

The Latvian parliament ratifies the government=s oil exploration deal with U.S.-based Amoco and Sweden=s OPAB. The agreement covers exploration of the offshore Dalders prospect, which is located in territory claimed by neighboring Lithuania. Lithuania=s government previously has stated that it will not respect the Amoco/OPAB agreement. If the dispute between the two countries is resolved, Amoco/OPAB plan to drill an exploratory well in early 1997. Exploration and appraisal costs could reach between $7-$30 million.

October 27

The London-press reports that the favored solution for disposal of the Brent Spar oil storage platform is to turn it into a tourist attraction within the Morecambe Bay wildlife resort in Cumbria, England. The 30 proposals now under review by the U.K. government mostly involve onshore disposal and recycling. However, a British company, Amec, proposes cutting the Spar in two and creating a wildlife viewing platform that would include glass elevators. The 1996, Shell=s planned offshore disposal sparked controversy after Greenpeace claims, which were later retracted, that the buoy contained toxic waste.

October 28

In the Falkland Islands, exploration licenses are awarded from the country=s first bidding round that was launched in October 1995. While 19 offshore tracts located north and south of the islands were on offer, companies only bid on seven northern concessions. Acreage was awarded to U.S.-based Murphy, Fina, International Petroleum Corporation, Clyde, Lasmo, and Sands Petroleum, among others. Neither British Gas nor Argentina=s YPF received acreage. The concessions offered in this bidding round are separate from the Special Area of Co-operation that recently was established between Britain and Argentina. This area lies to the west of the Falklands.

October 30

In Burma, a consortium comprising Texaco (50 per cent), Premier (30 per cent), and Nippon (20 per cent) signs a memorandum of understanding (MOU) to supply Thailand with gas from Burma=s offshore Yetagun field. Under the terms of the MOU, the Texaco-led consortium will supply Thailand with 200 Mmcf/d of gas starting in 1999. Although the 1.5-Tcf Yetagun field is located to the southeast of Unocal/Total=s Yadana development, the field=s onshore pipeline segment will run in the vicinity of the Yadana pipeline.

October 30

Ukraine and Russia agree to negotiate a long-term gas deal which would fix supplies and prices over a 20-year period. At present, Ukraine owes Russia $1.4 billion for past gas deliveries. A supply deal for 1997 also has not been reached. In 1997, Ukraine hopes to increase its purchases of Russian gas to 2.06, up from 1.85 Tcf in 1996.

October 30

Exxon confirms that it is in talks with state-owned Qatar General Petroleum Corporation concerning the application of new technology to convert natural gas to petroleum products. Exxon believes that technology developed in a successful 200-b/d a natural gas refinery project in Texas would work in Qatar, where a proposed $1-billion plant would be able produce between 50-100,000 b/d of middle distillate products. Under the proposal, Qatar=s 270-Tcf North field would supply between 0.5-1 Bcf/d of gas for use as feedstock. In the past, technological barriers and high costs have precluded the development of natural gas refineries.

November 1996

November 1

The $2.3-billion Gazoduc Maghreb-Europe (GME) pipeline begins its first gas deliveries between Algeria=s Hassi R=Mel gas field and Cordoba, Spain. By the end of 1998, Algeria is planning to boost the GME=s capacity to 935 million cubic feet per day (Mmcf/d), up from 695 Mmcf/d currently, in order to supply markets in Spain, Portugal, Morocco, France, and Germany.

November 4

Italy=s state-owned ENI discloses that it signed a $3-billion natural gas production and supply deal with Libya in June 1996. The deal covers development of the offshore NC-41 block and the onshore Wafra field. ENI plans to bring the areas online by 2000, with gas production flowing to Italy via a proposed underwater pipeline. Reserves in the two areas are estimated at 600 million barrels of oil equivalent. Production potential at the NC-41 block and Wafra field is estimated at 575 Mmcf/d and 195 Mmcf/d, respectively. The United States government recently implemented additional sanctions against Libya which bar new investments of more than $40 million by non-U.S. companies. However, ENI representatives state in a MEES interview that U.S. sanctions will not have an adverse material effect on the project.

November 4

Turkish Prime Minister Necmetten Erbakan states that Iranian natural gas shipments to Turkey will begin Awithin a year at the latest under the recently signed 22-year, $20-billion gas supply deal between the two countries. The agreement, which calls for delivery of up to 350 Mmcf/d of Iranian gas to Turkey, has met with U.S. criticism for providing financial support to the Iranian economy. Turkey=s demand for natural gas is expected to quintuple by 2010 to 2.9 billion cubic feet per year (Bcf/d).

November 5

British Petroleum Exploration (BPX) requests Colombia=s Prosecutor General to investigate the source of recent media reports that accuse the company of violating human rights and supporting paramilitary groups in eastern Colombia. In October 1995, several London newspapers reported the accusations from an unofficial study purportedly conducted by the Colombian presidency=s human rights office. BPX strongly denies the allegations.

November 6

In Greece, government anti-terrorism units arrest 16 Greenpeace activists who block entry to Motor Oil=s 100,000 b/d coastal refinery at Ayioi Theodori. The Greenpeace protest is in response to a large spill that occurred during loading operations in August and added to already high pollution levels in the Saronic Gulf. Motor Oil is owned jointly by Saudi Aramco and Greece=s Vardinoyannis.

November 7

An illegal oil depot explodes in eastern Thailand, killing 11 and injuring 17 people. The depot, located in an unpopulated area near Rayong, served as a storage area for smuggled shipments of refined products from Singapore.

November 8

In Baku, officials from Azerbaijan, Ukraine, and Kazakstan meet to discuss a possible export and refining deal for crude oil from Kazakstan=s Tengiz oil field. Under Ukrainian proposals, crude from Tengiz would be shipped by tanker across the Caspian through Duybend, Azerbaijan to Batumi, Georgia. From there, the oil would be carried by tanker to Ukraine=s 80,000-b/d refinery at Odessa. An alternate method under consideration would be use rail transport through the Caucasus through Russia to Ukraine.

November 8

Namibian Minister for Trade and Industry Hidipo Hamutenya states that a well recently drilled by Royal Dutch/Shell proves that gas reserves at the offshore Kudu field are in excess of 3 trillion cubic feet (Tcf) and could be as large as 7 Tcf. Shell has declined to comment on the field=s size. Both Namibia and Shell are hoping that Kudu=s reserves are sizable enough to justify a $670-million project to build a 250-Mmcf/d pipeline to export gas to landfall at Orangemund and eventually on to South Africa.

November 12

The Algerian daily newspaper Le Matin reports that state-owned Sonatrach and British Petroleum (BP) are discussing the feasibility of a third Algerian gas export pipeline to Europe. Currently, Algeria exports natural gas to Europe through the newly-inaugurated GME, the Transmed pipeline linking Algeria to Italy, and via liquefied natural gas (LNG) tankers. BP is seeking an outlet for future production from its gas fields in the In Salah region, which contains proven reserves of at least 7 Tcf. Through 2000, Italy and France are expected to remain the largest purchasers of Algerian gas, with 47 per cent and 20 per cent shares of Algeria=s gas exports, respectively. Through 2020, the European Commission has forecast that Algerian exports will not exceed a 25 per cent share of the European gas market.

November 14

In Abuja, Nigeria, officials from Benin, Togo, Ghana, and Nigeria meet to discuss a proposed $260-million West Africa gas project which would supply Nigerian natural gas to regional markets.

Chevron and Royal Dutch/Shell are in competition to construct the export pipeline associated with the project. The gas project would be aimed at lessening the region=s dependence on oil imports and hydroelectric power. While Nigeria has insisted that its gas reserves be used, Ivory Coast and Ghana have proposed using gas resources located in their offshore Foxtrot and Tano fields, respectively.

November 14

In response to the slow pace of the Middle East peace process, Israel is cut out of two separate gas export pipeline proposals. In the first project, Italy=s ENI decides to invest $2 billion in the next 30 months to build a 960-Mmcf/d underwater pipeline to carry gas from Egypt to Turkey. ENI=s original peace pipeline would have run between Port Said, Egypt to Haifa, Israel, via the Gaza strip. In the second project, U.S.-based Amoco signs an agreement with the governments of Egypt and Turkey to build a $2-$4 billion LNG export plant to transport Egyptian gas to Turkey. The LNG plan replaces an earlier moco proposal to build an Egypt-Israel gas export pipeline that would have competed with ENI=s peace pipeline.

November 14

European Union (EU) trade counselor Bernd Langeheine states that the EU has legislation ready that would block implementation of U.S. sanctions on European companies which invest in expropriated U.S. property in Cuba or in Iran and Libya=s oil sectors. The official says that if the United States invokes Article 21 (the national security exemption) under the General Agreement on Tariffs and Trade (GATT), the EU would easily win any subsequent challenge brought before the World Trade Organization

November 17

Iraq=s National Assembly approves a proposal by China National Oil and Gas Exploration and Development Corporation to develop the small al-Ahdab oil field, which is located near the giant East Baghdad field. While several French and Russian oil companies have conducted negotiations for the development of oil fields in Iraq, the deal for al-Ahdab is the first that has actually been signed. Existing U.N. sanctions forbid foreign investment in Iraq=s oil sector.

November 18

In Kazakstan, the Caspian Pipeline Consortium (CPC) approves a $1.5-billion plan to build an export pipeline linking Chevron=s 10-billion barrel Tengiz oil field in western Kazakstan with Novorossiysk, Russia=s oil export terminal on the Black Sea. Under the plan, a 560,000-b/d pipeline would be built by 1999, with an expansion to 1.37 million b/d occurring at a later date. Until recently, Chevron restricted output at the field to 60,000 b/d because of lack of space on existing Russian export pipelines. Current production is around 120,000 b/d and soon will be lifted to 180,000 b/d. In 1996, Chevron has tested alternative methods of exporting Tengiz crude, including the use of a swap arrangement with Iran and tanker and rail shipments across the Caspian and through Georgia, respectively. After a corporate restructuring anticipated for April 1997, CPC will include the governments of Russia, Kazakstan, Oman, Chevron, Mobil, Oryx, British Gas, Agip, Lukoil, and Rosneft, among others.

November 25

An Exxon-led consortium signs a memorandum of understanding (MOU) with the Chadian government to develop an estimated 900 million barrels of oil reserves in the Doba basin, located near Chad=s southern border with the Central African Republic and Cameroon. The proposed $3-billion development will include drilling 300 new wells and constructing a 650-mile, 30-inch pipeline across southern Chad to Kribi, Cameroon on the Atlantic coast. Peak production potential from the area is projected at 200,000-250,000 b/d. The development consortium comprises Exxon (through its affiliate Esso Exploration & Production Chad)(40 per cent), Societe Shell Tchadienne de Recherches et d=Exploitation (40 per cent), and Elf Hydrocarbures Tchad (20 per cent).

November 26

The United Nations and Iraq announce a final agreement on the long-awaited deal that would allow Iraq to sell up to $2 billion of oil every six months in exchange for food and other humanitarian supplies. Implementation of the deal recently has been delayed because of Iraq=s invasion of Kurdish safe haven areas in September and the U.S. review of the pricing formula in October. Iraqi oil exports under U.N. Resolution 986, which would be between 500,000-550,000 b/d with current oil prices, are expected to begin in December 1996, after metering equipment is installed on the Kirkuk-Ceyhan pipeline and U.N.-approved observers are positioned in Iraq to monitor oil flows.

November 27

Russia=s Interfax news agency reports that a $1.5-billion oil development deal will be signed between the State-Owned Oil Company of Azerbaijan (SOCAR), Amoco, Unocal, Saudi Arabia=s Delta Oil, and Japan=s Itochu in mid-December 1996. The deal will include the development of the Caspian Sea=s Dan Ulduzu and Ashrafi oil fields, which are estimated to contain combined reserves of 730 million barrel and 3 Tcf.

November 28

The Prime Ministers of Russia and Poland inaugurate the first 65-mile segment of the proposed 3,100-mile Yamal gas export pipeline. In August 1995, the Polish government gave approval for the $2.5-billion, 400-mile Polish segment of the Yamal pipeline, which has an initial capacity of 475 Mmcf/d. Ultimate Yamal pipeline gas shipments to Europe are expected to reach over 5.5 Bcf/d by 2010. This gas is targeted for export in 1998 to European markets via seven parallel gas pipelines running from Yamal to Belarus, Poland, and Frankfurt/Oder, Germany. However, continuous permafrost conditions and environmental concerns have raised development costs, and total project investment May exceed $12 billion. Russian gas exports now satisfy about 25 per cent of European gas demand.

November 29

France=s nation-wide truckers= strike enters its twelfth day, causing up to 5,000 gas stations either to run out of or to run low on gasoline stocks. Consequently, half of France=s 90 departments have put gasoline rationing measures in place. Striking truckers also set up roughly 250 roadblocks on French highways, which lead to the disruption of agricultural shipments from neighboring countries. In order to end the strike, the government proposes meet truckers= demands to lower the early retirement age to 55.

November 29

In Vienna, the Organization of Petroleum Exporting Countries (OPEC) agrees to roll-over its current quota of 25.33 million b/d. This quota was established at the Organization=s last meeting in June 1996 and allows room for Iraq to produce 1.2 million b/d. Without Iraqi exports under U.N Resolution 986, independent analysts have estimated that OPEC member countries are producing between 1.2-1.5 million b/d in excess of the aggregate quota. The majority of this overproduction comes from Venezuela.

December 1996

December 1

The London press reports that the al-Yamamah oil-for-arms deal between British Petroleum (BP), Royal Dutch/Shell, and Saudi Arabia will not be renewed in 1997. Under the arrangement, BP and Shell earn roughly $30 million per year in commission from their combined sales of 400,000 barrels per day (b/d) of Saudi oil. During the past 10 years, earnings from the oil sales are deposited in a Bank of England account to pay British Aerospace for weapons purchases by Saudi Arabia.

December 2

In northeastern India, a bomb blast damages two regional crude oil pipelines. The main line and a loop line have a combined throughput of 100,000 b/d. The United Liberation Front of Assam is believed responsible for the blast. The disruption of crude flows does not affect refinery operations in Assam. However, the 13,000-b/d Digboi refinery is closed after a small, unrelated fire in one of the plant=s two crude distillation units.

December 3

After bad weather causes high swells, Petroleos Mexicanos (Pemex) shuts its three largest ports on the Gulf of Mexico. The temporary closures of the ports at Cayo Arcas, Pajaritos, and Dos Bocas add pressure to already high world oil prices and aid in boosting crude oil futures for West Texas Intermediate to $24.80/barrel. During the first 10 months of 1996, Mexican oil production averaged 2.85 million b/d. Roughly 1.5 million b/d of this output is exported through Pemex=s ports on the Gulf of Mexico. Mexico currently is the third largest source of U.S. oil imports, behind only Venezuela and Saudi Arabia.

December 3

Petroleos de Venezuela (PdVSA) decides to postpone the multi-billion dollar Cristobal Colon liquefied natural gas (LNG) project for at least five years. The project would have involved a consortium of companies comprising PdVSA=s subsidiary Lagoven (33 per cent), Exxon (29 per cent), Shell (30 per cent), and Mitsubishi (8 per cent). The proposed LNG plant would have produced 1 billion cubic feet per day (Bcf/d) of gas from the 11 trillion cubic feet (Tcf) of reserves in Venezuela=s Paria Peninsula. The decision to postpone the project is made because of currently low world gas prices which make the project uneconomical.

December 5

Near Urkutsk in eastern Siberia, a distillation unit under repair at the 464,000-b/d Angarsk refinery explodes, killing three and hospitalizing four workers.

December 5

Saga Petroleum, Norway=s largest independent oil company, purchases Kuwait Petroleum Company-owned Santa Fe Exploration for $1.23 billion. Santa Fe has stakes in six producing North Sea oil fields, including Miller, Alba, and Gryphon. It also has a nine per cent share in the Britannia gas field development. The acquisition will raise Saga=s North Sea oil production from current levels of 50,000 b/d to 190,000 b/d.

December 6

Chevron and Mobil sign separate agreements to restructure the Caspian Pipeline Consortium (CPC) contract. Both of the companies agree to exercise their respective rights to boost their equity interest in CPC. After the contract revisions, the largest stakes in CPC will be held by Russia (24 per cent) and Chevron (15 per cent). Mobil also will hold a 7.5 per cent interest in the $2-billion project to link Kazakstan=s Tengiz oil field to an export terminal near Russia=s Black Sea port at Novorossiysk.

December 6

Canada-based Arakis Energy states that a consortium has been formed to build a $1-billion export pipeline linking oil fields in southern Sudan to Port Sudan on the Red Sea coast. The consortium will comprise China National Petroleum Corporation (CNPC)(40 per cent), Malaysia=s Petronas Carigali (30 per cent), Arakis (25 per cent), and a Sudanese government-owned company (5 per cent). Assuming financing can be obtained, Arakis hopes to begin initial exports of 150,000 b/d by 1999.

December 6

Following a week-long strike by 400 employees of Shell Gas (Lanka) Ltd. last month in Sri Lanka, the company says that is has agreed to a wage supplement plan which will ensure continuous operations for the next three years. Prior to the deal, the strike had severely hurt liquefied petroleum gas (LPG) production and distribution in Sri Lanka.

December 9

The United Nations approves U.N. Resolution 986, which allows Iraq to export up to $1 billion worth of oil every 90 days for an initial six month period. Income from the oil sales will be used in part for food and other humanitarian supplies. Implementation of the deal recently has been delayed because of Iraq=s invasion of Kurdish safe haven areas in September and the U.S. review of the pricing formula in October.

December 10

Turkey=s semi-official Anatolia news agency announces that Energy Minister Recai Kutan will soon sign a LNG purchase agreement with Yemen=s Prime Minister Abdulaziz Abdulghani. The deal adds Turkey to the list of customers for the $4.2-billion LNG plant proposed by a consortium comprising Total, U.S.-based Hunt Oil, Exxon, South Korea=s Yukong, and the Yemeni government. The LNG plant will utilize reserves from the 16-Tcf Marib and al-Jawf regions. The area=s current oil production is around 170,000 b/d.

December 14

Azerbaijan and a consortium of foreign oil companies sign a $2-billion deal to develop the Dan Ulduzu and Ashrafi oil and gas fields in the Caspian Sea. The two fields hold estimated combined reserves of 1.1 billion barrels of oil and at least 1 Tcf of gas. The consortium comprises the state-owned oil company of Azerbaijan (SOCAR)(20 per cent), Amoco (30 per cent), Japan=s Itochu (20 per cent), and a partnership between Unocal and Saudi Arabia=s Delta (30 per cent).

December 16

The U.S. Department of State releases the guidelines for non-U.S. companies to follow in order to comply with the Iran and Libya Sanctions Act, which penalizes foreign firms that invest in the oil and gas sectors of those two countries. Since the Act was passed in August 1996, several pertinent developments have occurred, including investments by Total and Petronas in Iran=s offshore oil sector, ENI=s planned investment in Libyan gas field development, and Turkey=s recent signature of a long-term, $22-billion gas purchase agreement with Iran.

December 16

A consortium comprising Texaco (30 per cent), Exxon (30 per cent), Amoco (20 per cent), and Norsk Hydro (20 per cent) sign a protocol production sharing agreement (PSA) for development of an estimated 2.2 billion barrels of recoverable reserves in 11 area fields in Timan-Pechora's Varandey region near the Barents Sea coast. A final agreement is expected to be signed in early 1997. Current plans foresee investments of $45 billion over the 50-year project life. Although stalled in mid-1996, talks that might allow Rosneft and another Russian partner to acquire a 20 per cent stake in the project reportedly are continuing.

December 17

A U.N. panel rules that Kuwait Oil Company will receive $610 million, almost two-thirds of the $951 million it claims to have spent while extinguishing oil fires left by the retreating Iraqi army in 1991. In addition, the United Nations has approved 862,000 claims amounting to $3.2 billion for people forced to leave Kuwait because of the Iraqi occupation.

December 17

Following the resignation of PetroPeru head Armando Echeandia, PetroPeru officials announce that key parts of the company, including the 62,000-b/d Talara refinery, will remain under government control. Originally one of the first targets for privatization, the Talara refinery has had its sale date pushed back three times. Until this announcement, roughly 60 per cent of its shares were slated for auction in April 1997. While chief of PetroPeru, Echeandia had been a supporter of the company=s privatization efforts.

December 18

Nigeria=s Liquefied Natural Gas (NLNG) Company sues ENEL, Italy=s state electricity utility, for withdrawing from a $15-billion contract to purchase Nigerian LNG for a 20-year period. NLNG is planning to build a $3.8-billion, 5.2-million ton per year (Mmt/y) LNG plant at Finima, on Bonny Island. NLNG consortium members are NNPC (49 per cent), Shell (25.6 per cent), Elf (15 per cent), and Agip (10.4 per cent). Prior to its decision not to build a LNG receiving terminal at Montalto di Castro on the Tuscany coast for economic and environmental reasons, ENEL had agreed to purchase 2.75 Mmt/y, or about half of NLNG=s currently contracted future LNG sales.

December 18

During a press conference, Iranian Deputy Foreign Minister Abbas Maleki states that Iran supports Athe free flow of oil through the Strait of Hormuz, but reserves the option of closing off the shipping route if it is threatened. Iran recently has admitted to deploying anti-aircraft and anti-ship missiles on Abu Musa, an island strategically located near the Strait of Hormuz=s shipping lanes.

December 19

After recently signing a large 20-year gas supply contract, Turkey and Iran announce plans to begin feasibility studies for the construction of a pipeline to carry Iranian crude oil for export from Iskenderun on Turkey=s Mediterranean coast. The two countries also agree to renew their annual $600-million oil sales agreement, which expires in April 1997.

December 20

Clashes between the Front for the Liberation of the Enclave of Cabinda (FLEC) and Angolan government troops continue as the respective forces attempt to capture territory previously held by the

National Union for the Total Independence of Angola (UNITA). According to a FLEC spokesman, a succession of recent clashes has resulted in more than five dozen deaths and has injured more than 100 combatants. About 60 per cent of Angola=s production of 700,000 b/d of oil occurs offshore Cabinda.

December 23

After two years of discussions, Qatar Liquefied Gas Company (Qatargas) signs a 12-year, $570-million loan deal with four international banks. The loan will be used to help finance a $6-billion, 6-Mmt/y LNG plant, which will consist of three LNG trains supplied by gas from the 270-Tcf North Field. Japan=s Chubu Electric has contracted to receive 4 Mmt/y of the project=s anticipated LNG production. The Qatargas consortium partners are Qatar General Petroleum Corporation (QGPC)(65 per cent), Mobil (10 per cent), Total (10 per cent), and a group of Japanese companies (15 per cent). The Qatargas loan agreement follows a recent $1.2-billion loan deal obtained by competitor Ras Laffan Liquefied Natural Gas Company (Rasgas), which also is planning to build a LNG plant in Qatar.

December 30

The United Nations announces that a total of 21 contracts have been approved for the limited Iraqi oil sales under U.N. Resolution 986. The approved contracts will allow for 43.68 million barrels of oil to be exported in the first 90 days of the sale. At present, exports of 26.37 million barrels have been approved for the second 90-day period of the sale, which allows Iraq to sell up to $1 billion worth of oil every 90 days for an initial 6-month period. In mid-December 1996, Iraq restarted the Kirkuk-Ceyhan pipeline, which is expected to carry up to 450,000 b/d of oil under the sales agreements approved so far under U.N. Resolution 986. Iraq=s remaining oil exports will flow through the Mina al-Bakr terminal.

95

Monthly Energy Chronology 1997

The following information contains a listing of the major energy events that occurred in 1997.

January 1997

January 1

A Turkish newspaper reports an agreement between Turkey and Iran to build a pipeline to carry Iranian oil from the Caspian Sea to Turkey's Mediterranean port of Ceyhan. The agreement also includes a pipeline to carry natural gas from Turkmenistan through Iran to Turkey. No details are provided. This is the latest in a series of Turkish-Iranian initiatives (including a $20 billion natural gas deal and several trade agreements) which have been criticized by the United States due to Iran's links to terrorism.

January 2

A Russian tanker carrying 19,000 tons of heavy oil breaks apart and sinks in rough seas in the Sea of Japan. Japanese helicopters sight an oil slick about 1 mile long and 90 yards wide located about 375 miles west of Tokyo. The tanker, Nakhodka, was en route from Shanghai, China to Russia's Kamchatka Peninsula.

January 3

President Clinton suspends for another six months a U.S. law, the Helms-Burton Act, that would allow Americans to sue foreigners doing business in Cuba on property confiscated from Americans.

January 3

The New York Mercantile Exchange posts its first longterm, forward contracts for crude oil, covering delivery as far out as December 2003. There are no takers.

January 6

Saudi Arabia reports a discovery of new reserves of super light crude oil in the central region, near Riyadh. According to Oil Minister Ali Naimi, this is the seventeenth such discovery in the area. Further assessments to determine potential reserves are planned.

January 6

Iraq informs its customers that it will reduce its contractual crude oil sales volumes in order to stay within the $1 billion limit for the first 90 days of the United Nations' oil-for-food agreement.

January 7

The United Nations approves three more contracts for the sale of Iraqi oil, bringing to 24 the total number of contracts approved so far under the oil-for-food agreement.

January 7

A Salomon Brothers' survey of 228 oil and natural gas companies worldwide indicates exploration and development spending will increase to $81.4 billion in 1997, up 14.7 per cent from $71 billion in 1996. This is described as the "strongest outlook for oil service demand" in the 15 years of the survey. Of the total, 62 per cent of the investment is planned outside North America, 28 per cent in the United States, and 10 per cent in Canada.

January 7

Algeria's Prime Minister, Ahmed Ouyahya, claims his country's proven oil reserves have recovered to their 1971 level (9.84 billion barrels, according to Organization of Petroleum Exporting Countries statistical sources).

January 7

A senior official of the China National Petroleum Company, China's state onshore oil and gas producer, says China plans to double its natural gas production capacity to almost 1.1 trillion cubic feet per year by 2005. To meet this goal, China has set a target to add 35 trillion cubic feet of verified onshore reserves by 2000. Primary areas of exploration are the Shaanxi-Gansu-Ningxia region, the Tarim Basin, and Sichuan province–whose combined total proven reserves are estimated to be at least 20.8 trillion cubic feet.

January 10

Russia's Gazprom approves a $2.5-billion project to build the world's deepest underwater natural gas pipeline (about 6,900 feet below the surface) linking Russia and Turkey via the Black Sea. A feasibility study will be completed in the second quarter of 1997.

January 10

Freezing weather disrupts shipments of petroleum products on Europe's inland waterways. Major areas affected include the Dutch canal system and waterways in Belgium, France, and the Netherlands which connect to the Rhine River.

January 15

U.S. Deputy Treasury Secretary Lawrence Summers announces that Mexico's oil receipts are now free of U.S. claims, as a result of its early repayment of $5 billion in debt to the United States and the International Monetary Fund. The U.S. loan had been guaranteed by the country's oil revenues.

January 16

Qatar Liquefied Gas Co. (Qatargas) starts up a second production line, with output capacity of 2.0 million tons per year of liquefied natural gas.

January 16

Oil from a wrecked Russian tanker in the Sea of Japan penetrates oil fences set up around two nuclear power plants belonging to Japan's Kansai Electric Power and Hokuriku Electric Power. A preliminary investigation indicates the tanker, which broke up and sank in the Sea of Japan on January 2, collided with a "halfsubmerged object".

January 18

Iraq agrees to export 25 million barrels of crude oil and 7 million barrels of petroleum products to Jordan in 1997. The total, which equates to 88,000 barrels per day, is 7 per cent more than in 1996. Jordan relies entirely on Iraqi crude oil, which is shipped under a special exemption from United Nations' sanctions against Iraq. Part of the oil is paid for at a reduced price ($19.10 per barrel in 1997, 25 per cent higher than in 1996) and the rest of the oil goes toward reducing Iraq's $1.3 billion debt to Jordan and paying for Jordanian exports of food and medicine to Iraq (slated to increase by 17 per cent to $255 million in 1997).

January 21

Yemen signs a final agreement endorsing the participation of a foreign consortium, Gas General, in its $4.2-million project to produce 5 million tons per year of liquefied natural gas. Participants include Total (36 per cent), Hunt Oil (15.3 per cent), Exxon (14.3 per cent), Yukong (8.4 per cent), and the Yemen government (26 per cent).

January 21

Japan's Prime Minister Ryutaro Hashimoto concedes that his government misjudged the seriousness of the January 2, 1997 oil spill in the Sea of Japan and announces an all-out effort to prevent further damage. Cleanup has been complicated by heavy seas and the nature of the oil, a specially refined distillate destined for use in Siberia, which did not coagulate in cold waters as expected. Fishermen and environmentalists describe the spill as a major ecological disaster and Japan's worst oil spill in 20 years.

January 21

Statoil, Norway's state-owned energy company, announces an agreement to supply 5.3 trillion cubic feet of natural gas to Italy under a 25-year contract between Norway's Gas Negotiating Committee (representing Statoil, Norsk Hydro, and Saga) and SNAM, a unit of Ente Nazionale Idrocarburi (ENI). The contract, valued at nearly $14 billion (at current prices), provides for initial delivery of 71 billion cubic feet in 2000. Annual contract volumes would increase to 212 billion cubic feet in 2002, and could be increased an additional 71 billion cubic feet in each subsequent year. Norway is currently the second largest source of natural gas for Europe (after Russia's Gazprom).

January 22

President Fidel Ramos orders full deregulation of the Philippine oil industry effective February 8, 1997, nearly a month ahead of the established deadline.

January 22

Libya officially notifies the United Nations that it will resume international flights in violation of United Nations' sanctions.

January 22

A senior Qatari official announces plans to increase Qatar's combined crude oil and condensate output capacity to 910,000 barrels per day by 2000. Current capacity is about 510,000 barrels per day.

January 23

An explosion and fire on the IraqTurkey oil pipeline in southern Turkey damages the unused portion of the line, but the line currently in use (carrying an estimated 440,000 barrels per day) is undamaged. Kurdish guerillas are blamed for the attack.

January 24

The U.S. Minerals Management Service (MMS), part of the Department of the Interior, proposes a new regulation that would revise how federal royalties are calculated on oil produced from federal lands. Under the proposed regulation, royalties would be based on a monthly average of futures prices at the New York Mercantile Exchange, except for oil produced in California or Alaska (which would be based on prices for Alaska North Slope crude oil). MMS would adjust the market prices to account for differences in quality and location. Currently, royalties are based on the posted (wellhead) prices of the oil.

January 28

The official Iraqi News Agency (INA) reports that Iraq has exported a total of 11.5 million barrels of crude oil from its Persian Gulf terminal of Mina al Bakr since it began selling oil under the United Nations' oil-for-food deal in December 1996.

January 29

The Clinton Administration refutes news reports that Iraq is threatening its neighbors, but restates the U.S. willingness to act if Iraq does become aggressive. The statement is in reaction to speculation following reports that Saddam Hussein's wife has been placed under house arrest and his son risks losing a leg to gangrene in the wake of a previous assassination attempt. March light, sweet crude oil futures prices settle sharply higher on the New York Mercantile Exchange (up 57 cents, to \$24.47 per barrel).

January 29

Amoco and Shell Deepwater Development (Shell) announce plans to jointly develop a deepwater Gulf of Mexico oil and natural gas discovery known as the Marlin prospect. The prospect is located in 3,240 feet of water and will cost \$500 million to develop. It is expected to produce 250 million cubic feet of natural gas per day and 40,000 barrels of oil per day by 1999. Amoco owns 75 per cent of the prospect and Shell owns the remaining 25 per cent.

January 29

Leftist rebels are suspected of being responsible for an explosion on a section of Colombia's main oil pipeline that causes a spill of about 5,000 barrels of crude oil and forces the first suspension of pumping on the pipeline this year.

January 30

United Nations' Secretary General Kofi Annan states he expects Iraqi oil exports under the oilforfood deal to continue through the year after an anticipated renewal of a sixmonth agreement at midyear.

January 30

Venezuela's stateowned oil company, Petroleos de Venezuela S.A. (PDVSA), announces that 259 companies have qualified to participate in its third auction of operating contracts for mature and marginal oil fields, scheduled for May 1997. According to PDVSA's estimates, the fields included in the 20 contracts hold reserves of between 2 billion and 3 billion barrels and output is expected to reach 300,000 barrels per day by 2006.

January 30

A United Nations' spokesman announces that disbursements from the escrow account set up under the Iraqi oilforfood plan have begun. The fund is expected to total $239 million by the end of the day.

February 1997

February 5

The Energy Policy Committee of the U.S.-Russian Joint Commission on Economic and Technological Cooperation (the Gore-Chernomyrdin Commission) meets in Washington to review progress in U.S.-Russian cooperation, including improvements in the foreign investment climate in the Russian oil sector. Other topics of discussion include nuclear nonproliferation; nuclear reactor safety and peaceful uses of nuclear technologies; the development of power projects, including for Russia's northern territories; and progress in energy efficiency.

February 5

Japan's Ministry of Finance announces plans to cut import tariffs on crude oil and most petroleum products from April 1, 1997, in a phased process that will reduce the country's crude oil import tariff rate to zero in April 2002.

February 6

President Clinton formally unveils a fiscal 1998 budget plan, which includes a proposal to raise $1.35 billion over the next 5 years with the sale or lease of crude oil from the Strategic Petroleum Reserve.

February 6

The United Nations reports that it so far has received a total of $343 million in proceeds from Iraq's oilforfood sale.

February 10

Lukoil, Russia's largest oil company, and the U.S. energy company Arco enter a $5 billion, 18-year joint venture to prospect for and develop oil in the former Soviet Union. Lukoil's share is 54 per cent and Arco's is 46 per cent. The venture, called Lukarko, plans investment of $400 million in 1997. Initial projects include a 12-per cent stake in the Caspian Pipeline Consortium and a 5-per cent share of the Tengiz field in Kazakstan.

February 10

South Africa's deputyminister of mineral and energy affairs indicates that South Africa will complete the restructuring of its oil industry by the end of 1997.

February 11

Woodside Petroleum receives approval from the Australian government to proceed with the $1 billion Laminaria development in the Timor Sea. The project will employ the world's biggest floating production, storage, and offloading facilities to produce 170,000 barrels per day of oil beginning in 1999. Crude oil reserves are estimated at 130-250 million barrels. Woodside's partners in the development are Broken Hill Proprietary (25 per cent) and Shell (25 per cent).

February 11

Ecuador's Congress names Fabian Alarcon president, ending a leadership struggle which began on February 6 when Congress voted to oust former president Abdala Bucaram for "mental incapacity."

February 11

In a special report, the International Energy Agency (IEA) predicts major increases in oil and natural gas production capacity in Algeria, Libya, and Egypt. Between 1995 and 2000, the IEA reports that Algeria is projected to increase its crude oil production capacity by 25 per cent (to at least 1.0 million barrels per day) and its natural gas production capacity by 58 per cent (to 3 trillion cubic feet annually); Libya will increase its crude oil production capacity by 18 per cent (to 1.65 million barrels per day) and maintain natural gas production capacity at 667 billion cubic feet annually; and Egypt will increase its crude oil production capacity by 11 per cent (to 1.0 million barrels per day) and its natural gas capacity by 76 per cent (to 1.2 trillion cubic feet annually). Natural gas liquids production will also increase in all three countries. In addition, the IEA projects that Algeria's natural gas export capacity will reach 2.9 trillion cubic feet and Libya's 159 billion cubic feet, representing increases of 123 per cent and 41 per cent respectively, compared with 1995 capacities.

February 12

A U.S. admiral headquartered in Bahrain reports that tankers are smuggling tens of thousands of tons of fuel oil out of Iraq in violation of United Nations' sanctions by reportedly skirting the shoals of Iran's coast, apparently with Iranian approval.

February 12

A spokesman for Ecopetrol, Colombia's national oil company, announces a suspension of drilling in the Coporo field, which had been hailed as one of the most promising in the country's history (with 700 million to 1.5 billion barrels of oil reserves). A company report indicates difficulties in extracting the oil due to low permeability and low pressure.

February 12

The Norwegian Oil Directorate announces it has upgraded its estimate of total extractable oil and natural gas reserves on Norway's continental shelf by 16 per cent, to 78.6 billion barrels of oil equivalent.

February 12

Petroleos Mexicanos (Pemex) announces plans to invest $5.92 billion in 1997 to boost Mexico's oil and natural gas production and upgrade operations (a 70-per cent increase from 1996). Pemex plans to increase production to 3.08 million barrels per day of crude oil and 4.55 billion cubic feet per day of natural gas.

February 13

Exxon Corp. files documents with the U.S. District Court in Anchorage, Alaska, appealing that court's $5.06 billion amended final judgment on the 1989 Exxon Valdez oil spill, which spilled 11 million gallons of crude oil into Prince William Sound.

February 13

The Congolese government announces plans to liberalize–and eventually privatize–the downstream activities of the country's petroleum sector. To participate, private sector companies will need to obtain approval from the Congolese government.

February 13

Canadian crude oil begins flowing through the new Express Pipeline to Caspar, Wyoming. Originating in Hardisty, Alberta, the pipeline interconnects at Casper with the Platte Pipe Line to transport oil to Wood River, Illinois and other points in the U.S. Midwest and Southeast. Commercial service on the 172,000-barrel-per-day pipeline is scheduled to start April 1.

February 14

Australia's Broken Hill Proprietary Co. receives formal approvals to proceed with the development plan and environmental assessments for developing the first two oil fields in the Timor Sea Zone of Cooperation, an area jointly administered by Indonesia and Australia. The two fields (Elang and Kakatua) contain estimated commercial oil reserves totaling 15 million barrels and are expected to begin producing in the third quarter of 1998.

February 17

Australia's Woodside Petroleum Ltd. announces a significant increase in natural gas and condensate reserves from acreage held by the North West Shelf venture. As of December 31, 1996, the total proved raw gas ultimate recovery is estimated at 24.4 trillion cubic feet, 60 per cent higher than the previous year's estimate, and the proved condensate ultimate recovery is estimated at 690 million barrels, an 18-per cent increase. The project is a joint venture shared equally among Woodside and its five partners: Broken Hill Proprietary, Chevron, Royal/Dutch Shell, British Petroleum, and Japan Australia LNG.

February 17

The Middle East Economic Survey reports that Iraq's State Oil Marketing Organization is increasing crude oil sales to reach its targeted $1.0 billion in sales for the first 90day period under the United Nations' oilforfood plan. The total value of contracts for the first 90 days is reported to be $800-$850 million, lower than expected due to lower oil prices (14 per cent less than when the sale was announced in mid-December 1996) and the failure of Russian companies to lift contracted volumes.

February 17

Raul Baca, Ecuador's new energy minister, suspends the bidding process announced by the previous government to build a new 246,000-barrel-per-day pipeline for heavy crude oil, due to "serious indications of illegalities."

February 18

A Colombian public workers' strike, in effect since February 11, ends when the government reaches an agreement with labor representatives, which reportedly includes a promise not to privatize state-held oil and telecommunications companies. The strike, which included 5,000 oil workers, had forced the shutdown of one of the state oil company Ecopetrol's gasoline plants in Barrancabermeja, the country's main hydrocarbons processing center.

February 18

Presidents Eduard Shevardnadze of Georgia and Geidar Aliev of Azerbaijan sign a package of 22 cooperation agreements, including: 1) an agreement for Azerbaijan to supply natural gas to Georgia, 2) an agreement on an oil pipeline from Azerbaijan across Georgia to the Black Sea port of Batumi, 3) an agreement calling for closer cooperation between the two countries' state oil companies, and 4) and an agreement under which Azerbaijan will ship 4.4 million barrels of oil by railroad through Georgia in 1997.

February 19

Conoco announces that the U.K. Department of Trade and Industry has approved development of the North Sea's Banff oil field, following a 6month production assessment which maintained a production rate of about 35,000 barrels per day. Discovered in 1991, the Banff field contains recoverable reserves of approximately 60 million barrels of oil (38 to 39 degree API gravity) and around 39 billion cubic feet of associated gas. Start-up is scheduled for June 1998. Conoco holds a 31.7-per cent share of the venture and will operate the field. Conoco's partners are: Enterprise Oil (27.9 per cent); Ranger Oil (26.2 per cent); Hardy Exploration & Production (12.4 per cent); and Santos Europe (1.8 per cent).

February 19

Deng Xiaoping, China's retired senior leader and architect of the country's economic reforms, dies at the age of 92. It is generally believed that his death will not have an immediate impact on China's economy, including its oil and petrochemical industries, as the retired leader had not played an active role in recent years.

February 20

U.S. Air Force officials announce the deployment of 30 warplanes and more than 1,000 airmen in Qatar to help train the emirate's Air Force and enforce the nofly zone over southern Iraq, which was established in 1991 after Iraq crushed a rebellion by Shiite Muslims following the Gulf War.

February 21

Cameroon's president Paul Biya officially inaugurates the Ebome offshore oil field, the country's second largest. The field began production on January 12 and is expected to produce up to 12,000 barrels per day.

February 21

China's official news service reports that China National Petroleum Corp. expects to produce 88,000 barrels per day of crude oil from Tarim Basin oil fields in the country's northwest in 1997, up from 63,000 barrels per day in 1996.

February 24

Romania announces plans to privatize 62 large state companies by the end of 1998, including oil refineries and energy distributors.

February 24

Russia oil giant AO Lukoil claims to have the largest proven oil reserves of any company in the world–10.77 billion barrels, according to an audit of its holdings. The estimate includes 7.927 billion barrels in Lukoil's main fields in Western Siberia and 2.85 billion barrels in European Russia. The company's natural gas reserves in European Russia are reportedly 1.0878 trillion cubic feet.

February 24

Qatar inaugurates the world's largest liquefied natural gas (LNG) exporting facility and formally launches Qatar Liquefied Gas Co., which will have total output capacity of 6 million tons per year of LNG. The facilities are part of a new $7.2 billion industrial zone which also includes a sea port with a capacity to handle 25-30 million tons of LNG annually. Qatar plans to build more gas liquefaction plants in the area to exploit its natural gas reserves of around 237 trillion cubic feet.

February 25

A senior oil official in Qatar reports that Qatar General Petroleum Corp. has approved a $5.0 billion, 5-year oil drilling program aimed at increasing oil production capacity to around 700,000 barrels per day (currently around 550,000 barrels per day). The goal is to spud 202 offshore wells, including 69 wells at the Al Shaheen oilfield. Plans include opening three new oil blocks and awarding development and productionsharing agreements to foreign oil companies.

February 25

Azerbaijan's parliament ratifies a $2 billion oil development project in the Caspian Sea–the country's fourth major oil exploration contract with foreign companies. The 25year productionsharing contract, first signed in December 1996, covers development of the Dan Ulduzu and Ashrafi oil fields. Partners are Amoco (30 per cent), Unocal (25.5 per cent), Itochu (20 per cent), Azerbaijan's state oil company (20 per cent), and Delta Nimir of Saudi Arabia (4.5 per cent).

February 25

Amoco Corp. signs an agreement with Shell Oil Co. to form a limited partnership called Altura Energy Ltd, which will operate the combined oil and natural gas producing assets of the two companies in the greater Permian Basin area of west Texas and southeast New Mexico. The new company (64 per cent Amoco, 36 per cent Shell), which begins operations on March 1, will be the largest oil producer in Texas and the third largest independent oil producer in the United States, according to Amoco officials. In 1996, Altura's 6,300 wells produced roughly 170,000 barrels per day of oil and 220 million cubic feet per day of natural gas.

February 27

Russia's State Property Committee approves a plan to privatize AO Rosneft, the country's last major oil company still entirely in state hands. Rosneft was among the last of the integrated oil companies the government formed from onceindependent production, refining, and marketing enterprises. In 1996, the company produced 261,000 barrels per day of crude oil and refined 95,000 barrels per day of crude oil.

February 28

Turkey's national pipeline company Botas receives 14 bids for the construction of a 162-mile natural gas pipeline from the Iranian border to the Turkish city of Erzurum. Iran will also build a 172mile natural gas pipeline from its Western city of Tabriz to the Turkish border. Each pipeline is expected to cost $200 million. Turkey has agreed to buy 10 billion cubic feet of natural gas annually from Iran beginning in 1998, with volume increasing to 33 billion cubic feet annually by 2005.

February 28

Russian gas producer and distributor RAO Gazprom agrees to the terms of an 8-year, $2.5 billion syndicated loan from 19 international banks. The loan, which is backed by revenues from Gazprom's gas exports to Europe, will pay for the construction of Gazprom's planned YamalEurope pipeline to western and central Europe.

February 28

The United Nations announces the approval of the 35th contract for the sale of crude oil under Iraq's oilforfood program. The contract is for delivery during the second 90-day period of the sale, which begins March 10. To date, about $900 million of oil has been contracted for the second 90day period, during which Iraq is allowed to sell up to $1 billion worth of crude oil.

March 1997

March 1

Sudan signs contracts with national oil companies from China and Malaysia and a private Canadian company (Arakis) to develop its oil reserves. The contracts also cover construction of a 50,000-barrel-per-day oil refinery in Khartoum and a 900mile oil pipeline to transport up to 250,000 barrels per day to Port Sudan on the Red Sea. Sudan hopes to increase crude oil output to 150,000 barrels per day by mid1999 (current production is estimated between 10,000 and 15,000 barrels per day). A civil war which erupted in 1983 hampered previous oil industry development in Sudan, which currently consumes about 50,000 barrels per day of refined products.

March 3

In its regular 60day review, the United Nations Security Council votes to maintain sanctions on Iraq. This is the 36th review since sanctions were first imposed in 1990.

March 3

Lyondell Petrochemical Co. and Citgo Petroleum Corp. announce the completion of a $1.1 billion upgrade of their joint-venture LCR refinery in Houston to process more than 215,000 barrels per day of very heavy crude oil (17 degree API gravity) from Venezuela and produce higher valued products such as reformulated gasoline and lowsulfur diesel. Affiliates of Venezuela=s state oil company (Citgo's parent company) supply up to 230,000 barrels per day of crude oil under a long-term contract and Citgo purchases and markets most of the refined product output. For Venezuela, the venture secures guaranteed, longterm markets for its heavy crude oil.

March 3

A Venezuelan Navy spokesman announces that a major oil spill (25,000-50,000 barrels) off Venezuela's western coast will take at least one month to clean up. The spill occurred when the Greek tanker Nissos Amorgos ran aground February 28 and began leaking its cargo of 476,000 barrels of heavy crude oil. Maraven, a unit of stateowned oil company Petroleos de Venezuela, reports the tanker was leased by Italy based oil company Agip.

March 3

Mexico's state oil company Petroleos Mexicanos (Pemex) establishes two new petrochemical subsidiaries covering the Cangrejera and Morelos complexes. Establishment of subsidiaries is the first step in the process of privatizing Mexico's petrochemical industry. To date, Pemex has established subsidiaries for 6 of its 10 petrochemical complexes (subsidiaries for Camargo, Cosoleacaque, Escolin and Tula were announced in late January). Competitive bidding for the sale of 49 per cent of each subsidiary is scheduled to begin in the second half of 1997.

March 4

Mexico's state oil company Petroleos Mexicanos (Pemex) reports the Campeche Sound (in the Gulf of Mexico) has hydrocarbon potential of 74 billion barrels of crude petroleum equivalent, proven reserves of 17.1 billion barrels, and cumulative production of 11.8 billion barrels. Pemex plans to publish estimates for Mexico's Southern Region in 1998 and its Northern Region in 1999. At the end of 1995, Pemex claimed that proven hydrocarbon reserves (crude oil and natural gas) for all of Mexico stood at 62.06 billion barrels of crude petroleum equivalent.

March 5

The United Nations approves the 36th contract for the sale of Iraqi oil and announces that the $1.07 billion limit for the first 90day period of Iraq's oil-for-food program has been "more or less" met. The $1.07 billion includes $70 million in pipeline fees to Turkey.

March 6

The North Sea's Captain Field, operated by a unit of Texaco, begins production. Located about 90 miles northeast of Aberdeen, Scotland, the field holds estimated reserves of 350 million barrels of oil and 53 billion cubic feet of natural gas. It is expected to reach peak production of 67,000 barrels per day by mid1997.

March 6

The Minerals Management Service of the U.S. Department of Interior reports a recordbreaking sale of offshore oil and natural gas leases in the Central Gulf of Mexico, attracting more than $824 million in high bids from 81 companies (compared with last year's $520.9 million in high bids from 78 companies). The high bid on each block must be evaluated to ensure fair market value before a lease is awarded; each block is approximately nine square miles.

March 7

Indonesia's stateowned oil and gas company Pertamina announces that it has secured a $1.13 billion syndicated loan to finance the development of its eighth liquefied natural gas (LNG) plant, to be located in Bontang, East Kalimantan. The plant, with annual production capacity of 2.95 million tons, is scheduled to begin operating at the end of 1999 and will supply LNG to the Korean Gas Company and the Chinese Petroleum Corporation under a 20year contract.

March 7

Spain's Cabinet of Ministers approves the public offering of the government's final 10 per cent stake in Repsol, the country's largest oil and energy group. The sale is expected to begin in April. The government has been reducing its stake in stages since privatization began in 1989.

March 8

Qatar's alKhaleej offshore oilfield begins production at an initial rate of 20,000 barrels per day, which will gradually increase to around 30,000 per day by November 1997. The field is being developed by a consortium including France's Elf Aquitaine and Italy's Agip under a productionsharing agreement with Qatar General Petroleum Corporation. Qatar plans to increase its oil output to around 700,000 barrels per day by 2000, from 550,000 barrels per day currently.

March 10

Exxon announces a 'significant discovery' in the Gulf of Mexico, called Hoover, which May contain developable reserves of more than 100 million barrels of oil equivalent.

March 11

Turkey's Minister of Energy Recai Kutan reports that Turkey will go ahead with its plans to import Iranian natural gas, despite objections from the United States. Turkey plans to begin operating gas power plants by 2000, when the country's annual demand for natural gas is projected to exceed 950 billion cubic feet.

March 11

Royal Dutch/Shell begins drilling its first appraisal well at Peru's giant Camisea field, now thought to be the largest natural gas field in South America. The field, discovered in 1980, could contain as much as 11 trillion cubic feet of natural gas and 600 million barrels of condensate.

March 12

By a vote of 99 to 1, Federico Peña is confirmed by the Senate and later sworn in as the eighth U.S. Secretary of Energy.

March 12

Turkey's BOTAS pipeline company reports that it has received 29.1 million barrels of oil from Iraq since December 14, 1996 at its Yumurtalik pumping station on the Iraq-Turkey pipeline, of which 8.4 million barrels were used for domestic consumption and 20.7 million barrels were shipped abroad by 27 tankers.

March 13

In his annual state of the union address to Congress, Venezuelan President Rafael Caldera reiterates his administration's determination not to use oil windfall tax revenue on current spending and notes that his administration has created a special account for these revenues in the Central Bank of Venezuela. Separately, Venezuela's Finance Ministry reports the country produced approximately 3.154 million barrels of oil per day (and exported 2.723 million barrels per day) in 1996–well above its current OPEC production quota of 2.359 million barrels per day. The government expects both production and exports to increase in 1997 (to 3.25 million barrels per day and 2.876 million barrels per day, respectively).

March 14

Officials at Japan's Ministry of International Trade and Industry predict a slowdown in the country's petroleum product demand growth over the next 5 years, to an average annual rate of less than 1 per cent (compared with 2.1 per cent annually over the past 5 years). Declines in demand for the electric power sector (as new coal, nuclear, and liquefied natural gas plants come onstream) are expected to partially offset continuing increases in other oil-consuming sectors.

March 14

Following the signing of a final agreement on the maritime boundary between Australia and Indonesia, Australia's Resources Minister Warwick Parer announces that his country will move to release exploration blocks for petroleum exploration at the earliest opportunity. He cites recent petroleum discoveries in the Timor Sea which previously had not been considered for petroleum exploration due to uncertainty over which nation had jurisdiction.

March 14

The prime ministers of three former Soviet republics in Central Asia Kyrgyzstan, Kazakstan and Uzbekistan sign a series of agreements aimed at forming an economic union.

March 15

Vietnam accuses China of violating its waters in the South China Sea and demands the immediate removal of a Chinese exploratory oil rig which allegedly has been operating between the Vietnamese coast and China's Hainan island since the beginning of March.

March 15

Saudi Arabian Oil Company (Saudi Aramco) reports that production from the Shaybah oil field will begin in June 1998, earlier than its initial plan which called for production of 500,000 barrels per day by December 1999. The field is estimated to contain 7 billion barrels of premiumgrade extra light crude oil with 40-42 degree API gravity. The $2.5 billion development program for Shaybah is the largest undertaking by Aramco in recent years.

March 17

Mexico President Ernesto Zedillo formally presents the country's 1997 energy sector program, under which the government will invest 69 billion pesos (about $8.7 billion at current exchange rates)–an increase of 41 per cent in real terms over the amount invested in 1996 and the highest level of investment in real terms since 1990. The president also reiterates his support for private sector participation–but only up to a point. Mexico has invited private companies to participate in the transport and distribution of natural gas, electricity generation, and minority ownership in the petrochemicals industry, but has refused to open up its oil production to private investors.

March 17

Ecuador authorizes Arco to extract oil from the Villano field in block 10 of the country's Amazon region. Initial production is expected to be 30,000 barrels per day of oil. Arco also plans to build an 80,000-barrel-per-day pipeline that will connect the field with Ecuador's main crosscountry oil pipeline by the second quarter of 1999.

March 18

Independent oil experts suggest that Mexico may have overstated the size of its petroleum reserves by as much as 30 per cent since the 1970s, and indicate that Mexico's latest reserves estimates are more in line with outside estimates such as the U.S. Central Intelligence Agency's. This is based on an analysis of a statistical report, released earlier this month, in which Mexico reduced its estimates of proven reserves in the Bay of Campeche to 15 billion barrels of crude oil and 9.7 billion cubic feet of natural gas (from previous estimates of 23.5 billion barrels of crude oil and 11.6 billion cubic feet of natural gas in 1996). The report was the first Mexican reserves report to be audited by independent experts. The Bay of Campeche currently accounts for about 80 per cent of Mexico's oil and natural gas production. Reserve studies of other major petroleum areas (including the Chincontepec and Tabasco-Chiapas fields) are not expected to be completed until 1999.

March 18

Mobil announces the start-up of oil production from the second production facility on the Wandoo field (the Wandoo B platform) on Australia's North West Shelf, which will boost the field's production to 40,000 barrels per day by May 1997 (current production is 8,000 barrels per day). Australia's Bureau of Agricultural and Resource Economics reports several other projects are expected to approach full production capacity in the next fiscal (July 1997 June 1998), boosting the country's total production by 5.7 per cent. New production will be concentrated in the Gippsland Basin offshore southeastern Australia. Other projects expected to come on stream after 1998 include the Elang/Kakatua and the Laminaria/Corallina fields in the Timor Sea, which together could add a further 170,000 barrels per day of production.

March 18

Iraq grants Russia most favored nation status to receive Iraqi oil exports in exchange for humanitarian goods. Of the first 37 contracts approved by the United Nations in the oil-for-food sale, 7 went to Russian companies representing almost 20 per cent of the volume of oil in the sale.

March 18

The United Nations approves the 38th and 39th contracts for the sale of Iraqi oil under Resolution 986, the Iraqi oil fo rfood sale, bringing the total dollar value of contracts approved to date to about $1.9 billion of oil (out of a total value of about $2.14 billion allowed over the first 6 months of the oil sale period). According to a Western diplomat, Iraq had shipped 55.9 million barrels of crude oil worth about $1.11 billion under the agreement through March 14, including 4.3 million barrels worth about $77.5 million since the second 90day period began on March 10. Most of the oil (about 63 per cent) has been shipped through the Iraq-Turkey pipeline from Kirkuk to Yumurtalik.

March 18

An explosion damages Colombia's Cano LimonCovenas oil pipeline near the Venezuelan border, the second attack on the country's oil facilities in less than two days. On March 17, an explosion affected three pipelines used to carry oil, natural gas, and petroleum products to Colombia's largest refinery in Barrancabermeja. British Petroleum asks Colombian President Ernesto Samper to beef up security near the Cusiana field, the country's largest (with estimated reserves of 1.6 billion barrels of crude oil). Earlier this month, Colombia's state oil company Ecopetrol reported that the Cano LimonCovenas pipeline was attacked 12 times in January and February, and the VasconiaCovenas pipeline was attacked once in February. The attackers reportedly are members of the National Liberation Army, a proMarxist group that opposes the government's oil policies.

March 18

India's oil minister T.R. Baalu announces a new oil policy aimed at increasing investment in the country's oil sector. The new policy allows–for the first time–oil exploration ventures that are 100-per cent private and foreign-owned. The previous policy required companies to form joint ventures that gave staterun partners up to 40-per cent equity when they bid for acreage.

March 19

The Czech government announces a 20-year agreement to purchase 1.9 trillion cubic feet of natural gas from a Norwegian consortium of natural gas exporters (Norsk Hydro, Saga Petroleum, and Statoil). Shipments will begin May 1 at the rate of 49 million cubic feet per day, increasing to 290 million cubic feet per day (106 billion cubic feet annually). Currently, Russia's Gazprom is the only supplier of natural gas to the Czech Republic. The contract represents the first time Norwegian natural gas will enter a market in central Europe. The agreement is part of the Czech government's strategy to diversify its imports of natural gas.

March 19

Texaco announces that it has received approval from the United Kingdom (U.K.) Department of Trade and Industry to develop the Galley oil and gas field in Block 15/23a of the U.K. North Sea. Initial production will begin in the first quarter of 1998 and reach 43,000 barrels per day of oil equivalent.

March 20

China's official news service reports that China hopes to build an annual natural gas production base of about 1.1 trillion cubic feet by 2005 (compared with current production of about 600 billion cubic feet) and double its known onshore reserves of natural gas by 2000 (to about 70.6 trillion cubic feet). Exploration and development work will focus on gasrich regions such as the ShaanGanNing Basin in centralwestern China (with proven natural gas reserves currently estimated at about 8.1 trillion cubic feet), the Tarim Basin in the far west (with estimated proven reserves of about 5.6 trillion cubic feet), and Sichuan province in the southwest (with estimated proven reserves of 7.1 trillion cubic feet). In addition, another 1.8 trillion cubic feet of natural gas was recently identified in the eastern part of Qinghai province, just east of Xinjiang.

March 20

Yemen's Oil Minister Mohammad Said alAttar reports that he expects Yemen's 1998 oil output to rise to 420,000 barrels per day from a current average of 360,000 barrels per day.

March 20

Colombia's Mines and Energy Minister Rodrigo Villamizar reports that his country's oil output will top 800,000 barrels per day by the end of 1997.

March 21

Great Lakes Gas Transmission announces plans to double its capacity to transport Canadian natural gas to pipeline interconnections in the United States no later than the winter of 1999-2000, at a cost of $2.5 billion. The project, which involves about 1,000 miles of pipeline and associated gas compression facilities along the entire length of the company's pipeline system, will allow shipment of an additional 2 billion cubic feet per day of natural gas primarily to markets in the northeastern United States and eastern Canada.

March 22

Iraqi Oil Minister Amer Rashid announces the establishment of a new Iraq/Russian oil company which will work independently of Iraq's national oil company, and reports that other agreements would be signed with France and China. Russia and France were Iraq's main arms suppliers before the Gulf War.

March 24

Royal Dutch/Shell reports that local protesters have detained 127 of its staff operating in western Nigeria and taken over six flow stations, disrupting 100,000 barrels per day of crude oil production at six fields. The company plans to maintain oil production at approximately 900,000 barrels per day by increasing production from other fields in the eastern region.

March 24

Presidents Geidar Aliev of Azerbaijan and Leonid Kuchma of Ukraine sign 17 economic, cultural, and military cooperation agreements, including a memorandum on cooperation in the oil and gas industries under which Ukraine will build platforms for a future Caspian Sea drilling project.

March 25

The United Arab Emirates appoints Obeid bin Saif alNasiri oil minister, replacing Rakadh Bin Salem who had served as acting oil minister since May 1995.

March 27

Trinidad and Tobago signs a production sharing agreement with British Gas, Agip, and Deminex for the North Coast Marine Area, a new hydrocarbon province off Trinidad's northern coast. The consortium plans to drill two exploration wells in the next 6 months. The area includes four undeveloped gas fields (Hibiscus, Orchid, Iris and Poinsettia) with potential reserves of more than 3 trillion cubic feet of natural gas.

March 27

Shell reports it is losing 210,000 barrels per day of production due to protests at its oil installations in Nigeria (twice as much as originally estimated) but that exports are not being affected because the company has large crude oil stockpiles in the country.

March 30

The stateowned Dubai Natural Gas Company opens a new methyl tertiary butyl ether (MTBE) plant with a capacity of 500,000 metric tons a year. The company's chairman, Hussain Alsayegh, reports that most of the output from the $250 million facility will be exported to the United States.

April 1997

April 1

A Shell spokesman confirms the company will declare force majeure at its Nigerian Bonny terminal due to local protests which disrupted 210,000 barrels per day of the company's oil production. Although the protests have ended and production is returning to normal, the backlog is temporarily delaying loadings by 3 days.

April 1

Enron begins producing natural gas from the Tapti field offshore India. Production is expected to reach 150 million cubic feet per day by the end of 1997 and continue for 25 years.

April 2

The U.S. Department of Energy announces plans to offer two types of interest in the sale of the government portion of the giant Elk Hills oil and gas field near Bakersfield, California: a single "operating working interest" (representing about three quarters of the government's interest) and separate "non-operating working interests" (each representing 2 per cent of the government's interests). The sale, scheduled to be completed by February 10, 1998, is one of the largest divestitures of federal property in the nation's history. Current ownership is divided between the U.S. government and Chevron.

April 2

The Potential Gas Committee of the American Gas Association estimates that 1996 U.S. natural gas reserves had risen 6 per cent from 1994–to 1,039 trillion cubic feet (representing about 55 times the current yearly U.S. production levels). The estimate includes: 165.1 trillion cubic feet of proven gas reserves as estimated by the U.S. Department of Energy; 190.9 trillion cubic feet of probable gas reserves from current fields; 240.7 trillion cubic feet of possible gas reserves from new fields; 309.8 trillion cubic feet of speculative gas reserves from frontier fields; 13.4 trillion cubic feet of probable coalbedmethane reserves; 36.3 trillion cubic feet of possible coalbed-methane reserves; and 82.6 trillion cubic feet of speculative coalbed-methane reserves. The committee also presents a higher estimate (1,232 trillion cubic feet) calculated using the statistically aggregated mean values for reserves rather that its traditional methodology.

April 2

EssoProduction Malaysia announces the start-up of production and sales from the Lawit gas field (150 miles offshore Malaysia), whose production is expected to reach about 450 million cubic feet per day upon completion of development drilling later this year. The company plans to install compression facilities in 1998 to achieve design capacity of 700 million cubic feet a day.

April 3

Colombia Gas System announces plans to build a $600 million natural gas pipeline to carry Canadian natural gas to U.S. markets in the northeast. The 380-mile Millennium Pipeline would carry at least 650 million cubic feet per day beginning in early 1999.

April 3

TransCanada PipeLines and Northern States Power Co. announce plans to build a $1 billion pipeline to pump natural gas from Alberta, Canada to the U.S. Midwest. The 800-mile Viking Voyageur Gas Transmission project would carry up to 1.2 billion cubic feet per day of natural gas through Minnesota, Wisconsin, and Illinois.

April 7

Express Pipeline begins shipments of Canadian crude oil from Hardisty, Alberta to the Platte Pipe Line in Casper, Wyoming. The 785-mile pipeline gives Canadian producers access to U.S. markets in the Rocky Mountain and Midwest regions. Initial design capacity is 172,000 barrels per day.

April 8

Nigeria's military government approves the first gasoline imports since September 1996, in response to a 2-week-old fuels shortage. A dusk-to-dawn curfew in response to communal troubles near the country's Warri refinery is impeding fuel deliveries.

April 8

Turkmenistan President Saparmurat Niyazov names Deputy Prime Minister Batyr Sardjaev as the new oil and gas minister.

April 9

Romanian Prime Minister Victor Ciorbea announces plans to close 10 large state companies in the next few weeks, including two oil refineries–Petromidia Navodari in the Black Sea port of Constanza and the Darmanesti Rafinaria in Bacau. The closures are expected to reduce 1997 refining capacity by about 47 per cent, to about 363,000 barrels per day.

April 11

China and Vietnam conclude three days of expert-level talks with no resolution over rights to potentially oilrich waters claimed by both countries in the South China Sea. Earlier in the week, a Chinese oil rig left an offshore area between Vietnam's north coast and China's Hainan Island after completing one month of exploration and drilling.

April 14

Australia's minister for resources and energy, Warwick Parer, announces the release of 34 new offshore oil and gas exploration areas–including 11 areas adjacent to the Zone of Cooperation between Australia and Indonesia in the Timor Sea and 6 areas in the Browse Basin offshore Western Australia. The minister notes that this action "is in keeping with the government's strategy to encourage more intensive exploration in Australia's vast continental shelf."

April 14

An Iraqi Oil Ministry official reports that Iraq expects to earn more than $80 billion from its contract with Russia for the development of the West Qurna oil field in southern Iraq. The contract, which was initialed in March and approved by Iraq's National Assembly on April 13, calls for 560 wells which will produce 4.4 billion barrels over 23 years. According to the official, the part of the field being developed with Russia has 11.5 billion barrels in reserves and the entire West Qurna field has reserves of 38 billion barrels. The official states that production will begin "soon" (initially about 250,000 barrels per day, increasing to 600,000 barrels per day).

April 17

Royal Dutch/Shell Group estimates Nigeria's proven oil reserves rose to 24 billion barrels at the end of 1996 (8.3 per cent higher than the 22 billion barrels estimated in 1995) and reports that Nigerian production could reach 3 million barrels per day by 2002 (and 4 million barrels per day by 2010) with an annual investment of $4$6 billion.

April 18

An annual ranking of the world's top oil companies by the Price Waterhouse World Energy Group and Petroleum Intelligence Weekly indicates stateowned oil companies continue to dominate the upstream sector of the world oil industry and are beginning to catch up to the major publicly traded companies in the downstream sector. Based on 1995 data, state-owned companies accounted for 17 of the top 20 companies in terms of oil reserves, 8 of the top 10 companies in terms of oil production, and 4 of the top 10 companies in terms of refining capacity. However, 4 stateowned companies (Petroleos de Venezuela S.A., Saudi Aramco, Indonesia's Pertamina, and Kuwait's KPC) increased their refining capacity by more than 1 million barrels per day in the 1993–1995 period, accounting for about half of the net global increase in refining capacity during that time.

April 19

A senior Iranian oil industry official reports that Iran and Russia signed three agreements on cooperation in the oil and gas industries: 1) an agreement whereby Russia's Gazprom would invest directly in joint operations in the development of Iranian gas fields, production capacity, refining, liquidation, and transport of natural gas and associated extraction industries in the Persian Gulf and other areas; 2) an agreement between a newly formed Russian consortium and the National Iranian Oil company (NIOC) for drilling in Iran's continental shelf; and 3) an agreement between NIOC and the oil company of Tatarstan for cooperation in preextraction and postextraction industries.

April 21

Occidental Petroleum reports that Colombia's main oil pipeline (Cano Limon-Covenas) was bombed 21 times in the first quarter of 1997 (compared with a total of 45 attacks in all of 1996) and that the leftist National Liberation Army (ELN) is growing increasingly active in the region near its operations. Occidental operates the Cano Limon field in partnership with Ecopetrol (Colombia's national oil company) and Shell. As a result of the repeated attacks, Occidental plans to spend $20 million on security in 1997, including a cooperation contract with the Colombian armed forces.

April 22

U.S. Secretary of State Madeline Albright announces sanctions to bar new U.S. investment in Burma (also known as Myanmar) because of the continued oppression of opposition activists by the country's military rulers. About 20 U.S. companies have direct investment or employees in Burma, and the eight largest investors are oil or mining companies: Atlantic Richfield, Baker Hughes, East Asia Gold, Halliburton, McDermott International, Newmont Mining, Texaco Inc., and Unocal. U.S. companies have a total of $240 million in investment in Burma. Unocal, the single largest investor, is a partner in a $1.2 billion joint venture led by Total of France to develop several offshore blocks with significant natural gas reserves.

April 23

Pakistan's government approves a plan to privatize 13 stateowned companies (including its electrical and natural gas holdings) over the next 6 months.

April 24

Russian President Boris Yeltsin names reformist First Deputy Prime Minister Boris Nemtsov to the additional post of minister of fuel and energy. Nemtsov, who has called for radical liberalization of the energy sector, replaces Pyotr Rodionov, who resigned in March.

April 24

Norwegian and British authorities agree on a natural gas treaty covering the Frigg pipeline running from the North Sea to St. Fergus in northern Scotland, reopening the way for exports of Norwegian gas to the United Kingdom and Ireland. Disagreement between the two governments on the use of the pipeline had previously stopped Norway from making new gas sales. The Frigg pipeline currently transports 230 billion cubic feet of gas per year, but capacity can be boosted to 424 billion cubic feet per year. In addition, the two governments reached a framework agreement on the use of connector pipelines to facilitate the use of new oil and gas pipelines which tie installations together on both sides of the British and Norwegian gas sectors.

April 25

Shell Nigeria declares force majeure on loadings of Forcados crude oil, effective April 29-May 20, due to the loss of about 90,000 barrels per day from the usual daily production of 450,000-460,000 barrels per day of Forcados crude oil production. The disruption is the result of an escalation of fighting between local communities in the Warri region of Nigeria.

April 25

Russian President Boris Yeltsin signs a decree in support of the Caspian Pipeline Consortium (CPC). The decree gives CPC a 16-year special exemption from rules mandating it must convert foreignexchange revenues into rubles as part of a package of support for the venture–on condition that the company reinvest its revenues in the project. The CPC is a $2 billion joint project of the governments of Russia, Kazakstan, and Oman and a group of private oil companies set up to build a 932-mile pipeline (planned capacity is 1.5 million barrels per day) from Kazak oilfields on the eastern shore of the Caspian Sea across southern Russia to an export terminal near the Black Sea port of Novorossisk.

April 28

The Middle East Economic Survey reports that Iraq's State Oil Marketing Organization (SOMO) is continuing to hold negotiations with international oil companies, despite the fact that the 50 sales contracts approved so far have reached the $2 billion target for the first 180 days under the oilforfood agreement with the United Nations. According to United Nations statistics, Iraq has sold 121.164 million barrels (averaging 673,000 barrels per day for the 6month period ending June 7) worth a projected $2.142 billion (including $142 million in pipeline fees for Turkey).

April 28

Russian President Boris Yeltsin signs a decree ordering sweeping reforms in the country's natural gas, electric, rail and telecommunications sectors aimed at reducing rates and stimulating investment. At a news conference announcing the decree, First Deputy Prime Minister Boris Nemtsov explains that the state will retain control over these key natural monopolies and will not break up their national supply networks. With respect to natural gas, the government will retain its 40-per cent equity stake in the monopoly RAO Gazprom indefinitely and work to strengthen Gazprom's international position and boost its share price. Gazprom will lose its monopoly right to develop new gas deposits (which instead will be allocated at tenders open to competitors), will offer equal access and competitive rates on its national pipeline network to all producers (giving oil and other companies the opportunity to compete in the gas business), and will be required to ensure transparency of its finances with detailed annual financial reports.

April 28

Turkey signs a $12 billion agreement to boost its imports of Russian natural gas (from the current level of 212 billion cubic feet) by 18 billion cubic feet in 1997, 71 billion cubic feet in 1998, 106 billion cubic feet in 1999, and 159 billion cubic feet in 2000. By the year 2001, Turkey's total gas purchases from Russia will reach 494 billion cubic feet per year. The gas will be shipped through the existing pipeline, which will be expanded by two new joint ventures which will invest $1.5 billion to add compressor stations and new stretches of pipe to double the capacity.

April 29

The U.S. Department of Energy begins soliciting public comment regarding the future of the Strategic Petroleum Reserve (SPR), which will be incorporated into a Statement of Administration Policy on the SPR to be issued in September 1997.

April 29

In a report to Congress, the U.S. State Department estimates that the Caspian Sea basin contains proven recoverable oil reserves of 15.6 billion barrels and possible reserves of an additional 163 billion barrels (compared with more generally used estimates of 27.5 billion barrels of proven oil reserves and an additional 40-60 billion barrels of estimated reserves). According to the report, Kazakstan has the largest reserves in the region (10 billion barrels proven, 85 billion barrels possible), followed by Turkmenistan (1.5 billion barrels proven, 32 billion barrels possible), Azerbaijan (3.6 billion barrels proven, 27 billion barrels possible), Iran (no proven reserves, 12 billion barrels possible) and Russia (no proven reserves, 5 billion barrels possible).

April 30

For the fourth year in a row, three Organization of Petroleum Exporting Countries (Iran, Iraq, and Libya) are named on the U.S. State Department's list of the top sponsors of international terrorism (other countries on the list are Cuba, North Korea, Sudan and Syria). Appearance on the list automatically triggers U.S. government sanctions, including: a ban on military sales and aid; controls over export of dualuse items that can be used for either military or civilian purposes; prohibitions on economic aid, and miscellaneous financial and other sanctions.

April 30

Shell Nigeria reports that four out of the five flow stations that had shut in 80,000 barrels per day of Forcados crude oil production in the Western Delta area of Nigeria since April 25 have been reopened. The Warri River flow station is the only one that is still shutin.

May 1997

May 1

In its regular 60day review, the United Nations Security Council votes to maintain sanctions on Iraq. This is the 37th review since sanctions were first imposed in 1990. This vote, however, does not affect the humanitarian oil sales.

May 2

Shell Nigeria announces the release of 15 contractor staff held hostage for 12 days by a local community in Nigeria's Western Delta and the reopening of flow stations that had been disrupted by unrelated civil unrest in the area since April 25. The company expects output of Forcados crude oil, which initially was disrupted by 80,000 barrels per day, to return to normal within a few days.

May 2

South Korea Development Corporation buys a 15-per cent stake in the North Sea's Captain field from Texaco Inc. The purchase will provide a steady supply of high-sulfur crude oil that will be refined into fuel oil for Korean power generation and will enable the country to develop a strategic fuel reserve.

May 2

British Petroleum announces a new oil discovery, the Liberty field, off the north coast of Alaska. Recoverable reserves are estimated at 120 million barrels.

May 4

Chinese deputy Vice Premier for trade Li Lanqing and Iranian vicepresident Mohammad Hashemi sign a new trade protocol under which exports of Iranian crude oil to China would be increased from the current level of 70,000 barrels per day to 100,000 barrels per day (200,000 barrels per day by 1999). China also accepts an offer from Iran to equip a Chinese oil refinery to handle high-sulfur Iranian crude oil.

May 5

The New York Mercantile Exchange (NYMEX) adds Colombia's Cusiana crude oil as a deliverable for its light, sweet crude oil futures contract, effective with the July 1997 contract.

May 6

The Energy Information Administration releases the International Energy Outlook 1997 which forecasts worldwide demand for crude oil and natural gas will grow steadily through 2015. Crude oil demand is projected to average 77.8 million barrels per day by 2000 and 104.6 million barrels per day by 2015. Natural gas demand is projected to reach 98.1 trillion cubic feet by 2000 and 144.5 trillion cubic feet by 2015. This represents an average annual demand increase of 2.1 per cent for crude oil and 3.1 per cent for natural gas between 1995 and 2015.

May 6

The director of exploration and production for Petrobras, Brazil's stateowned oil company, reports that the company plans to spend $3 billion per year to increase its production to 1.6 million barrels per day by 2001 (from the current level of 900,000 barrels per day and a projected level of 1.2 million barrels per day in 1998).

May 7

Mexico's energy regulatory commission (CRE) awards the concession to distribute natural gas in the northern state of Sonora to a U.S.Mexico consortium, including U.S. company KN Energy Inc.

May 8

Elf Petroleum Nigeria begins construction of its Obite Gas Project, which will supply 253 million cubic feet per day of natural gas to the Nigeria liquefied natural gas plant being constructed at Bonny. The project, which is scheduled for completion in 1999, also involves the construction of an 87-mile pipeline to Bonny and production of about 18,000 barrels per day of condensates.

May 9

Colombia's national oil company, Ecopetrol, reports that guerrillas have attacked the country's largest oil pipeline (Cano LimonCovenas) 24 times so far in 1997, spilling a total of 78,000 barrels of crude oil and reducing production by 500,000 barrels. Colombian armed forces blame the National Liberation Army for 460 separate attacks since the pipeline started operating in 1986.

May 10

General Binford Peay, head of the U.S. Central Command, warns that U.S. forces will respond if Iran tries to make good on a threat to close the Persian Gulf's Strait of Hormuz if it felt threatened by the United States. Iran issued the threat earlier in May, following reports that U.S. officials were considering missile strikes against Iran if it was found responsible for a truck bombing that killed 19 American servicemen in Saudi Arabia in 1996. In addition, Iran held military exercises in the Gulf to demonstrate its naval capability. The United States has 18,550 Air Force, Army, Navy, and Marine personnel on 14 ships or at installations in Saudi Arabia, Kuwait, Qatar, and Bahrain. The Strait of Hormuz is the gateway to about onefifth of the world's oil supply.

May 10

Qatar Liquefied Gas Company (Qatargas) signs a contract with Spain's Enagas for the sale of 420,000 tons of liquefied natural gas (LNG)–32,300 tons per month for delivery over a 13-month period, beginning September 1997. This will be the company's first LNG sale to the European market. Current capacity is 6 million tons of LNG per year, with most of the output committed to Japan's Chubu Electric Power Company under a 25year contract.

May 10

Turkey and Iraq sign a preliminary agreement to build an 807-mile pipeline to carry Iraqi natural gas to Turkey's Mediterranean port of Ceyhan. The project must secure $2.5 billion in financing prior to implementation.

May 10

Libya's Oil Minister reports that his country has lost about $3 billion in oil revenues since the United Nations imposed sanctions in 1992. The sanctions (which limit diplomatic contacts, ban arms sales, and prohibit air traffic in and out of Libya) are intended to force the surrender of two Libyan suspects wanted in connection with the 1988 bombing of Pan Am Flight 103 over Lockerbie, Scotland, in which 270 people were killed.

May 13

Russian President Boris Yeltsin signs a decree substantially increasing the state's role in managing natural gas giant RAO Gazprom. The order names First Deputy Prime Minister Boris Nemtsov to head a special commission charged with setting government policy at the 40-per cent stateowned company (Russia's largest) and calls on the government to draft new regulations on foreign investment in Gazprom shares.

May 14

Yemen's Oil Minister, Mohammed alAttar, reports that his country's oil production capacity is expected to increase to 500,000-550,000 barrels per day by 2000.

May 14

The leaders of Iran, Turkey, and Turkmenistan sign an agreement under which 1.059 trillion cubic feet per year of Turkmen natural gas would be exported to Turkey, via Iran, for shipment to European markets.

May 15

Turkish Energy Minister Recai Kutan reports that Turkey's state pipeline company, Botas, has started construction work for the first phase of a pipeline that will transport Iranian gas into Turkey. He adds that Iran has also begun building its part of the pipeline under the $20 billion, 22year gas project announced by the two countries in August 1996.

May 15

The managing director of Shell Exploration and Production Namibia reports that the company has applied to Namibia's Ministry of Mines and Energy to have the Kudu gas field declared an official petroleum field. This marks the first step in developing the field, located off the country's southern Atlantic coast, which is believed to contain sufficient natural gas to power a planned 750-megawatt power plant in Namibia for at least 20 years.

May 15

The Canadian Gas Potential Committee estimates that Canada has 570 trillion cubic feet of discovered and undiscovered natural gas in conventional and unconventional reservoirs (including coalbed methane, tight gas reservoirs, and shale gas). The reserves represent approximately 50 years of supply from conventional sources and another 50 years from unconventional sources. Most of the natural gas is located in the Western Canada Sedimentary Basin.

May 15

Tosco Corporation announces the reopening of its Marcus Hook refinery in Trainer, Pennsylvania, after a $100 million refurbishment program. Tosco bought the refinery in 1995 from British Petroleum, and shut it down in January 1996 after failing to reach an agreement with union workers over a new labor contract. In 1996, Tosco said it was cutting the refinery's crude processing capacity to 150,000 barrels per day from the previous level of 190,000 barrels per day.

May 16

A final agreement creating the Caspian Pipeline Consortium (CPC) is signed by project participants: Russia (24 per cent), Kazakstan (19 per cent), Chevron Corp. (15 per cent), AO Lukoil/Arco Corp. (12.5 per cent), Mobil Corp. (7.5 per cent), AO Rosneft/Shell Corp. (7.5 per cent), Oman (7 per cent), Agip SpA (2 per cent), British Gas PLC (2 per cent), Oryx Corp. (1.75 per cent), and Kazakstan Pipeline Ventures, a joint venture of Kazakstan's state oil company and Amoco Corp. (1.75 per cent). The Russian government plans to transfer its stake to two Russian oil companies, AO Lukoil and AO Rosneft. CPC plans to begin building a 932-mile pipeline to transport crude oil from the Caspian region to Russia's Black Sea coast in 1998 and begin shipping around 558,000 barrels per day of oil in 1999 (planned peak capacity is 1.4 million barrels per day).

May 16

USX Corp. and Ashland Inc. announce plans to combine most of their downstream oil businesses, creating one of the largest U.S. refiners and continuing an industry trend to consolidate gas station operations. This would be accomplished by a joint venture (for which Ashland Petroleum Co. has signed a letter of intent with Marathon Oil Co., a unit of USX) involving 5,400 gas stations and 6 per cent of U.S. refining capacity.

May 20

President Clinton signs an executive order barring new U.S. investment in Burma (also known as Myanmar), effective May 21 and renewable annually. U.S. companies have invested about $250 million in Burma, primarily in the oil and gas sector. The biggest U.S. investor is Unocal, which is building (with France's Total) a $1.2 billion pipeline from Burma's Yadana natural gas field to an electric power plant in Thailand.

May 21

The U.S. Department of Energy authorizes its financial advisors to begin contacting prospective bidders for the Elk Hills Naval Petroleum Reserve, one of the largest producing fields in the lower 48 states. Bids must be received by October 1, 1997, and the sale is expected to close by February 10, 1998.

May 22

A Mobil Oil executive reports that his company has relocated three rigs outside Nigeria, and May relocate a fourth, because of the Nigerian government's failure to maintain payments for its share of joint venture production costs. Most of Nigeria's oil production (currently 2.2 million barrels per day) comes from joint ventures with international oil groups led by Shell, Mobil, and Chevron.

May 23

The International Energy Agency (IEA) issues a communique reaffirming that oil security remains a serious concern, particularly given the prospect of increasing import dependence and the increasing concentration of remaining oil reserves in the Middle East. The IEA states that spare crude oil production capacity is now lower than before the 1990 Gulf crisis, and the potential for further fuel switching has diminished. The communique was issued following a meeting of the IEA's Board of Governors, at which Hungary was welcomed as the organization's 24th member.

May 24

Mohammad Khatami, a moderate cleric, is declared the winner in Iran's presidential election. He will succeed President Hashemi Rafsanjani, who steps down in August after two, fouryear terms.

May 27

Former Iraqi Oil Minister Issam AlChalabi estimates Iraq needs $5 billion of outside investment and two to three years for its oil industry to restore production to the level prior to the imposition of United Nations sanctions (3.8-4.2 million barrels per day). He also indicates that it would take 5 years and $30-50 billion to achieve production capacity of 5.5 million barrels per day.

May 28

The United Kingdom is reported to have added at least 161,000 barrels per day of new offshore oil production in 1997. New oil field startups could add up to 190,000 barrels per day more by the end of the third quarter of 1997, and another 84,000 barrels per day at the beginning of the fourth quarter.

May 28

West Africa magazine reports that a widespread, complicated series of interethnic claims and conflicts stretching across state and local government boundaries in Nigeria's Niger Delta leaves many onshore and offshore oil and gas projects vulnerable. This area produces about 37 per cent of Nigeria's onshore oil production.

May 28

Russian President Boris Yeltsin signs a decree designed to limit foreign investment in natural gas giant RAO Gazprom.

May 28

Royal Dutch/Shell Group (RD) reports that the Cornea oil structure, located in the Browse Basin offshore northwestern Australia, could contain up to 2.67 billion barrels of oil.

May 30

Iraq's Oil Minister, Amer Mohammed Rasheed, reports that the full allotment of $2.14 billion in Iraqi oil (including $140 million for maintaining oil pipelines) has been exported under the United Nations' 6-month oilforfood program. Oil sales totaled 120 million barrels under 51 contracts with international companies. The United Nations is expected to renew the agreement, which began in December 1996 and expires in early June.

June 1997

June 1

Oman and Iran agree to draw up a joint development plan for the Bukha/Hengam oil field which is jointly owned by the two countries.

June

Shell Oil Company and Mobil Corporation combine their California exploration and production units into a joint venture company, Aera Energy LLC (58.6 per cent Shell, 41.4 per cent Mobil). With proven reserves of about 1 billion barrels of oil equivalent and production of about 250,000 barrels per day, the venture is now the largest oil producer in California.

June 2

Amoco Corporation announces plans to sell about 15 per cent of its domestic oil and natural gas reserves to allow the company to focus future efforts on six regions: Trinidad (where it is expanding a liquefied natural gas plant); Venezuela; the Eastern Mediterranean (especially Egypt's Nile Delta); the Caspian Sea; heavy oil projects in Alberta, Canada; and deep-water projects in the Gulf of Mexico. The reserves–located in Wyoming, Colorado, New Mexico, Oklahoma, and along the coast of the Gulf of Mexico–total about 450 million barrels of oil equivalent (about two-thirds natural gas) and provide about 10 per cent of the company's domestic production.

June 3

Japan's Minister of International Trade and Industry says Japan needs to diversify the sources of its oil supplies further to improve its energy security.

June 4

In a unanimous vote, the United Nations Security Council renews for another 180-day period its "oil for food" initiative with Iraq. Under the resolution, Iraq May sell $2 billion worth of oil to buy food, medicine and other necessities to alleviate civilian suffering under the sanctions imposed when it invaded Kuwait in 1990.

June 4

China signs separate agreements with Kazakstan and Iraq to secure oil supplies into the next century. The agreements cover a $4.3 billion investment in Kazakstan's Aktyubinsk oil enterprise over the next 20 years, a planned oil pipeline from Kazakstan to China, and the development of the Ahdab oil field in southern Iraq (1 billion barrels at a cost of $1.2 billion).

June 5

Russia's Deputy Prime Minister Alfred Kokh announces plans to sell an additional 15 per cent share of state-owned oil producer AO Lukoil.

June 6

Australian Prime Minister John Howard announces that his country will phase out the use of lead in automobile gasoline by 2010.

June 6

Venezuela completes a 5-day auction of 20 marginal oil fields in its third round of sales to foreign investors, earning about $2 billion (twice as much as expected). Successful bidders receive a 20-year production contract with certain minimum investment levels and must pay set royalties and fees for each barrel of oil produced. They include U.S. companies Atlantic Richfield, Chevron, Pennzoil, and Phillips Petroleum; British company Lasmo; and the China National Petroleum Company.

June 10

Chevron Corporation signs a production-sharing agreement to perform seismic tests and drill wells deep beneath China's second largest oil field (Shengli). This is the first onshore exploration pact in China for Chevron, which already has offshore leases in China's Bohai Gulf and the South China Sea.

June 10

The National Iranian Oil Company agrees to spend $250 million to upgrade a large refinery in southern China to increase its capacity for processing Iranian crude oil.

June 10

Kazakstan and Azerbaijan agree to build a 1600-mile pipeline along the floor of the Caspian Sea to ship 315-350 million barrels per year of Kazak oil via Turkmenistan to Baku in Azerbaijan. The pipeline would continue overland across Georgia and Turkey to the Mediterranean Sea. Construction is slated to begin in 2000 and be completed within 3 years.

June 11

Canadian Prime Minister Jean Chretien names Ralph Goodale the country's new natural resources minister.

June 13

Triton Energy reports that it expects combined production from Colombia's Cupiagua and Cusiana fields to reach 500,000 barrels per day in early 1998 (currently 185,000 barrels per day). The company had planned to achieve this goal by the end of 1997, but has recently encountered problems in construction and drilling operations in the Cupiagua field.

June 18

Turkey and Ukraine sign a deal to build a pipeline from Turkey's Mediterranean coast to its Black Sea coast for oil shipments to Ukraine.

June 19

The U.S. Supreme Court rules in favor of the Federal Government in an 18-year-old boundary dispute with Alaska over control of offshore areas along the Arctic coast. The decision effectively prevents development of the area's oil and natural gas reserves, as the Federal Government has said it would not permit drilling off the coast of wildlife refuges.

June 19

Exxon appeals the $5.3 billion verdict awarded in 1994 to fishermen, native Alaskans, and others harmed by the 1989 Exxon Valdez oil spill.

June 20

Talisman Energy receives permission to develop the Ross oil field in the United Kingdom sector of the North Sea. The field, which contains estimated reserves of 60-100 million barrels of oil and 20-30 billion cubic feet of natural gas, is expected to begin producing in September 1998 and reach peak production of 40,000 barrels per day in 1999.

June 20

Russian President Boris Yeltsin orders cuts of up to 40 per cent in natural gas prices for businesses that pay their overdue Gazprom bills in cash by the end of the year. High customer debt (estimated at $12.15 billion) makes it difficult for the company to pay its national tax arrears.

June 24

Russia's State Duma (lower house) approves a long-awaited law that would allow production-sharing agreements for development of major natural resource deposits, including five oil fields (Samotlor, Krasnoleninsk, Romashkinskoye, Prirazlomnoye, North Sakhalin), the Kuranakhskoye gold field, and the Yakovlevskoye iron-ore deposit. Investment in these projects is expected to total $16 billion. The legislation now goes to the upper house for consideration, which is expected to consider the measure this fall.

June 25

Partners in Australia's Stuart Oil Shale project announce that they have obtained financing to proceed with the first phase of development for what could become the world's first commercial project extracting crude oil from oil shale. If successful, the project would produce 85,000 barrels per day within 10 years. Partners are Canada's Suncor (50 per cent) and two Australian companies (Southern Pacific and Central Pacific, 25 per cent each).

June 26

The Organization of Petroleum Exporting Countries (OPEC) extends its current output ceiling of 25.033 million barrels per day of crude oil through the end of 1997. The official communique indicates quota adherence will be "closely monitored" and announces that the next meeting would take place November 26 in Jakarta, Indonesia. OPEC Secretary General Rilwanu Lukman tells a news conference after the meeting that OPEC's target price remains at $21.00 per barrel. Individual quotas remain as follows (in millions of barrels per day): Saudi Arabia, 8.0; Iran, 3.6; Iraq, 1.2; Venezuela, 2.359; Nigeria, 1.865; Indonesia, 1.33; Kuwait, 2.0; Libya, 1.39; United Arab Emirates, 2.161; Algeria, 0.75; and Qatar, 0.378.

June 26

Iranian Oil Minister Gholamreza Aghazadeh reports that U.S. sanctions to deter Iran's ability to attract outside capital "haven't worked at all" because creative financing methods allow foreign companies to avoid punishment under the Iran and Libya Sanctions Act (which was passed in 1996 and applies to investments of $40 million or more). He cites the use of buyback contracts under which companies pay for development rights in oil rather than in dollars, and direct financing by foreign banks (which are not covered by the sanctions law).

June 26

The government of Portugal confirms the sale of a 27.5 per cent stake in petroleum company Petrogal to Saudi Aramco. Under an agreement scheduled to be finalized by the end of 1997, Saudi Aramco would supply Petrogal's sour crude requirements under a long-term contract. Petrogal operates two refineries (with capacity totaling 300,000 barrels per day) and supplies 50 per cent of the country's petroleum products. Portugal's government retains a 45 per cent stake (down from 55 per cent) and Petrocontrol, a private consortium of shareholders, retains a 27.5 per cent stake (down from 45 per cent). A leading Portuguese newspaper estimates the reference value of the Saudi share at $1.4-1.5 billion.

June 27

International Energy Agency head Robert Priddle says he "regrets" the previous day's action by the U.S. House Appropriations Committee approving a $209 million sale of crude oil from the U.S. Strategic Petroleum Reserve. Priddle notes that U.S. strategic stocks have already dropped from a 90-day supply to less than 70 days currently, following earlier sales of 30 million barrels of crude oil.

June 27

China's Premier Li Peng and Russia's Prime Minister Viktor Chernomyrdin sign a series of agreements to boost trade and economic relations between their countries. Included is a planned $4-5 billion project involving the development of Russian natural gas reserves in the Irkutsk region of western Siberia to supply up to 1 trillion cubic feet per year via pipeline to China's coast.

June 30

Iran's top military commander says his country does not intend to start a war with the United States, but promises to turn the Persian Gulf into a slaughterhouse if attacked. The pledge follows a series of U.S. warnings about Iran's potential to use missiles to close the Strait of Hormuz to oil tankers and other shipping.

June 30

Nigeria's government authorizes the Nigerian National Petroleum Company to swap 100,000 barrels per day of crude oil for petroleum product imports, reversing a year-earlier decision banning the practice due to widespread abuses. The action comes one week after the government authorized importing 33 cargoes of oil products to ease fuel shortages associated with problems at domestic refineries.

July 1997

July 1

The Russian government begins offering many of its remaining oil-industry holdings in a series of auctions and investment tenders expected to raise $780 million by mid-December 1997. The privatizations, which are open to foreign bidders, include Russia's stakes in AO Vostsibneftegaz (38 per cent), AO Vostochnaya Neftyanaya (51 per cent), AO Sibur (36.28 per cent), AO Tyumenskaya Neftyanaya (48.68 per cent), AO Komitek (27.1 per cent), and AO Norsi-Oil (45.45 per cent).

July 1

The British colony of Hong Kong reverts to Chinese rule as a Special Administrative Region. Chinese Prime Minister Li Peng declares Beijing's rule of law will protect foreign interests in post-handover Hong Kong. China has promised laissez-faire oversight under the principle of "one country, two systems."

July 5

China National Petroleum Corporation (CNPC) reports plans to intensify exploration in eight basins that contain 80 per cent of the country's onshore oil and gas reserves while continuing its efforts to secure overseas supplies. According to CNPC, the company has verified 25.6 billion barrels of crude oil reserves and 22.6 trillion cubic feet of natural gas reserves in the past 6 years.

July 6

The government of India approves the sale of shares in four state-run companies, including Gas Authority of India Ltd. and Indian Oil Corp.

July 9

Corpoven (a unit of Venezuela's state-owned oil company) signs a formal agreement with three U.S.-based companies (Atlantic Richfield, Phillips Petroleum, and Texaco) creating a $3.5 billion joint venture to extract and upgrade 197,000 barrels per day of extra-heavy crude oil in the Hamaca Zone of Venezuela's Orinoco Belt. Drilling is expected to start in late 1998, with oil production beginning in 1999 at an initial rate of 35,000-40,000 barrels per day. The first of two upgrading facilities, designed to produce a grade similar to Alaskan North Slope crude oil, is scheduled to be completed by 2002.

July 11

The German cabinet approves a supplementary budget that includes proposals to sell part of the government's strategic oil reserves to meet European Monetary Union criteria for a budget deficit of 3 per cent of gross domestic product. Germany plans to sell about $224 million worth of reserves in 1997 (roughly 15 million barrels of oil), and additional volumes in 1998 and 1999.

July 11

A tripartite agreement on the transportation of Caspian Sea crude oil via the Baku-Grozny-Novorossiysk pipeline is signed in Baku by Russian Oil and Energy Minister Boris Nemtsov and representatives of Azerbaijan's state oil company and the Chechen oil company. The pipeline, which will pump oil from Azerbaijan across Chechnya to a Russian export terminal in Novorossiisk, is expected to be operational within one month.

July 15

The president of the Colombian Petroleum Association declares that repeated attacks on Colombia's oil installations have led to a "critical" situation for private oil companies operating in the country. This is in reaction to three attacks in the last 10 days, which forced state-owned oil company Ecopetrol and field operator Occidental Petroleum Corporation to close down production and declare force majeure for all export commitments in the Cano Limon field. The Cano Limon-Covenas pipeline has been attacked 470 times since it began operations in 1986, causing damage in excess of $1.5 billion; however, this is the longest time field operations have been suspended due to the attacks.

July 15

Guatemala awards oil exploration contracts in its northern Peten province to four foreign firms (including U.S.-based Oil Technology Service) and one Guatemalan company. Minister of Energy and Mines Leonel Lopez Rodas expects company investments of $128 million in nine potential fields in 1997 and 1998 will make Guatemala self-sufficient in energy supply. Guatemala previously awarded 23 drilling contracts to private firms.

July 16

The Brazilian Senate approves the creation of a regulatory body (the National Petroleum Agency) required for implementation of a 1995 constitutional amendment allowing private domestic and foreign enterprises to compete with federally controlled oil conglomerate Petroleo Brasileiro SA (Petrobras). The legislation also creates a National Energy Policy Council.

July 16

Tengizchevroil, operator of Kazakstan's Tengiz field, awards a $250 million contract for construction of a new oil and gas processing plant that will increase production capacity by 30 per cent (from 160,000 barrels per day currently to 240,000 barrels per day by the end of 1999).

July 17

Texaco, the government of Saudi Arabia, and the U.S. unit of Royal Dutch/Shell agree to merge their U.S. East Coast and Gulf Coast refining and marketing businesses in a joint venture company pooling four refineries (with total capacity of 823,000 barrels per day) and 14,717 gasoline stations currently owned by Shell and Star Enterprise (a joint venture between Texaco and Saudi Aramco). Under the agreement, Shell would own 35 per cent of the new company, and Texaco and Saudi Arabia would hold 32.5 per cent each.

July 19

Iran opens an oil jetty for petroleum product exports at Nowshahr port on the Caspian Sea. According to Iran's managing director of Ports and Shipping Organization, the country plans to increase its total port capacity to 4 million barrels per day by the year 2022 (currently 800,000 barrels per day).

July 22

The first shipments of oil produced from Kazakstan's Tengiz field arrive at terminals on the Black Sea in Novorossiysk (Russia) and Batumi (Georgia) for subsequent export through the Bosphoros Strait. Volumes total between 100,000 and 150,000 barrels per day.

July 23

A senior official with Royal Dutch/Shell reports that the company plans to spend at least $10 billion in Latin America by the year 2005, primarily in four countries: Peru, Venezuela, Brazil and Argentina. Major projects include Peru's Camisea fields and the Bolivia-Brazil natural gas pipeline.

July 23

The U.S. State Department rules that Turkey's August 1996 agreement to purchase $23 billion worth of natural gas from Iran over a 20-year period does not violate the Iran and Libya Sanctions Act. In a May 1997 memorandum of understanding with Iran and Turkmenistan, Turkey modified the original arrangement so that the natural gas will be purchased from Turkmenistan rather than Iran.

July 23

Officials from Turkmenistan, Pakistan, Unocal, and Delta (a Saudi oil company) sign an agreement to build a natural gas pipeline from Turkmenistan across Afghanistan to Pakistan (871 miles). Under the agreement, a consortium will be formed by October 1997, construction will begin by December 1998, and the project will be completed by 2001 at an estimated cost of $2.0-2.7 billion. The pipeline will carry up to 706 billion cubic feet of natural gas annually from Turkmenistan's largest natural gas fields, at Daulatabad.

July 24

The government of Chile approves CMS Energy Corp.'s $650 million Atacama project, which will transport natural gas via pipeline from northern Argentina across the Andes Mountains to two new natural gas-fired electric generating units at Mejillones in northern Chile. The project is being built in conjunction with Chile's Empresa Nacional de Electricidad SA. Construction of the pipeline will begin in the fourth quarter of 1997 and commercial operations will begin by early 1999.

July 25

Presidents Gonzalo Sanchez de Lozada of Bolivia and Fernando Henrique Cardoso of Brazil sign construction contracts for the longest natural gas pipeline in South America–a 1900-mile pipeline from Rio Grande, Bolivia to three Brazilian cities (Sao Paolo, Curitiba, and Porto Alegre). The 32-inch diameter pipeline, estimated to cost about $2 billion, is scheduled to begin operating in 1999 at an initial rate of 283 million cubic feet per day, increasing to 565 million cubic feet per day after 7 years. Major contractors include Enron, Bechtel and Royal Dutch/Shell.

July 26

Iran's President Hashemi Rafsanjani formally inaugurates the 232,000-barrel-per-day Bandar Abbas refinery–the country's eighth oil refinery. The facility, which will run Iranian Heavy crude oil, has partially started up and is scheduled to be fully operational in March 1998. Iran hopes to become selfsufficient in gasoline, kerosene, jet fuel, and diesel, and begin exporting petroleum products at that time. Domestic petroleum demand averages about 1.1 million barrels per day. The country's refining capacity (excluding Bandar Abbas) totals about 1.2 million barrels per day.

July 28

Egypt's Petroleum Minister Hamdy el-Banbi announces that the private sector will be allowed to launch distribution networks for natural gas as part of the government's plan to substitute natural gas for oil.

July 28

Azerbaijan's President Heydar Aliyev, on his first visit to the United States, repeats his support for Turkey's port of Ceyhan on the Mediterranean Sea to be the main export route for his country's Caspian Sea crude oil.

July 29

State-owned oil company Petroleos de Venezuela (PdVSA) signs agreements turning over operating contracts for 17 marginal crude oil fields to 17 national and international consortiums, the winners in the country's third round of auctions for 20 marginal oil field contracts held in June 1997. The auction raised $2.08 billion. Two fields received no bids and the winning bid on one field (Mata) was later revoked by PdVSA.

July 30

The U.S. Federal Energy Regulatory Commission approves several natural gas pipeline projects for delivery of Canadian natural gas to the United States: 1) the U.S. portion of Alliance Pipeline Ltd.'s new $3.6 billion pipeline from British Columbia and Alberta to the Midwest (conditional upon Canadian regulatory approval); 2) expansion and extension of Northern Border Pipeline from the Canadian border to the Chicago area; and 3) a portion of the Maritimes & Northeast Pipeline from Sable Island, Nova Scotia, to markets in New England.

July 31

Russian state-owned oil company AO Rosneft announces it is pulling out of a $1 billion oil deal (signed July 4) to develop Azerbaijan's Kyapaz field, in the Caspian Sea. The deal has been challenged by neighboring Turkmenistan, which claims the field as its own (under the name Serdar).

August 1997

August 1

Azerbaijan signs production-sharing contracts with Chevron Corporation, Exxon Corporation, Mobil Corporation, and Amoco Corporation. Officials from Azerbaijan value the Caspian Sea projects at more than $10 billion. Mobil will develop the Oguz exploration area, Chevron the Zeynabdin Tagiyev area, Exxon the Nakhchivan area, and Amoco the Inam Prospect.

August 1

United Nations Secretary-General Kofi Annan approves Iraq's request to add the Iraqi-Syrian border as an entry point for food and humanitarian aid under the oil-for-food program. The request provides further evidence of re-established official ties between the two countries. Ties were severed after Syria supported Tehran during the Iran-Iraq war, and this new link between Iraq and Syria follows the reopening in mid-June of three border points that were shut in 1980.

August 1

Mobil Corporation announces a favorable $78.3 million judgement in litigation with the federal government. The lawsuit involved Mobil's offshore North Carolina federal Outer Continental Shelf oil and gas leases. The company said that the judgement recognizes Mobil's right to terminate five Outer Continental Shelf leases, and recover the purchase price. Mobil sued the federal government in 1992 for the government's failure to allow development of certain oil and gas leases offshore Alaska, Florida, and North Carolina. The Alaska and Florida parts of the litigation were settled in 1995.

August 1

Russia annuls a $1 billion oil deal with Azerbaijan in support of Turkmenistan's protest that the deal is illegal. On July 4 Russia's Rosneft and Lukoil signed a deal with Azerbaijan's Socar to develop the Kyapaz field at the edge of Azerbaijan's zone in the Caspian Sea. The Turkmen government protested claiming the field, which Turkmenistan calls Serdar, as its own.

August 2

Shell Netherlands B.V., a unit of Royal Dutch/Shell Group, unexpectedly shuts down part of Europe's second largest oil refinery. A company spokesperson indicated that the unit at Shell's Pernis refinery near Rotterdam could be closed all month, but did not give a reason for the shutdown.

August 4

Mohammad Khatami, a cleric viewed as moderate, is sworn in as Iran's new president.

August 4

In Colombia, Occidental Petroleum, a California-based international oil company, and Ecopetrol, Colombia's national oil company, declare force majeure on all oil exports from the Cano Limon field. The declaration comes after a series of attacks dating back to July 30 knocked out a major oil pipeline transporting oil from the field to the Caribbean port of Covenas. The pipeline has been attacked 45 times this year which is equal to the total number of attacks for 1996. Responsibility for the attacks has not been determined, but leftist guerrillas from the National Liberation Army are usually blamed for such attacks. The force majeure declaration does not apply to the oil contained in the 2 million barrel storage facility at Covenas.

August 4

United Nations (U.N.) Secretary-General Kofi Annan approves a revised Iraqi aid distribution plan under the U.N. oil-for-food program. The oil-for-food program allows Iraq to sell approximately $2 billion of crude oil every six months. Sale periods are divided into two 90-day intervals. The proceeds from the oil sales are paid into a U.N.-run escrow account from which humanitarian supplies are purchased for the Iraqi people. The program went into effect in December 1996 and was renewed for another six months on June 8, 1997. However, since late May 1997, Iraq has refused to sell any oil until the U.N. approved the revised aid distribution plan which is designed to speed up the procedure for approving shipments of humanitarian goods. The next step for Iraq is to gain U.N. approval of an oil pricing formula, and then the U.N. must approve purchase contracts from buyers. Following these two approvals, Iraq can resume its oil exports. Iraq has until September 5, 1997 to sell $1.07 billion of oil under the current 90-day period.

August 5

Mobil reveals that production at Nigeria's 100,000 barrel per day Oso condensate field has been completely shut-in since the previous week to repair faults in the gas reinjection wells. Bolts securing one offshore injection wellhead sheared, and as a precaution, Mobil decided to replace all the bolts securing offshore injection wellheads. Officials expect delays to Oso condensate loadings, but could not specify how many days' cargoes would be delayed.

August 6

Nigeria's government releases $145 million to the state petroleum corporation to begin repair work on its oil refineries. All four of the country's refineries were either closed or had production severely reduced in May due to maintenance problems. These problems have forced Nigeria to import oil to meet domestic needs, despite its position as the world's sixth-largest producer of crude oil. Repairs are expected to begin October 1 and take at least 60 days.

August 7

Romanian Prime Minister Victor Ciorbea announces the closure of three of the country's nine oil refineries. Petrotel, Vega, and Darmanesti will close immediately in order to reduce a surplus of oil refining capacity and cut government spending.

August 7

Exxon Corporation confirms that it has been chosen as sole foreign partner in a proposed multi-billion-dollar project to expand oil refining and build a major petrochemical plant in Fujian province, China. The joint project, involving Exxon and Fujian Petrochemical, includes increasing oil refining capacity at Xiaocuo from 80,000 barrels per day to 240,000 barrels per day and constructing a 600,000-metric-ton-per-year ethylene plant. In addition, the plant will include a 450,000-ton-per-year polyethylene plant and a 300,000-ton-per-year polypropylene plant. Exxon has a 50 per cent share in the joint venture.

August 7

Nova Corporation, a Canadian natural gas services and petrochemicals company, officially opens the GasAndes pipeline. The pipeline will initially transport 140 million cubic feet per day of natural gas from Argentina to Chile. The volume will increase to 250 million cubic feet per day by fall 1998. The new pipeline will supply three new gas-fired power plants in Chile that are set to open in October 1997, spring 1998 and fall 1998. The power plants are part of a program to improve the air quality in the Chilean capital, Santiago, by switching from coal-fired to natural gas-fired electrical generation. Nova holds a 56 per cent interest in the GasAndes project.

August 8

The United Nations approves a sale-price formula for Iraqi crude oil sales under the oil-for-food plan. The approval cleared the way for Iraq to resume limited oil exports immediately through the Turkish port of Ceyhan on the Mediterranean Sea and Iraq's Gulf port of Mina al-Bakr. The United Nations will also begin reviewing contracts for Iraqi crude oil purchases. Iraq has until September 5 to raise the $1.07 billion allowed under the existing 90 day oil-for-food plan window. Iraqi officials state they will boost exports to 2 million barrels per day to meet the sales target. However, industry experts say that Iraq's export capacity is untested beyond 1.4 million barrel per day.

August 8

Shell Nigeria announces that it expects to produce 7 per cent to 8 per cent less oil than had been originally planned for 1997. Officials cite a reduction in hard currency available to its joint venture partners as the reason for the decrease. A cutback in production would reduce output by approximately 70,000 barrels per day to 856,000 barrels per day.

August 11

Colombia's national oil company, Ecopetrol, lifts the force majeure declaration on all oil exports from the Cano Limon field. A company spokesman states that crews completed repairs to the Cano Limon-Covenas pipeline on Saturday August 9 and that normal operations have resumed.

August 12

A spokesman for Saga Petroleum ASA announces that oil production from the Snorre oil field in the Norwegian sector of the North Sea has been restarted and that it should return to full production within a day. A restart date for the Vigdis field, which was also shut down, has not been set. Production at the two fields was stopped on August 11 following the discovery of a production fault. Snorre produces 200,000 barrels per day while Vigdis has been gradually working towards 40,000 barrels per day following repairs to the gas compressor.

August 12

Exxon Corporation shuts down a catalytic cracking unit at its refinery in Baytown, Texas because of an unspecified operating problem. Exxon declined to say when the unit, which produces more than 100,000 barrels per day of gasoline components, is expected to return to service.

August 14

Stefan Balabanoff, acting head of economics and finance in the OPEC research division, predicts that OPEC's share of the world oil market will rise from a current level of about 40 per cent to 52 per cent by 2020. Balabanoff estimates that OPEC's production capacity must rise to 50 million barrels per day to meet the increased demand. From 1995 to 2000 he sees oil demand increasing by an average of 2.1 per cent per year and from 2000 to 2010 demand is forecast to rise by 1.4 per cent per year. Asian and Pacific countries are predicted to lead the increases in demand. In addition, Balabanoff forecasts growth for non-OPEC supply centered on the North Sea, non-OPEC developing countries, and the Commonwealth of Independent States.

August 18

Three international groups confirm that they plan to bid on Petroleos Mexicanos' (Pemex) Cantarell project in the Yucatan peninsula. Pemex, the Mexican state-owned oil monopoly is seeking to boost oil production by injecting nitrogen into the terrain of Campeche state and into the sea floor beneath Campeche Bay in the Gulf of Mexico. Dresser Industries Inc. of Dallas will submit a joint bid with U.S. companies Marathon Oil and Praxair Inc., Nova Gas International of Canada, and Bufete Industrial of Mexico. Shell Oil Company entered a joint bid with Enron Corporation and Air Products and Chemicals Inc., both from the U.S., and Japan's Mitsui and Company. The third group is led by BOC Group PLC of Britain. Industry officials estimate the price of the project at $950 million.

August 18

A spokesman for Saga Petroleum ASA reports that production at the Vigdis oil field in the Norwegian section of the North Sea was restarted despite the absence of a replacement gas seal. A temporary measure of transferring gas from the field to the Snorre field for processing enabled the company to restart Vigdis. The replacement seal is in production and scheduled to be completed by August 25.

August 20

The Iranian parliament, the Majlis, confirms all 22 of President Mohammad Khatami's nominations for his cabinet. The confirmations are seen as a strengthening of Khatami's mandate to move towards a less strict society and a weakening of the hard-liners' power base in parliament. One of the nominees has advocated direct talks with the United States.

August 20

Mexican President Ernesto Zedillo inaugurates the first U.S.-originated natural gas delivery to the Mexican state of Baja California. The new gas distribution system in Mexicali, the state capital just across the United States border, is the first under the Mexican government's natural gas privatization program. A binational consortium comprised of Enova International, Pacific Enterprises International and Proxima Gas, S.A. de C.V. will distribute the gas for the next 30 years. The deal is valued at $20 million over the first three years for the supply more than 10 billion cubic feet of gas.

August 20

The Saudi Arabian Oil Company (Aramco) shut down its 300,000-barrel-per-day oil refinery on the Persian Gulf for maintenance. Aramco expects the shutdown to continue until September 17. Reports suggest that the shutdown will result in the release of extra supplies of Arab Light and Extra Light crude oil for export in August and September.

August 20

YPF SA, an Argentine energy company, and Amoco Corporation sign an agreement to operate U.S. natural gas fields with estimated reserves of 989 billion cubic feet of gas. YPF reveals that the joint venture includes 2,000 productive wells in the Texas Panhandle and western Oklahoma, two processing plants with combined daily capacity of 230 million cubic feet of gas, a 932-mile distribution system and 889,560 acres of leased land for exploration. The companies have set up a 50-50 holding company called Crescendo Management LLC to operate the joint venture.

August 21

The United Nations approves the pricing formula for September crude oil exports under the United Nations/Iraq oil-for-food sale. Shipments of Iraqi Kirkuk crude oil headed for Europe will increase 20 cents per barrel, and shipments of Basrah Light crude oil to the Far East and United States, as well as Kirkuk crude oil to the Unites States, will increase 10 cents per barrel.

August 22

An analyst with Calgary's First Energy Capital Corporation reports that western Canada's natural gas production is declining for the first time in many years and that demand could outstrip supply soon. In addition, the analyst estimates that natural gas reserves in western Canada are declining between 17 per cent and 18 per cent per year. Ontario Hydro's recent decision to close seven of nineteen nuclear power plants could increase the demand for western Canadian natural gas thereby exacerbating the potential shortage.

August 26

Enterprise Oil announces that it has received approval from the British government for a development plan for the Pierce Field in the central North Sea. The company expects the first oil from the field in August 1998 with an initial rate of production of 20,000 barrels per day. Enterprise expects a continuous rate of 45,000 barrels per day shortly thereafter. In addition, Enterprise reveals that it completed a deal with British Petroleum which increases its stake in the Pierce Field to 75 per cent.

August 26

China plans to spend $600 million combining onshore and offshore exploration activities at its biggest offshore oil field, Suizhong 36-1 in Bohai Bay. The Bohai Oil Company will build facilities for oil and gas processing, storage, and marketing as well as a new oil berth. Suizhong 36-1 covers 18.5 square miles and has proven oil reserves of 2.1 billion barrels.

August 26

The leader of Norway's oil workers union, Terje Nustad, said that the union plans to expand a strike already affecting five oil rigs on the Norwegian continental shelf. The expansion will start after September 10 and will include the West Epsilon rig, the Transocean rig, and the Poly Saga rig. Approximately 340 oil workers went on strike August 24 after rejecting a 7.8 per cent raise offered by the Norwegian Shipowners' Association. The strike has not slowed daily oil production because the five affected rigs are mainly used for drilling new wells. However, if the strike is expanded, the annual maintenance stoppage at Statfjord B in the North Sea could be extended. Moreover, the restart of Norne field production, scheduled for September, could be delayed.

August 27

Consulting firm Arthur Anderson reports that world oil companies increased spending on U.S. exploration and development to $21.6 billion in 1996, an increase of 24 per cent over 1995. Major oil companies spent $11.1 billion in the U.S., a 13 per cent increase, while independent companies spent $10.5 billion, a 37 per cent increase. Royal Dutch/Shell Group, Texaco Inc., and Amoco Corporation led the major companies in 1996 U.S. exploration expenditures. Independent spending leaders were Union Pacific Resources Group Inc., Burlington Resources Inc., and Enron Oil & Gas Company.

August 27

The United Nations Security Council announces that it will discuss a possible rollover of Iraqi oil sales from the current 90-day period to the second 90-day period under the renewed oil-for-food program. The existing six month period of the oil-for-food program began on June 8, 1997, but Iraq stalled on the oil sale until mid-August 1997. The delay makes it impossible for Iraq to sell $1.07 billion of oil during the first 90-day period which is set to end on September 5, 1997. If the rollover gains approval, it could prevent Iraq from losing an estimated $400 million in sales because of the late start.

August 28

Russian officials annul a high-profile deal with Exxon Corporation to develop huge oil deposits in Russia's far north. Late last year Exxon's affiliate, Exxon Arkhangelsk, beat out several international consortia for a 50 per cent stake in oil fields in the Timan-Pechora province. Russian officials cite five undisclosed violations committed by Exxon during the tender process as the reason for scrapping the agreement. Some analysts suggest that the cancellation of this $1.5 billion investment project could shake investor confidence in Russia's energy sector and May deter private investment in Russia.

August 29

A spokesman for Enterprise Oil confirms that production at Enterprise's North Sea Nelson oil field is to be halted at midnight while an unexploded World War II bomb is removed from close to the Nelson pipeline. The work May take as many as five days to complete; however, Enterprise hopes to restart production on September 1 Normal production from the field is 160,000 barrels per day.

August 29

Russia and Turkey sign an agreement to build an underwater pipeline across the Black Sea to transport Russian natural gas to Turkey. The proposed 225-mile pipeline would run from the Russian port city of Tuapse to the Turkish port city of Samsun, and initially carry 280 billion cubic feet of gas.

The amount will reportedly double by 2003. Construction of the pipeline is scheduled to start next year and is expected to be complete by 2000.

September 1997

September 1

South Korea's state-run Daehan Oil Pipeline Corporation (DOPCO) dedicates a 593-mile oil pipeline and a large oil storage terminal in Songnam, 20 miles south of Seoul. DOPCO has invested about $890 million over seven years to complete the two projects. The storage terminal is capable of storing 1.97 million barrels of petroleum products and loading 186,000 barrels of oil per day. The pipeline will transport oil products from five domestic refineries including Yukong Ltd. and Hyundai Oil Refining Company.

September 1

Saga Petroleum, a Norwegian upstream oil company, reports that production stoppages at the North Sea Vigdis Field, caused by gas compressor problems, have reduced this year's output target from 25 million barrels to 16 million barrels.

September 1

Shell Nigeria resumes operations at Nembe Creek flow stations following repairs of a pipeline leak. Nembe Creek flow stations are a major trunk line feeding Nigeria's Bonny terminal. Shell officials suspect that the line had been sabotaged to force Shell to pay compensation to the local communities for other pipeline leaks. Shell confirms that negotiations are under way with the local communities.

September 3

Japanese and Russian governors sign an oil and gas development agreement at a meeting on economic cooperation between three Russian Far East regions and Japan's northernmost prefecture. The agreement targets development of the Sakhalin continental shelf and encourages private companies to participate in offshore oil and gas drilling projects. Commercial production is expected as early as 1999.

September 3

After months of negotiations, Chechnya and Russia reach a compromise on transporting Caspian Sea oil across Chechnya. The last issue holding up the agreement was disagreement over tariff rates for the use of the existing Chechen pipeline.

September 3

Workers from Colombia's oil, banking, and judicial sectors launch a 24-hour strike to protest government plans to privatize state-held assets. The oil workers are also protesting the detention of 17 oil workers, on charges of bombing Colombia's oil installations and talks between Ecopetrol and British Petroleum to renegotiate a contract to develop the Piedemonte field.

September 5

Amoco Corporation announces an agreement to merge its oil and natural gas business in South America's southern region with Bridas Corporation of Argentina. The deal will create what will be Argentina's second-largest oil and natural gas company with more than $3 billion in assets. The new company is currently unnamed with Amoco holding a 60 per cent interest and Bridas the remaining 40 per cent. The venture will focus on exploration and production in Argentina, Bolivia, Brazil, Paraguay, Chile, Peru, and Uruguay.

September 5

Mobil Oil Canada and Chevron Canada Resources enter a strategic alliance for exploration and development of 29 million acres in the Grand Banks area, offshore Newfoundland. The companies indicate the alliance enables them to move more quickly on proposed exploration and development plans.

September 8

YPF SA, Argentina's largest energy company, confirms that its board approved a previously announced agreement with Amoco Corporation to operate U.S. natural gas fields with an estimated 989 billion cubic feet of gas reserves. The venture will be called Crescendo Resources LP, with YPF holding an indirect 64 per cent interest and Amoco the remaining 36 per cent.

September 8

China National Oil Development Corporation, a unit of state-owned China National Petroleum Corporation (CNPC), and Pan-China Resources Ltd., a unit of Canada's Ivanhoe Capital Corporation, sign a production-sharing contract to develop oil reserves in China's Hebei province. The two firms will develop the 34.9-square-mile Kongnan block of the Dagang oil field, located 62 miles south of Tianjin. The contract is CNPC's latest effort to stabilize production in China's eastern oil fields.

September 10

Norway's state-owned Statoil AS and United Kingdom's (U.K.) Enterprise Oil sign a deal to develop and produce the Pierce oil field located in the U.K. sector of the North Sea. According to Statoil, the deal is worth around $280 million and will last for a minimum of 5 years. The Pierce field has recoverable reserves of 84 million barrels of oil. When it comes on stream in the fall of 1998, it is expected to produce 45,000 barrels per day.

September 11

China completes a natural gas pipeline from Shaanxi province to Beijing. The 534-mile pipeline is the county's longest land pipeline. The project cost about $475 million and will supply gas from the Changqing field in northwest China's Ordos Basin. The pipeline is expected to supply Beijing with 10.6 billion cubic feet of gas in 1998 and eventually deliver 212 billion cubic feet of gas annually.

September 11

Israel's Prime Minister Benjamin Netanyahu orders a freeze on a major natural gas deal with Russia, which includes the construction of an underwater pipeline to transport the gas from Turkey across the Mediterranean to the Israeli coast. Prime Minister Netanyahu cites suspicions that Russia is helping Iran develop ballistic missiles as the reason for the freeze.

September 11

Colombia's President Ernesto Samper inaugurates a natural gas pipeline stretching from the southwestern city of Cali to the city of Mariquita. The 211-mile pipeline cost $312 million and has a maximum capacity of 200 million cubic feet of gas per day. The line will transport gas to 3 million people in about 47 municipalities.

September 12

The United Nations Security Council passes a resolution that allows Iraq to reach the $2.14 billion oil sales limit under its oil-for-food program by December 5. The current 6-month oil sales window, running from June 8 to December 5, will be split into a 120-day segment and a 60-day segment instead of two 90-day segments. During each segment Iraq can sell $1.07 billion worth of oil. The resolution should enable Iraq to make up for lost revenues during a delay in the start of oil sales during the first two months of the current six month sale period.

September 15

U.S.-based Bechtel, Peru's Cosapi, and Brazil-based Odebrecht win a $2 billion contract for the design and construction of a natural gas pipeline stretching from Peru's Camisea gas fields to the capital city, Lima. The field is being developed by a joint venture between Royal/Dutch Shell and Mobil Corporation.

September 16

Amoco Exploration Company announces a major North Sea drilling program designed to increase reserves and production from its central and northern United Kingdom (U.K.) oil and natural gas fields. The cost of the program is expected to exceed $160 million and extend over a three-to-five year period. The program will initially concentrate on additional development drilling in Amoco's Arbroath field and then continue on to other fields in the central and northern U.K. sector. The program will also utilize new drilling rigs built to operate in harsh conditions.

September 17

One day after sending money to Chechnya for repairs to a war-torn pipeline, a Russian commission announces a decision to construct a new pipeline that would bypass Chechnya. The new 176-mile pipeline will run along the Chechen border in the Dagestan republic to the Stavropol region, and then on to Russia's oil terminal at the Black Sea port of Novorossiysk. First Deputy Prime Minister Boris Nemtsov estimates the cost of the project at $220 million.

September 22

Texaco Inc. confirms that is has agreed to sell all of its assets in Burma to the United Kingdom based Premier Oil PLC for $260 million. Texaco holds a 42.9 per cent interest in the Yetagun natural gas field and a 50 per cent interest in Block M-10, a nearby exploration area. Premier will reportedly sell most of the interests to Petronas, Malaysia's state-owned oil and gas company.

September 23

Greenpeace begins legal proceedings against the United Kingdom Department of Trade & Industry in an attempt to block oil and gas exploration west of the Shetland Islands and Scotland in the Atlantic Frontier. Greenpeace is asking for a full judicial review of operating licenses awarded in April's 17th Offshore Licensing Round. The environmental group contends that the European Habitats Directive requires special consideration be given to reef habitats before licenses are granted. If Greenpeace is successful, a full judicial hearing could be held in November.

September 24

A four-week-old Norwegian oil strike intensifies as the Norwegian Shipowners' Association threatens to lock out oil workers from 13 floating oil rigs beginning on October 8. The strike began on August 24 and has mainly affected exploration drilling rather production. However, on September

22 the strike forced the closure of the 35,000-barrel-per-day North Sea Yme field. The oil workers union called the strike after rejecting a 7.8 per cent pay increase offered by the Shipowners' Association.

September 25

Dominique Simondon, chief of conceptional engineering for Elf Aquitaine on the Girassol project, reveals that reserve estimates for Elf's Angolan block 17 oil discoveries have been increased to 3 to 4 billion barrels. Initial reserve estimates of 1 to 3 billion barrels had already made it one of the largest finds offshore west Africa. The Girassol field is due to begin producing oil in the second half of 2000 and is expected to have a production capacity of 200,000 barrels per day.

September 25

Russia's State Property Committee approves a privatization plan for AO Rosneft. The approval clears the way for the sale of the Russian government's last major holding in the oil industry. The sale will take place during the fourth quarter of 1997 and the first half of 1998 and will be open to foreign and domestic investors.

September 26

Kazakstan and China sign a $4.3 billion deal to explore the Aktyubinsk oil fields in western Kazakstan. The deal comes two days after the China National Petroleum Corporation and the Kazak Ministry of Energy and Natural Resources signed a contract for joint exploration of the Uzen oil field, which includes the construction of a $3.5 billion, 1,863-mile pipeline from western Kazakstan to China. Work on the pipeline is expected to begin in 1998 and be completed by 2005.

September 29

France's Total SA confirms that it reached a $2 billion agreement with the National Iranian Oil Company (NIOC) to develop part of the South Pars natural gas field. Other companies that are part of the deal include Russia's Gazprom and Malaysia's Petronas. The South Pars gas field has estimated reserves of 300 trillion cubic feet and is located in the Persian Gulf adjacent to Qatar's North Dome natural gas field. Under the terms of the agreement, Total and its partners will be responsible for the initial development and production of the field until 2002. At the completion of this work, NIOC will take over as production operator. In a related story, the French government warns the United States not to apply the "D'Amato law" to the agreement. The law allows the President of the United States to impose sanctions on foreign companies investing more than $40 million per year in Libya or Iran.

September 30

Brazil's Petrobras begins oil production in two major fields in the Campos Basin. The Barracuda and Caratinga fields together contain reserves of one billion barrels of oil equivalent and 530 billion cubic feet of natural gas. The fields are located 50 miles off the coast of Rio de Janeiro.

October 1997

October 1

Russian oil company AO Sidanko's petition to settle a lengthy legal battle with the Russian government and state-owned AO Rosneft for control of the Siberian oil production company AO Purneftegaz is accepted by the Supreme Arbitration Court. The end of the lawsuit clears the final hurdle for the privatization of AO Rosneft.

October 1

South Korea's state-owned Korea Petroleum Development Corporation (Pedco) agrees to participate in a consortium to develop Venezuela's Onado oil field. Pedco will hold a 12 per cent interest in the development project with the remaining 88 per cent split among CGC of Argentina, Carmanah Resources Limited of Canada, and Distral of Venezuela. The Onado field contains 162 million barrels of oil and is expected to produce 52,000 barrels per day beginning in January 1998.

October 2

In a joint statement, Norway's Saga Petroleum and state-owned Statoil announce a deal to jointly develop the Haltenbanken South oil and gas area in the Norwegian Sea. The deal combines Haltenbanken's four fields, Kristin, Lavrans, Tyrihans, and Trestakk, into one unit and should speed up the development of the area. Combined reserves are estimated at around 7 trillion cubic feet of natural gas and 630 million barrels of crude oil and condensate. Production is expected to start in 2001.

October 2

The Venezuelan Congress approves a proposed $2.5 billion joint venture between Lagoven, a unit of state-owned Petroleos de Venezuela, U.S.-based Mobil, and Germany-based Veba Oel AG to extract and upgrade extra heavy crude oil from the Cerro Negro zone of the Orinoco belt. The project is projected to last 35 years with peak production of 120,000 barrels per day reached by 2001.

October 5

Japan Gasoline Corporation (JGC) announces that it has won a contract to design, build, and equip an oil refinery in Pakistan. JGC estimates the total cost of the project to be around $600 million. The refinery will include a crude oil distillation unit and a hydrocracker and will have a processing capacity of 100,000 barrels of crude oil per day, making it Pakistan's largest refinery. It is designed to help meet Pakistan's domestic demand for oil products and is expected to be complete by the end of 2000.

October 6

Japan's Itochu Corporation and Japan Gasoline Corporation secure an oil refinery construction contract from the Georgian government. The contract is estimated to be worth between $200 and $300 million, with construction beginning in 1998 and start-up scheduled for 2001. The refinery will be built in Supsa on the Black Sea and will process 60,000 barrels of crude oil per day. It will supply gasoline, gas oil, and heavy fuel oil for domestic consumption in Georgia and export to Ukraine, Turkey, and other neighboring countries.

October 6

The Malaysian government confirms that it has approved a proposal submitted by Japan's Chiyoda Corporation to build a 118-mile crude oil pipeline across the Malay peninsula. The $2 billion pipeline will have the capacity to carry 2 million barrels per day from the Malaysian town of Alor Setar to the Thai port of Sai Buri. Thailand still needs to approve the project before it can go forward, but, Chiyoda is expecting that to occur soon.

October 6

The U.S. Department of Energy announces that it will sell its 78 per cent share of the Elk Hills Naval Petroleum Reserve in California to Occidental Petroleum for $3.6 billion. Currently, Elk Hills produces 60,000 barrels per day of oil, down from a high of 150,000 barrels per day in 1981. The U.S. Congress and the U.S. Department of Justice have 31 days and 60 days, respectively, to review the sale.

October 8

The United Nation's Iraq Sanctions Committee votes to provide funds to Turkey to purchase spare parts for the Kirkuk-to-Yumurtalik pipeline that carries Iraqi crude oil to the Turkish port of Ceyhan. The Committee said that the funds are for medium-term maintenance of the pipeline which is to carry at least half of Iraq's oil exports under the current six-month phase of the oil-for-food program.

October 9

Turkey's state-owned oil company, Turkiye Petrolleri AO (TPOA), signs a $750 million deal with Kazakhstan to develop oil reserves in the northwestern Aktyubinsk region near the Caspian Sea. The contract is expected to last at least 29 years, and Kazakhstan could earn $1.8 billion in revenues. TPOA and Amoco will set up a 50-50 joint venture to finance and implement the project. The first four years of the contract are set aside for exploration and the following 25 years for production. Turkish officials estimate that the region contains about 327 million barrels of oil.

October 9

Norway's state-owned oil company, Statoil AS, reports that so far about 60 million barrels of recoverable oil have been found in eight exploration wells drilled this year. In the North Sea, new discoveries were made in the Beta West structure on the Yme field, in the Glitne oil field north of the Sleipner area, and in the Statfjord field. By the end of the year, Statoil will drill four more exploration wells. Its greatest expectations are for the first deep water well on the Vena Dome in the Voering plateau in the Norwegian Sea.

October 11

Pilipinas Shell Petroleum Corporation, a unit of Royal Dutch/Shell, and Occidental Philippines Incorporated, a unit of Occidental Petroleum Corporation of the U.S., announce a $5 billion joint venture to distribute natural gas from the Malampaya/Camago natural gas field off the western Philippine island of Pauline. The Philippines has depended heavily on imported coal and oil for its energy needs and has begun to develop its domestic energy resources to reduce this dependence. The Filipino government will retain 60 per cent of the revenues generated by the project which should translate into an estimated $8 billion. The remaining 40 per cent goes to the Shell/Occidental partnership. Shell and Occidental stand to earn $500 million per year in revenues with investments fully recovered in seven to eight years.

October 11

Nine days after signing a $2 billion contract with the National Iranian Oil Company, France's Total SA begins Phases Two and Three of the development plan for Iran's South Pars natural gas field. The field contains an estimated 280 trillion cubic feet of natural gas and 7.5 billion barrels of natural gas liquids. Phases Two and Three are designed to produce two billion cubic feet of natural gas per day over 30 years. The deal has the support of the European Union, but has drawn harsh criticism from the Unites States.

October 11

Egypt and Libya announce a $1 billion project that includes the construction of a 386-mile oil pipeline between the two countries. The line will transport up to 150,000 barrels of oil per day, carrying Libyan crude oil from the border town of Tobruk to Egyptian refineries.

October 13

The Gulf Cooperation Council (GCC), whose membership includes Bahrain, Kuwait, Oman, Qatar, Saudi Arabia, and United Arab Emirates, opens a two-day conference with the European Union in Manama, Bahrain, focusing on advanced oil and gas technologies. Fayez al-Sadah, general manager of Bahrain National Oil Company, opened the conference by stating that world crude oil demand is expected to rise from today's level of 72 million barrels per day to 96 million barrels per day by 2010 and that the GCC states are ready to meet any increases in world oil demand.

October 14

Shell U.K. Exploration and Production reports that production has started on the Central North Sea Kingfisher condensate field, four weeks ahead of schedule and significantly under budget. The field's reserves are estimated at 56 million barrels of crude oil and condensate and 368 billion cubic feet of gas. Kingfisher is located about 155 miles northeast of Aberdeen. A second production well should be flowing by the end of 1997 and four more wells are expected on line by early 1999.

October 14

At a joint news conference, officials from Turkmenistan and Iran announce an agreement to set up a joint consortium to develop offshore oil and gas fields located on their joint border in the Caspian Sea. The two countries also agreed to further efforts toward determining the legal status of the Caspian region in order to provide more guarantees to international companies interested in long-term work in the Caspian.

October 20

Two of Mexico's main Gulf Coast crude oil export ports reopen after being closed for five days due to poor weather conditions. The port of Pajaritos usually exports around 400,000 barrels of oil per day and the port of Dos Bocas exports around 300,000 barrels per day. These two ports, along with the port of Cayo Arcas, handle the vast majority of exports from state-owned Petroleos Mexicanos.

October 20

The Maritime and Port Authority of Singapore confirms that a 166,500-barrel oil spill from a collision of two oil tankers has been contained in a region southwest of Singapore. On October 15, an empty, very large crude carrier and an oil tanker collided in the Singapore Strait. The oil tanker was carrying about 800,000 barrels of marine fuel oil before the collision. Neither ship was in danger of sinking and clean-up operations began immediately. However, clean-up efforts have been hampered by thick haze from forest fires in nearby Indonesia. The individuals in charge of the two vessels have been charged with reckless navigation. The spill has been described as Singapore's worst.

October 21

The U.S. Department of Energy announces a new method of producing electricity from gasoline that yields twice as much useful energy as today's car engines with 90 per cent less pollution. This marks a significant step toward developing an electric car without having to frequently recharge heavy batteries. The new fuel cell makes an electric current by combining hydrogen and oxygen into water. Energy Secretary Federico Pena said that cars using this new technology could be on the road by 2010.

October 21

Thailand's Songkla Petroleum & Chemical Limited signs a $2.2 billion contract with a German joint venture company, Kruppe Uhde GmbH & Thyssen Rheinstahl Technik GmbH, to design and build a crude oil refinery, a petrochemical complex, and a 150-megawatt power plant. The project is part of Thailand's Southern Seaboard Development program and will be constructed in Songkla province in southern Thailand. The oil refinery will have a capacity of at least 125,000 barrels per day.

October 21

Qatar General Petroleum Corporation announces that it will move forward with its plans to build a gas pipeline from its North Dome natural gas field to Jebel Ali in the United Arab Emirates. The proposed pipeline will have a capacity of 800 million cubic feet per day and is scheduled to be completed in three years. The project is being developed by a five-member consortium led by Atlantic Richfield of the U.S.

October 22

Kazakhstan's Ambassador to Rome, Olxas Suleimenov, announces that Italy's Ente Nazionale Idrocarburi (ENI), Texaco, BG PLC, and Kazak Gas will invest $9 billion over 40 years to develop Kazakhstan's giant Karachaganak oil and gas field. ENI and BG will each hold a 32 per cent stake in the field, Texaco 20 per cent and Kazak Gas 16 per cent. The Karachaganak field contains reserves of 17.66 trillion cubic feet of gas and 2.2 billion barrels of crude oil and condensates.

October 23

Oilwatch Alaska, an oil industry watchdog group, is calling for an antitrust investigation of the companies that control the Trans-Alaska oil pipeline. The group charges that high shipping charges are strangling competition for development of the North Slope oil fields. Oilwatch also accuses the companies of overcharging for oil shipments through the pipeline, which allegedly cuts into profits and costs the state of Alaska billions of dollars in royalties and taxes. The 800-mile pipeline carries 20 per cent of US domestic oil production. The companies that control the pipeline include: British Petroleum 50 per cent, Arco 22 per cent, Exxon 20 per cent, and Mobil 3.2 per cent. Amerada Hess, Philips Petroleum, and Unocal each own less than 2 per cent.

October 23

The Afghanistan Taliban Islamic movement and Turkmenistan have reached an agreement on setting up a tripartite commission to build a natural gas pipeline from Turkmenistan to Pakistan via Afghanistan. Officials from Turkmenistan and Pakistan signed an agreement in July 1997 to build the pipeline which will carry up to 706 billion cubic feet per year of Turkmen gas.

October 23

In response to Iraq's recent refusal to comply with United Nations (U.N.) arms inspections, the U.N. Security Council approves a resolution condemning Iraq's actions. Although the measure does not add new sanctions on Iraq, it expresses the Council's intentions of adopting travel restrictions on Iraqi military and intelligence officials if Iraq continues to inhibit U.N. arms inspections. The resolution passed by a 10-0 vote with five abstentions.

October 25

The State Oil Company of Azerbaijan opens an oil pipeline stretching northwest from Baku on the Caspian Sea, through a 93-mile segment in Chechnya, and on to the Russian port of Novorossiysk on the Black Sea. The ceremony marks the first time Azerbaijan has exported its own oil to Western markets in 65 years.

October 27

Unocal Corporation organizes a multinational consortium to build a natural gas pipeline from Turkmenistan to Pakistan, and possibly India, via Afghanistan. The CentGas pipeline will cost about $2 billion and will transport two billion cubic feet of gas per day 790 miles from the Turkmen border to Multan, Pakistan. The pipeline is expected to take two years to complete. The consortium has also proposed a $600 million, 400-mile extension to New Delhi. Unocal will act as project leader, retaining a 46.5 per cent interest in the project. The rest of the consortium consists of Saudi Arabia's Delta Oil (15 per cent), the Turkmen government (7 per cent), Japan's Itochu Oil Exploration (6.5 per cent), Indonesia Petroleum (6.5 per cent), Korea's Hyundai Engineering & Construction (5 per cent), and Pakistan's Crescent Group (3.5 per cent). A 10 per cent share has been reserved for Russia's Gazprom, which is expected to sign onto the project soon.

October 27

The Tehran Times newspaper reports that Russia and Iran have signed a memorandum of understanding for Russia to build an oil refinery and a petrochemical complex in the Chahbahar free trade and industrial zone in southeast Iran.

October 27

Under pressure from the U.S. Congress, the U.S. Export-Import Bank is expected to suspend $186 million in loan guarantees to 21 oil service companies, that were expecting to sell oil well drilling and measuring equipment to Russia's Gazprom over the next year. The U.S. Congress is calling for the suspension of the loans because of Gazprom's ties to Iran. Gazprom recently signed onto a $2 billion contract to develop Iran's South Pars natural gas field, a deal that May violate the Iran and Libya Sanctions Act. The Export-Import Bank is due to vote on the loan guarantees by the end of 1997.

October 28

The U.S. Senate passes the $13.8 billion Interior Appropriates bill that includes a provision directing the Department of Energy to sell $207.5 million of oil from the Strategic Petroleum Reserve. Revenues from the sale will pay for the Reserve's operations over the next year. The bill also extends for another year the current moratorium on offshore oil and gas drilling in parts of the Gulf of Mexico and Alaska. The U.S. House of Representatives has already passed its version of the bill and President Clinton is expected to sign it.

October 29

A consortium of South Korean, British, and Italian firms has discovered an oil field in Libya with estimated recoverable reserves of one billion barrels. The field is located 466 miles south of Tripoli on the Murzuk NC-174 block. A contract to explore the field was signed in 1990 by the consortium and Libya's National Oil Company Under the terms of the agreement, the Libyan company will retain 75 per cent of the oil produced and the rest will be equally split among the consortium's members.

October 29

Iraq's Revolution Command Council, the country's main decision making body, announces that it will no longer allow U.S. citizens and U.S. aircraft to serve with the United Nations (U.N.) arms inspection teams. The council's statement gives U.S. citizens working with the inspection teams one week to leave Iraq. Iraq has also asked the U.N. to stop flights by American reconnaissance aircraft monitoring its compliance with U.N. resolutions requiring the elimination of weapons of mass destruction. In response to this statement, the U.N. Security Council unanimously approves a statement condemning Iraq's threats to expel the Americans.

October 31

Chevron Corporation announces that its Kuito Field (formally known as Block 14) off the coast of Angola, May contain at least one billion barrels of oil. This estimate makes the field one of Chevron's largest finds in several decades. Production is expected to begin by early 1999. Chevron holds a 31 per cent interest in the field followed by France's Total, Italy's Agip, and the Angolan government each with 20 per cent share, and by Portugal's Petrogal with about a 9 per cent share.

November 1997

November 3

An official from Russian oil drilling company Zarubezhneft denies an Iraqi news report that it will begin developing Iraq's 600,000-barrel-per-day West Qurna oil field. Zarubezhneft is part of a consortium led by Russia's Lukoil that signed a 23-year production sharing agreement with Iraq, valued at $3.8 billion, to develop the field. Zarubezhneft's Deputy General Director Yuri Agababov indicates that there have been negotiations but reiterates that work in Iraq will not start until United Nations sanctions are lifted.

November 3

Dutch and Belgian oil worker unions call for a boycott of Nigerian crude oil as part of a campaign to secure the release of jailed Nigerian oil union leaders Milton Dabibi and Frank Korkori.

November 4

The U.S. Department of the Interior rejects a petition from environmental groups seeking to halt crude oil and natural gas drilling in the Beaufort Sea off the coast of Alaska's Arctic National Wildlife Refuge. The petition expressed concern for potential damage to the ecosystem of the refuge from pollution and possible oil spills. However, the Interior Department states that Atlantic Richfield Company's plan for exploration in the Beaufort Sea has adequate safeguards to protect the area.

November 4

Turkmenistan withdraws three of eleven blocks from the first round of an offshore oil and gas tender opened on September 1, 1997. The three blocks, Lachin, Burgut, and Yelbars, lie on the Turkmen-Iranian Caspian Sea border-zone, calling ownership into question. The three blocks contain estimated reserves of 17.1 trillion cubic feet of natural gas and 5.4 billion barrels of oil.

November 5

Texaco North Sea U.K. Limited announces that natural gas and condensate production has begun in its United Kingdom North Sea Erskine field. This follows a November 3, 1997 announcement that first production had been delayed again from the original start-up date of October 14, 1997 due to final testing on a pipeline that will carry gas and condensate to a London platform operated by Amoco Exploration Company. Peak production is expected to reach 120 million cubic feet of gas per day and 29,600 barrels per day of condensate by October 1998. Estimated recoverable reserves total more than 330 billion cubic feet of natural gas and over 75 million barrels of condensate.

November 5

Turkmenistan resumes natural gas shipments to Ukraine following the signing of a new agreement on paying for the fuel. Under the agreement, Turkmenistan will supply 99 billion cubic feet of gas through the end of 1997 and 706 billion cubic feet in 1998. Ukraine is required to pay for at least half of the shipments in cash and the remainder through barter and other mechanisms. Turkmenistan cut off gas shipments to Ukraine in April 1997 to force repayment of a $300 million debt from previous gas shipments.

November 7

Mexico's state-owned Petroleos Mexicanos announces that it will spend $3.5 billion over the next three years to build coker units at three oil refineries in Salina Cruz, Ciudad Madero, and Minatitlan. The project will begin in 1998 and enable the refineries to produce more high quality products, such as diesel fuel and gasoline, from heavy Mexican crude oil. The upgrade is designed to meet rising domestic demand for oil, but should also permit increased exports.

November 8

Azerbaijan's parliament ratifies an exploration agreement signed between Azerbaijan's state oil company and Mobil Corporation on August 1, 1997. The $2 billion deal is for the exploration of the Oguz oil field located about 50 miles southeast of the Apsheron Peninsula in the Caspian Sea. Each company holds a 50 per cent stake in the project, and reserves are estimated at 733 million barrels of oil.

November 11

Russian President Boris Yeltsin and Chinese President Jiang Zemin sign an agreement to build a $12 billion pipeline to carry natural gas from eastern Siberia to China and then on to Japan and South Korea. The pipeline is expected to bring 353 billion cubic feet of gas to China and make available an additional 353 billion cubic feet to Japan and South Korea over a 30-year period. South Korea and Japan have yet to sign off on the deal, but the two countries are expected to provide equipment and financing.

November 11

Russian gas company Gazprom postpones a bond offering to raise up to $3 billion for a natural gas exploration project in Iran. Gazprom cites uncertain market conditions as the reason for the delay. The lead underwriter for Gazprom is Goldman, Sachs, an American investment bank, which May be in violation of the 1996 Iran-Libya Sanctions Act for its role in the bond offering. Gazprom and its partners, France's Total and Malaysian oil company Petronas, signed a $2 billion contract last month with Iran to begin exploration of Iran's South Pars gas field.

November 12

BG Exploration and Production and its partners Royal Dutch/Shell Group and Edison International sign a natural gas sales agreement with Egyptian General Petroleum Company to supply gas from the Rosetta Concession, offshore Egypt's Nile Delta. Gas deliveries will begin in January 2000 and reach a maximum of 250 million cubic feet of gas per day. The agreement is scheduled to last for a minimum of 20 years.

November 12

The United Nations (U.N.) Security Council unanimously approves a resolution to condemn Iraq for its decision to expel Americans from U.N. weapons inspection teams. The Council's measure imposes a travel ban on Iraqi officials who obstruct the inspection teams and expresses the firm intention to take further unspecified measures if Iraq continues to defy the U.N.

November 13

Australian crude oil and natural gas producer, Santos Limited, reports that it has made one of the largest natural gas discoveries in Australia at its Barrolka field in southwest Queensland state. Barrolka's reserves are estimated at 575 billion cubic feet. Santos holds a 58.86 per cent interest in the field.

November 14

Coastal Corporation signs a memorandum of understanding with Maraven, a unit of Venezuela's state-owned Petroleos de Venezuela, for a 35-year joint venture to produce, refine, and market extra heavy crude oil from the Zuata region of Venezuela's Orinoco Belt. The crude oil will be transported via pipeline to the Jose upgrading complex near Puerto La Cruz and then on to Coastal's 100,000-barrel-per-day refinery in Corpus Christi, Texas. The agreement calls for the joint venture to purchase the Corpus Christi refinery and install a new delayed coking unit and an additional hydrotreatment unit. The deal is still subject to final approval by the Venezuelan Congress and by Coastal's board.

November 17

Royal Dutch/Shell announces that it will invest $1 billion in a joint venture with Russia's Gazprom to produce crude oil, natural gas liquids, and natural gas in Russia and elsewhere. Shell and Gazprom will share equal ownership in the partnership. In a separate memorandum of understanding, Shell, Gazprom, and Russian oil company, Lukoil, announce that they will submit a joint bid for a stake in Russian oil company, Rosneft, when it becomes available for privatization in the near future.

November 18

Mobil Oil Qatar, a unit of Mobil Corporation, signs a memorandum of understanding with state-owned Qatar General Petroleum Corporation to evaluate development of Qatar's North Field. The giant field contains estimated reserves of more than 380 trillion cubic feet of natural gas and is thought to be the world's largest nonassociated natural gas field. Mobil says the project will eventually deliver one billion cubic feet of natural gas per day, primarily to the Qatari domestic market, more than 30,000 barrels per day of condensate, and more than 15,000 barrels per day of propane and butane.

November 18

Kazakhstan signs an agreement with a consortium that includes US-based Texaco, British Gas, Italy's Agip, and Russia's Lukoil to develop the Karachaganak crude oil and natural gas field in northwestern Kazakhstan. The field's reserves are estimated at more than 2 billion barrels of crude oil and condensate and about 20 trillion cubic feet of natural gas. The consortium will develop the field for the next 40 years, with Agip and British Gas each holding a 32.5 per cent interest in the field, Texaco 20 per cent and Lukoil 15 per cent.

November 18

Kazakhstan signs an agreement with a consortium that includes Mobil Corporation, British Petroleum, Royal Dutch/Shell, Norway's Statoil, France's Total, British Gas, and Italy's Agip to explore an offshore area in the North Caspian Sea known as Kazakhstancaspishelf. The agreement allows the consortium to split production from 12 blocks selected by the consortium and also calls for additional seismic data. Drilling is expected to start in 1998.

November 18

Nizar Hamdoon, Iraq's United Nations (U.N.) representative, announces that Iraq will not continue the U.N.-sponsored oil-for-food sale unless the U.N. agrees to a specific date for lifting economic sanctions. Hamdoon indicates that Iraq will continue to participate in the current six-month phase of the oil-for-food sale, which runs through December 4, 1997.

November 20

Iraq's Revolution Command Council formally endorses an agreement, arranged by Russia, that enables United Nation's (U.N.) weapons inspection teams to resume operations in Iraq. The deal ends a three-week standoff between the U.N. and Iraq that began in late October 1997 after Iraq announced it would no longer allow U.S. citizens to serve on U.N. weapons' inspection teams.

November 20

Egypt and Libya sign an agreement to link the two countries' natural gas networks by constructing a pipeline between Tobruq, Libya and Alexandria, Egypt. The two countries also have agreed to form a joint company to manufacture pipelines and integrate their petrochemical industries.

November 21

Turkey's state-run petroleum company Turkiye Petrolleri Anonim Ortakligi will team with Amoco Corporation to explore and develop the Alibegamolla oil fields in Kazakhstan. The two companies plan to invest $850 million in the oil fields, located in the Temir region.

November 24

Exxon Services Venezuela, a unit of Exxon Corporation, agrees to form a joint venture with Corpoven, a unit of Venezuela's state-owned oil company Petroleos de Venezuela. The joint venture will develop extra heavy crude oil from the Hamaca area of the Orinoco belt. Although the project still requires the approval of Venezuela's Congress, Exxon indicates that its initial investment will be $800 million, and Corpoven foresees the companies investing $4.9 billion over 35 years. Exxon will hold a 70 per cent interest in the venture and Corpoven the remaining 30 per cent.

November 25

Canada's AEC Pipelines initiates an expansion of its Alberta Oil Sands Pipeline to accommodate extra output from the Syncrude Canada plant. The initial phase of the project involves a 30-inch looping of the existing 22-inch system and will increase pipeline capacity to 300,000 barrels per day. This phase is estimated to cost about $155 million and is expected to be completed as early as 1999.

AEC also reports that when the pipeline expansion is fully complete, the system will carry up to 950,000 barrels of crude oil per day from Ft. McMurray to Edmonton.

November 25

U.S. oil company Chevron Corporation and representatives from Australia's Aborigines finalize a land access agreement for a proposed $2 billion, 1,553-mile natural gas pipeline between Papua New Guinea and Australia's Queensland state. The agreement is a significant step toward final approval of the project, that is scheduled for October 1998.

November 26

Mexico's state-owned oil company Petroleos Mexicanos signs a $2.46 billion contract with a consortium consisting of South Korea's Sunkyong Engineering and Construction, Germany's Siemens, and Mexico's Tribasa to expand and modernize an oil refinery in northeastern Mexico. The project involves the construction of nine new units, the expansion and updating of ten existing units, and the construction of 754 miles of oil pipeline. Construction is expected to be complete within three years.

November 28

United Nations Secretary-General Kofi Annan decides not to give a recommendation on the amount of oil Iraq can sell in the next six month phase of the oil-for-food program. The current six month phase ends December 4, 1997. Annan will suggest improvements to the program, but will not specify a recommended amount of oil that Iraq will be allowed to export. Currently, Iraq is allowed to export $1.07 billion worth of oil every 90 days for two consecutive 90-day periods.

November 29

For the first time in four years, the Organization of Petroleum Exporting Countries (OPEC) agrees to an increase in its production ceiling. OPEC has raised the ceiling to 27.5 million barrels per day for the first half of 1998, effective January 1, 1998. The new ceiling represents a 10 per cent increase over the current ceiling. The new quotas are as follows: Saudi Arabia 8.76 million barrels per day (bbl/d), Iran 3.942 million bbl/d, Iraq 1.314 million bbl/d, Venezuela 2.583 million bbl/d, Nigeria 2.042 million bbl/d, Indonesia 1.456 million bbl/d, Kuwait 2.19 million bbl/d, Libya 1.522 million bbl/d, United Arab Emirates 2.366 million bbl/d, Algeria 0.909 million bbl/d, and Qatar 0.414 million bbl/d.

December 1997

December 2

Royal Dutch/Shell has been offered a lead role in building a natural gas pipeline from Turkmenistan to Turkey, according to Alan Parsley of Shell Exploration & Production. The 1,242-mile pipeline, running from Turkmenistan through Iran and into Turkey and European markets, will be designed to carry between 989 billion cubic feet and 1.06 trillion cubic feet of gas annually.

December 4

Canada's National Energy Board approves construction of eastern Canada's first offshore natural gas project and a pipeline to carry the gas to markets in eastern Canada and the U.S. Northeast. The combined cost for the two projects is estimated at $2.1 billion. The Board's approval removes some of the uncertainties surrounding the Sable Offshore Energy Project, which is scheduled to start producing 460 million cubic feet of natural gas per day from offshore Nova Scotia in late 1999.

December 4

Iraq's United Nations (U.N.) Ambassador Nizar Hamdoon warns that Iraq will not allow oil to flow during a third six-month phase of the U.N.'s oil-for-food sale until the U.N. approves an aid distribution plan. Despite the warning, the U.N. Security Council approves a third six-month phase following the end of the second six-month phase. Like the first two phases, the third phase allows Iraq to sell up to $1.07 billion of oil in each of two 90-day periods. However, the sales level may be increased by the Security Council in January 1998 after U.N. Secretary-General Kofi Annan reports on Iraq's needs.

December 5

Iraq stops pumping oil into the Iraqi-Turkish pipeline at the end of the second six-month phase of the United Nations (U.N.) oil-for-food program in spite of U.N. approval of a third six-month phase.

December 7

Japan's Isuzu Motors Inc. announces that it will begin producing trucks that run on condensed natural gas in August 1998. Isuzu plans to manufacture gas-powered trucks with a two ton load capacity. The company expects to produce 150 trucks in the first year of production, and another 300 trucks in the following year. Isuzu also expects demand to increase to more than 500 per year in 2000.

December 8

As a result of construction delays, Malaysian Refinery Company (MRC) has delayed the start of operations at its 100,000-barrel-per-day oil refinery in Malacca from January 1998 to late March 1998. MRC is a joint venture between Petronas, Malaysia's state oil company, Conoco of the United States, and Norwegian state oil company Statoil. Petronas holds a 45 per cent stake in the refinery, Conoco 40 per cent and Statoil 15 per cent.

December 9

Russian oil company AO Yukos announces that it has acquired a controlling interest in Russia's state owned Eastern Oil Company. Yukos won a 45 per cent stake in the Siberian oil producer for $775 million, adding to the 9 per cent that Yukos already held. The purchase secures Yukos's position as Russia's second largest oil producer behind AO Lukoil.

December 9

French oil company Elf Aquitaine confirms the discovery of a third oil field in Block 17 in the Gulf of Guinea off Angola. The oil field, named Dalia 2, has estimated reserves of 730 million barrels, which will boost Elf's world wide reserves by 15 per cent. Elf holds a 35 per cent stake, Exxon 20 per cent, British Petroleum 16.67 per cent, Norway's Statoil 13.33 per cent, Norsk-Hydro 10 per cent, and Belgium's Fina 5 per cent.

December 10

The Australia Gas Light Company has reached a final agreement with Native Title claimants, opening the way for construction of a $35 million natural gas pipeline in Australia's New South Wales state. The 140-mile pipeline will connect the towns of Dubbo and Marsden. Construction is set to commence in January 1998, with first gas deliveries expected in mid-1998. Under the agreement, the Aborigines win guarantees that selected cultural sites will be protected during construction and promises that they will be offered jobs.

December 10

Texaco, Brown & Root, and Syntroleum Corporation announce plans to develop a plant that will convert natural gas to synthetic crude oil. The partners state that it will be capable of turning gas into heavy and light synthetic crude which can then be processed into petroleum products. The proposed plant, the first of its kind, will have a capacity of 2,500 barrels per day and will be located outside the United States at a sight to be announced in early 1998.

December 11

Delegates from 150 industrial nations attending a United Nations climate conference in Kyoto, Japan reach agreement on a protocol to control heat-trapping greenhouse gases. The protocol, if ratified, would commit nations to roll back emissions of six greenhouse gases (carbon dioxide, methane, nitrous oxide, hydrofluorocarbons, perfluorocarbons, and sulphur hexafluoride) below 1990 levels. Under the protocol, the United States would be required to reduce its greenhouse gas emissions by 7 per cent below 1990 levels, while Europe and Japan would make cuts of 8 per cent and 9 per cent, respectively. Developing countries are exempt from the emissions ceilings for the time being.

December 16

Talal Kanaan, an Iraqi official, states that two new oil fields have been discovered in northern and western Iraq. The field in northern Iraq, called Ismail, is expected to add to Iraq's 112 billion barrels of reserves while the western field, located in the western desert, could open a new area for development.

December 17

A consortium led by United Kingdom-based Premier Oil PLC signs a $250 million contract to develop an onshore oil field in Albania. The deal involves the development of the Patos Merinze oil field where production of heavy crude is to be increased from 6,000 barrels per day to between 25,000 and 50,000 barrels per day within four years. The consortium consists of Premier with a 25 per cent stake, Germany's Preussag 17.5 per cent, International Finance Corporation 15 per cent, and Albpetrol, Albania's state owned oil company, 42.5 per cent.

December 17

Australia's Broken Hill Proprietary (BHP) announces that the Australian government has granted a production license to the company for the development of the deep water Blackback oil field in the Bass Strait. BHP states that it will spend about $71 million on the field, while BHP's 50 per cent partner, Esso Australia, claims that total expenditures will exceed $128 million. Production is expected to begin in the first half of 1999, with peak production reaching 18,000 barrels of oil per day in the first year. Esso Australia, a unit of Exxon Corporation, adds that the field will also yield 35.3 million cubic feet of natural gas per day.

December 18

Norway's state-owned oil company, Statoil, reports that its Smoerbukk South component of the Aasgard oil field will produce more oil than originally expected. Olav Strandenaes, staff engineer for Aasgard, states that a well at Smoerbukk South produced an oil flow of about 15,700 barrels per day but could produce between 30,000 and 37,000 barrels per day. Peak output from Aasgard is estimated to reach 200,000 barrels per day.

December 18

Russia's giant gas company Gazprom cancels a $750 million loan deal with the U.S. Export-Import Bank. The loan, signed in November 1994, guaranteed financing for purchases of equipment and services from American companies. Gazprom reports that it decided to turn down the loan guarantees before they could be withdrawn. The U.S. government has been considering sanctions against Gazprom for its participation in a $2 billion project to develop Iran's South Pars gas field.

December 19

In a deal valued at nearly $3.5 billion, KN Energy, a Colorado-based natural gas utility, confirms that it will purchase gas pipeline giant MidCon Corporation from Occidental Petroleum. MidCon, headquartered in Lombard, Illinois, primarily transports natural gas from the Gulf of Mexico to as far north as the Canadian Border. KN Energy's acquisition of MidCon complements its decision in August 1997 to purchase Interenergy Corporation, an energy company that processes and markets natural gas in the Rocky Mountains.

December 23

After changing one word, the United Nations (U.N.) Security Council agrees to a statement criticizing, but not condemning, Iraq for refusing to grant U.N. weapons inspectors full access to suspected weapons sites. Opposition from Russia and other council members prompted the wording change. The statement comes after chief weapons inspector Richard Butler told the Security Council that Iraq would not allow access to all suspected weapons sites, including Iraqi President Saddam Hussein's palaces and homes.

December 24

A spokesman for Shell Nigeria confirms that loadings of Nigerian Forcados crude oil cargos are to be delayed by 1-3 days until the end of January 1998. This announcement follows the earlier closures of 20 flow stations. Two stations, Tunu and Opukushi, had been closed since November 25, 1997, after villagers attacked and occupied the facilities. Those closures forced Shell Nigeria, on December 19, 1997, to declare force majeure on crude exports from its Forcados terminal from December 21, 1997 to January 11, 1998. The other 18 flow stations were shut down on December 21, 1997 because of a contractual dispute. All 20 flow stations have now begun to reopen, with output of Forcados crude returning to the normal 450,000-barrel-per-day rate, and Shell Nigeria continuing talks with local communities and its contract workers.

December 24

Japan's Arabian Oil Company (AOC) confirms that it will begin a feasibility study in early 1998 for the development of the Dorra gas field in the Neutral Zone, shared equally by Saudi Arabia and Kuwait. The study will last for about six months and, if it shows that the project is economically viable, AOC will begin commercial production of natural gas after 2000.

December 29

The Presidents of Iran and Turkmenistan open a 125-mile natural gas pipeline built to transport Turkmen gas to Iranian power plants. The pipeline cost $190 million and has an initial capacity of 4 million cubic feet per day. The line runs between the Korpedzhe field in Turkmenistan and the Iranian border village of Kurtkoy and it is the first pipeline to carry energy southward from Central Asia.

96

Monthly Energy Chronology 1998

The following information contains a listing of the major energy events that occurred in 1998.

January 1998

January 5

France's Total SA reports that it has successfully tested a new well in Block 14 offshore Cabinda, Angola. The new discovery, called Landana, produces 7,300 barrels per day of 36-degree API crude oil, encouraging Total and its partners to continue with three more exploration wells in 1998. Total holds 20 per cent of Block 14, with Chevron Corporation holding 31 per cent, Angola's national oil company Sonagol 20 per cent, Italy's Agip 20 per cent, and Petrogal 9 per cent.

January 7

Due to the continuing Asian economic crisis, South Korea's refiners have reportedly cut operations to around 80 per cent of capacity. The refiners have also had difficulty securing crude oil supplies for delivery in late January or February, which could cut operations to as low as 70 per cent-75 per cent of capacity.

January 8

Elf Exploration Angola, a unit of France's Elf Aquitaine, and its partner Socieded Nacional de Combustiveis de Angola begin production of 9,500 barrels of oil per day from the Oombo field, located offshore Angola. Elf Exploration operates the field and holds a 50 per cent interest.

January 9

Premier Oil PLC announces that it has made a significant gas discovery in the Natuna Sea, Block A, Indonesia. After four drill tests produced a combined flow rate of 43 million cubic feet of gas per day, Premier has suspended the Pelikan-1 as a future producer. The Pelikan-1 well is located about 31 miles from the Anoa oil production platform, and it follows a gas discovery at the Gajah Puteri well drilled in late 1997. The oil rig will now move to another location in the block to drill the Adik Anoa oil and gas prospect. Premier operates Block A, holding a 66.67 per cent interest, with Kuwait Foreign Exploration Company holding the remaining 33.33 per cent interest.

January 9

Shell Philippines Exploration, a unit of Royal Dutch/Shell and its partner Occidental Philippines, a unit of Occidental Petroleum Corporation, sign two gas supply agreements for the Malampaya natural gas-to-power project in the Philippines. The first agreement is with First Gas Holdings, a company 40 per cent-owned by British Gas. The agreement calls for 1.46 trillion cubic feet of natural gas to be sent over a 20-year period to two power plants that will be built by First Gas. Initial delivery is scheduled to begin January 1, 2002. The second agreement is with National Power Corporation of the Philippines. It is a 20-year agreement beginning January 1, 2002, to supply gas to a turbine power plant to be built at Ilijan, Batangas. Total financing for the gas-to-power project is around $4.5 billion which will cover the development of the offshore subsea gas production wells, the construction of a 317-mile offshore pipeline and shore terminal, and the construction of the three power plants.

January 12

The Albanian government awards oil exploration blocks to two international oil consortia, one led by Austria's OMV and the other led by an Albanian subsidiary of Occidental Petroleum. OMV and its United Kingdom partners Enterprise Oil and Clyde Gas have received two blocks in the south and one in central north Albania. Meanwhile, Occidental and its partner International Petroleum Corporation have received two blocks in the south and one in the north. Although the production-sharing agreements still need to be approved by the Albanian government, drilling is expected to begin in mid-1998.

January 12

After running aground in rough seas near the United Arab Emirates on January 7, 1998, a sunken barge has spilled about 28,000 barrels of fuel oil into the Persian Gulf,. The spill created a 5 mile by 2 mile oil slick along the coastline of the northern emirate of Umm Al Quwain. International environmental experts have teamed with the Emirates Coast Guard, the Federal Environment Authority and the Abu Dhabi National Oil Company in cleanup efforts.

January 13

Japan's Mitsubishi Petroleum Development Company begins oil production from a well in Block 3/91, located 25 miles off the coast of Angola. Daily output from the well is estimated at 9,500 barrels, with the concession expected to yield a total of around 15 million barrels of crude oil.

January 14

In a show of support for Richard Butler, chairman of the United Nations (U.N.) Special Commission on Iraq, the U.N. Security Council unanimously approves a statement deploring Iraq's recent actions to impede inspections by U.N. weapons monitors.

January 14

China National Offshore Oil Corporation and Norway's Statoil begin production from the Lufeng 22-1 oil field. Lufeng 22-1 is the deepest oil field in the South China Sea and is currently producing about 60,000 barrels per day.

January 14

Iraq begins exporting crude oil under the third phase of the United Nations sponsored oil-for-food program. First loadings are filling the million barrel tanker White Sea for French oil company Elf Aquitaine. Other companies expected to lift Iraqi crude in January include France's Total, Italy's Agip, Russia's Lukoil, and China's Sinochem.

January 15

Environmentalists hail the implementation of a 50-year moratorium on mining and oil exploration in the Antarctic. A protocol for the protection of the Antarctic was adopted by twenty-six countries in 1991, but it could not be implemented until Japan's ratification cleared the way last month. Antarctica contains 70 per cent of the world's fresh water, and the moratorium attempts to preserve the world's least polluted continent.

January 19

Elf Exploration Incorporated, a subsidiary of France's Elf Aquitaine announces the development of its Virgo discovery located in the Gulf of Mexico off the coast of Louisiana. The next step for Elf and its partners is to construct a drilling and production platform that will have a daily production capacity of 120 million cubic feet of gas and 15,000 barrels of oil. Elf Exploration holds a 64 per cent interest in the license with Coastal Oil and Gas 16.2 per cent, Pogo Producing Company 10.8 per cent, and Nippon Oil Exploration USA Limited 9 per cent.

January 19

Bohai Petroleum Company, a unit of China National Offshore Oil Corporation, begins production from the Boxi Oil field located in the Bohai Sea. The field's reserves are estimated at more than 37 million barrels of oil and about 353 billion cubic feet of natural gas.

January 19

Gazprom, Russia's giant natural gas company, signs a $340 million agreement with Poland designed to help repay Poland's $200 million debt to Gazprom and to finance the Jamal-Europa gas pipeline. Under the agreement, Poland will give Gazprom $175 million worth of goods and services. The other $165 million will finance the construction of the pipeline, which will cross Russia and Eastern Europe into Western Europe. The Polish section of the pipeline is expected to be completed later this year.

January 20

Russia's second and seventh largest oil concerns, AO Yukos and AO Sibneft, announce a merger to create a new company to be named AO Yuksi. In terms of reserves, Yuksi will surpass Royal Dutch/Shell Group as the world's largest private oil company and rank third behind Shell and Exxon Corporation in terms of production. Yukos's shareholders will take a 60 per cent share in the new company, with shareholders from Sibneft controlling the remaining 40 per cent.

January 21

Following a speech last week by Luis Giusti, president of state-owned Petroleos de Venezuela, in which he claimed that Saudi Arabia has been producing 9.12 million barrels of crude oil per day and that its overproduction has led to falling oil prices, Saudi Arabia defends itself stating that it is abiding by its new OPEC production quota of 8.76 million barrels per day. Giusti also said that Venezuela is currently producing 3.3 million barrels per day, which is about 25 per cent over its OPEC quota of 2.583 million barrels per day, and that production will reach 3.6 million barrels per day in 1998.

January 22

Statoil, Norway's state-owned energy company, reports that 10 drilling services contracts have been awarded to companies holding production sharing contracts in the Caspian Sea off the coast of Azerbaijan. The 10 contracts, worth $110 million, have been awarded to Azdri MI, Schlumberger, Geoservices, Oceaneering, Baker Hughes Intec, Franks International and Security DBS. The first exploration well in the Shah Deniz area, where Statoil holds a 25.5 per cent interest, is scheduled to be drilled in June 1998.

January 25

Toyota Motor Corporation announces that it will attempt to introduce its hybrid electric/gasoline-powered Prius compact simultaneously in the United States and Europe in the year 2000. This new hybrid system generates half the carbon dioxide of a gasoline engine and reduces emissions of other toxic substances such as nitrogen oxides. The Prius was introduced in Japan in December 1997, and sales have been strong, with 3,500 orders received in the first month.

January 26

The Algerian Energy Council adopts a series of measures designed to strengthen Algeria's oil and gas industry, including the opening of the refining, petrochemical, and fertilizer sectors to foreign investment.

January 26

Delegates from OPEC's Ministerial Monitoring Committee hold an emergency one-day meeting to discuss falling oil prices. The delegates agree that OPEC's crude oil output is above the 27.5 million barrels per day ceiling by at least 500,000 barrels per day, but less than 1 million barrels per day. The delegates also say that there is little OPEC can do to lift oil prices soon outside of recommending strict adherence to production quotas or possibly cutting production.

January 27

Su Shulin, Executive Deputy Director of the Daqing Petroleum Management Bureau, states that China's largest oil field, Daqing, will maintain annual crude production of around 100,000 barrels per day through 2010. Proven reserves in the field are said to be around 40 billion barrels.

January 28

Amoco Corporation announces a crude oil discovery at its Immortelle oil field, located 34 miles off the east coast of Trinidad. This represents Amoco's largest crude oil find in Trinidad in 25 years. Reserve estimates range between 40 and 70 million barrels, and as many as eight wells could be drilled by the end of the year. [Editor's note: EIA estimates Trinidad oil production for 1997 at 124,000 barrels per day.]

January 29

Hungarian oil and gas company MOL renews an agreement with Nafta Industria Srbia (NIS) to ship Russian natural gas to Serbia. Under the agreement, MOL will ship 99 billion cubic feet of Russian gas across Hungary from the border with Ukraine to the Hungarian-Serbian border in 1998. Annual gas shipments to Serbia will gradually increase over the course of the agreement eventually reaching 134 billion cubic feet in 2002. Over the next 20 years, MOL expects to ship a total of at least 2.6 trillion cubic feet of gas to Serbia. In addition, MOL is set to develop a pipeline system to deliver natural gas to Bosnia. MOL says it will invest around $77 million in the new system and expects to have it in service by the year 2000.

January 31

New York, Massachusetts, Vermont and Maine have decided to hold out for California's zero emission standard for cars rather than join the federal National Low Emission Vehicle program. Under the federal program, automakers will agree to begin producing cars that will significantly reduce the amount of smog-producing emissions if all 13 northeastern states will agree not to adopt California's tougher standard. The automakers say that the new cars would be 70 per cent more effective in eliminating emissions of hydrocarbons and 50 per cent more effective in reducing emissions of oxides of nitrogen. The cleaner cars could be available in the northeast as early as this year and nationally by 2000. Many federal officials and some automakers believe that the auto industry will agree to the new program even without participation from all 13 states. The automakers have until February 17 to decide whether they will go along with the National Low Emission Vehicle program.

February 1998

February 1

A U.S. federal judge denies a preliminary injunction requested by civil rights lawyers that would have removed California-based Unocal Corporation from a $1.2 billion pipeline project in Burma. The lawsuit alleges a variety of human rights violations by the Burmese government in connection with the project. The judge's ruling states that prohibiting Unocal's participation in the project would not stop the human rights abuses claimed in the lawsuit. President Clinton imposed trade and economic sanctions on Burma in April 1997, but the ban did not affect existing investments. Unocal holds a 28 per cent share in the pipeline project, which is being constructed from Burma to Thailand.

February 1

United Nations (U.N.) Secretary-General Kofi Annan calls for an increase in the amount of oil Iraq can sell under the U.N.-sponsored oil-for-food program. Annan recommends raising the sales limit from $2.14 billion every six months to $5.2 billion. The recommendation now moves to the U.N.'s 15-member Security Council for approval.

February 2

Western partners in the Caspian Pipeline Consortium have frozen funds for a $2 billion pipeline and are pushing for a shake-up in the Consortium's Russian-led management. The 900-mile pipeline will eventually transport 1.4 million barrels of oil per day from Kazakhstan's Tengiz oil field through Russia to the Black Sea port of Novorossiysk. Oil shipments were scheduled to begin in late 1999, but will likely be delayed at least one year as a result of this latest development. Continued funding of the project was contingent upon the Consortium obtaining Russian rights of way and various federal and local permits for the pipeline by the end of 1997. During a meeting in December 1997, the Western partners discovered that much of this work had not been completed.

February 3

Azerbaijan's state oil company Socar and U.S.-based Conoco announce an agreement to conduct a joint study of Azerbaijan's natural gas processing industries. The study will examine the commercial viability of gathering, processing, and transporting Azerbaijan's natural gas and natural gas liquids for export and domestic use. The study will be conducted over a period of six months.

February 3

Russia's RAO Gazprom has sold its 10 per cent stake in Centgas, a consortium set up to construct a $2 billion Turkmenistan-Afghanistan-Pakistan natural gas pipeline. U.S.-based Unocal Corporation acquired 7 per cent of Gazprom's shares, bringing its total stake in the Centgas to 54 per cent. The remaining 3 per cent went to Japanese-owned Indonesia Petroleum and Itochu, South Korea's Hyundai Engineering & Construction Company, and Pakistan's Crescent Group. The 800-mile pipeline will extend from the Daulatabad gas field in southeastern Turkmenistan through Afghanistan to Pakistan.

February 4

General Motors and Amoco Corporation announce a groundbreaking partnership to jointly develop clean-burning motor vehicle fuels. The partnership is the first of its kind between an automaker and an oil company. General Motors Vice Chairman Harry Pearce says that cleaner diesel fuels and cleaner-burning reformulated gasoline are necessary for the success of future hybrid vehicles, which will run on a combination of gasoline and electricity.

February 4

General Motors, Chrysler Corporation, and Ford Motor Company announce that they will participate in the Environmental Protection Agency's National Low Emission Vehicle Program. The Program requires automakers to build cars that will be 70 per cent cleaner than current models. In exchange for the voluntary participation of the car companies, eight Northeastern states and the District of Columbia gave up the right to adopt higher California emissions standards. The automakers will begin producing the cleaner cars in 1999 in the Northeast and two years later in the rest of the country. Ten foreign automakers have agreed to participate as well, including Honda, Toyota, Nissan, BMW, Mercedes, Saab, Subaru, Suzuki, Volvo, and Rolls Royce.

February 5

Following a ruling by a federal judge denying a request from environmentalists and Native Americans seeking to block the sale of the Elk Hills Naval Petroleum Reserve, the U.S. Department of Energy formally transfers ownership of the reserve to Occidental Petroleum Corporation. Occidental purchased a 78 per cent interest in the field for $3.65 billion. Chevron Corporation currently holds the remaining 22 per cent. Elk Hills contains 450 million barrels of proven oil reserves; however, officials from Occidental believe the reserve May contain one billion barrels of recoverable reserves.

February 5

Norway's state oil company Statoil AS and its partners will invest $400 million to expand the development of Venezuelan oil field LL 652. The investment will cover the first phase of the expansion designed to increase production from 10,000 barrels per day to 115,000 barrels per day over the next ten years. The field's recoverable reserves are estimated at 500 million barrels. Statoil and Chevron each hold a 30 per cent interest in the project with Phillips and Atlantic Richfield each holding 20 per cent.

February 5

The United Kingdom's Hardy Oil & Gas announces the discovery of a major new natural gas field in the South West Miano concession onshore Pakistan. The exploration well Sawan-1, operated by Austria's OMV, produced a flow rate of 58 million cubic feet per day. Reserves are estimated at approximately 918 billion cubic feet. Commercial development is expected to begin in 1998, and initial production is to start by the end of 1999. Hardy and Pakistan Petroleum Limited each hold a 30 per cent interest in the concession, with OMV holding a 25 per cent interest.

February 6

Iraq rejects key parts of United Nations (U.N.) Secretary-General Kofi Annan's proposal to increase the amount of oil Iraq is permitted to sell under the U.N.'s oil-for-food program from $2.14 billion to $5.2 billion. In a letter to Annan, Iraqi Foreign Minister Mohammed Saeed al-Sahhaf objects to additional funds to pay for U.N. monitoring, proposals to repair electric power stations in northern Iraq, and plans for U.N. agencies to target aid to specific groups such as the poor and children. Al-Sahhaf writes that the Iraqi government should deliver the aid and determine which power stations are repaired not the U.N.

February 9

In a letter to United Nations (U.N.) Secretary-General Kofi Annan, Iraqi Foreign Minister Mohammed Saeed al-Sahhaf informs the U.N. that Iraq can only export up to $4 billion of oil in six months. The letter is a response to Annan's recommendation to the U.N. to allow Iraq to increase the amount it can export under the oil-for-food program from $2.14 billion to $5.2 billion. In addition, al-Sahhaf writes that a larger share of the oil sales should go towards humanitarian aid, while the amounts funding U.N. programs and a compensation fund for Persian Gulf war victims should be reduced.

February 9

Georgian President Eduard Shevardnadze escapes an assassination attempt as he is headed home in Tbilisi, Georgia's capital. This represents the second assassination attempt against him in three years. Shevardnadze is not injured, but two bodyguards are dead and four others seriously wounded. Shevardnadze suggests that the attack could have been related to his country's attempt to be part of a pipeline route transporting Caspian Sea oil to the Black Sea.

February 10

The United Nations (U.N.) approves 34 oil contracts submitted by Iraq which will meet its $2 billion quota under the third phase of the U.N.-sponsored oil-for-food program. The 34 contracts total 151.8 million barrels of oil. Iraq exported 121.1 million barrels during the first phase and 127.3 million barrels during the second. The increase in volume is due to weakening oil prices on international markets.

February 10

Philippine President Fidel Ramos signs into law legislation deregulating the country's oil industry. It is the last major structural reform sought by the International Monetary Fund (IMF) before it allows the Philippines to withdraw from its current IMF borrowing program. The new law provides for a five-month transition period before full deregulation and a $73 million buffer fund that can be used to subsidize certain petroleum products.

February 10

Energy Africa Limited, a South African oil and gas exploration company, announces the discovery of oil off the coast of Congo about 10 miles southwest of the Nkossa oil field. The Bilondo Marine 1 well produced flow rates of 8,520 barrels of oil per day. The well also tapped into a different geological structure from Nkossa, proving the existence of a separate field. The discovery was made within the Haute Mer exploration permit which could contain estimated oil reserves of 500 million barrels. Energy Africa holds a 4 per cent interest in a consortium that includes Elf Aquitaine (51 per cent), Chevron (30 per cent), and Hydro Congo (15 per cent).

February 11

Russia's RAO Gazprom and Italy's ENI SpA sign a preliminary agreement for joint exploration and exploitation of the Astrakhan region in Russia. The Astrakhan fields are estimated to contain up to one billion barrels of oil equivalent. The deal is worth $3 billion and includes ENI taking a yet-to-be-determined stake in Gazprom. In addition, the two companies envision joint ventures in other regions of the world, including Europe, the Mediterranean, and China.

February 16

The United Nations (U.N.) Iraq Sanctions Committee approves a pricing plan for Iraqi oil to be shipped under the third phase of the U.N.-sponsored oil-for-food program. The approval clears the way for Iraq to export $1.07 billion of oil in each of two 90-day periods. The first 90-day period ends on March 5, 1998.

February 17

A barge carrying about 200,000 gallons of diesel fuel sinks in rough seas off the coast of the United Arab Emirates. The barge is suspected of transporting smuggled fuel from Iraq.

February 18

United Nations (U.N.) Secretary-General Kofi Annan receives unanimous support from the U.N. Security Council for his diplomatic trip to Iraq. Annan is scheduled to meet with President Saddam Hussein and other Iraqi leaders in an attempt to reach a diplomatic solution to the standoff between Iraq and the U.N. over weapons inspections.

February 18

The international environmental group Greenpeace praises a policy decision by the European Union (EU) Commission to ban the disposal of offshore oil and gas platforms at sea. The Commission also called on countries in the Ospar Convention to adopt the same policy. The Ospar Convention regulates marine pollution in the North East Atlantic from Gibraltar to the Arctic. The policy, which carries an exemption for large concrete platforms, still requires approval from EU foreign ministers. The ministers will address the policy at a Ministerial meeting on July 20-24, 1998.

February 19

The U.S. Justice Department intervenes in four lawsuits against DuPont Company's Conoco Unit, Royal Dutch/Shell Group's Shell Oil Company, Burlington Resources, and Amoco Corporation. The companies are accused of violating the False Claims Act by knowingly underpaying royalties to the US government and Indian nations for oil extracted from public and Indian lands. No estimate of the value of the underpayments has been released, but a Justice Department attorney says that, since 1988, the companies had undervalued hundreds of millions of barrels of oil taken from the Gulf of Mexico and western states.

February 20

The United Nations (U.N.) Security Council votes unanimously to more than double the amount of oil Iraq can export under the U.N. oil-for-food program. The Security Council's vote increases the amount Iraq can export from $2.14 billion to $5.26 billion over six months. Iraq maintains that it only has the capability to export up to $4 billion over a six-month period.

February 22

Sudan signs a $600 million agreement with three foreign companies to construct a 994-mile pipeline to carry oil from the western fields of Higlig to Bashair Port, near Port Sudan on the Red Sea. The pipeline will transport 150,000 barrels per day and is expected to be operational by June 1999. The Sudanese government's partners are China's CPECC Oil Company, Argentina's Techint, and Britain's Weir and Allan Diesel

February 22

United Nations (U.N.) Secretary-General Kofi Annan reaches an agreement with senior Iraqi officials over U.N. inspections of suspected Iraqi weapons sites. The deal includes opening eight Iraqi presidential compounds to weapons inspectors, one of the major points of contention between Iraq and the U.N. Annan will now present the agreement to the U.N. Security Council for approval.

February 23

Amerada Hess Corporation of the U.S. signs two production-sharing contracts with Malaysia's Petronas Carigali, marking Amerada Hess' entry into exploration activities in Malaysia. The contracts give Amerada Hess a 70 per cent stake in block PM304 offshore Terengganu and an 80 per cent stake in block SK306 offshore Sarawak. Amerada Hess has committed to spending $24.9 million and drilling five well on the two blocks.

February 23

Japan's Export-Import Bank signs a loan agreement worth approximately $80 million with Gas Authority of India Limited to help finance capacity expansion work on a natural gas pipeline linking Gujarat, Madhya, and Uttar provinces with the Bombay North oil and gas field. The loan is being co-financed by the Asian Development Bank. The loan will be used to boost the capacity of the pipeline from 636 million cubic feet per day to 1.2 billion cubic feet per day.

February 24

Turkmenistan and Shell Exploration B.V. sign a Memorandum of Understanding (MOU) providing Shell with the exclusive right to promote a project to transport Turkmen natural gas to Turkey. As part of the MOU, Shell will conduct a feasibility study on a proposed 1,200-mile gas pipeline connecting Turkmenistan and Turkey via Iranian territory.

February 26

Longtime competitors Halliburton Company and Dresser Industries announce plans to merge for $8.1 billion. The new company will retain the Halliburton name at its Dallas, Texas headquarters and will form the world's largest oil-drilling services company. The deal is still subject to approval by shareholders and regulatory agencies.

February 26

State oil company Petroleos de Venezuela (PdVSA) announces that three major Venezuelan oil joint ventures will be delayed 6-8 months due to budget cuts. Lower-than- expected oil prices have forced PdVSA to cut its investment plan for 1998 by approximately $1 billion. The three projects are the Cerro Negro joint venture with Mobil Corporation, the Total-led Sincor project in the Zuata region, and the Arco-led Petrolera Ameriven S.A. joint venture in the Hamaca zone. All three are extra-heavy crude upgrading projects in the Orinoco Belt region in southeast Venezuela.

March 1998

March 1

The Jordanian government announces that it has signed an agreement with the Royal Dutch/ Shell group allowing the company to extract crude from oil shales in Jordan. A recent study by Jordan's Natural Resources Authority indicates that there are 40 billion tons of oil shale reserves in the southwestern regions of the Kingdom. Jordan also announced that it has received offers from an international consortium to build a refinery in Aqaba, located 217 miles south of the Jordanian capital Amman. The refinery is expected to cost $2 billion, with a capacity of 250,000 barrels per day.

March 1

Iraqi officials and independent experts report that Iraq's oil production system has fallen into such disrepair that Iraq cannot take full advantage of a recent United Nations (U.N.) Security Council resolution that more than doubled U.N.-authorized Iraqi oil sales to $5.26 billion every six months. Iraq says that, under current conditions, it can sell only $4 billion worth of oil over the next six months, unless it is allowed to purchase more equipment. Experts say that Iraq's production system needs $800 million of repairs, and that these repairs could take a year or more to complete.

March 2

The United Nations Security Council unanimously approves a resolution endorsing Secretary Kofi Annan's agreement with Iraq on weapons inspections which includes the controversial presidential palaces. The resolution threatens any Iraqi violation of the agreement with the "severest consequences," but does not authorize an automatic military response.

March 3

The Independent Petroleum Association of America (IPAA) announces that it has filed a lawsuit against the U.S. Department of the Interior, challenging the validity of a natural gas transportation regulation that went into effect on February 1, 1998. The IPAA contends that under the new law, independent oil and gas producers will no longer be allowed to deduct the full cost of transporting gas to the marketplace, which effectively adds these costs to the wellhead price. This, in turn, allegedly will drive up the cost of gas and increase royalty payments required by the government for that production. IPAA claims that this situation will threaten future natural gas production on federal lands.

March 4

The Hungarian government announces that it will sell 1 per cent, or 984,000 ordinary shares, in state-owned oil and gas company MOL Rt. The stake May be increased to 2 per cent if demand from domestic small investors is greater than the 1 per cent offer. The offer is part of the government's plan to privatize MOL and reduce its ownership from 36.2 per cent to 25 per cent.

March 6

Ford Motor Company and Mobil Corporation form an alliance to develop fuel and vehicle technologies to increase efficiency and reduce emissions. This announcement follows a similar agreement between General Motors Corporation and Amoco Corporation. Ford and Mobil will focus on developing a direct-injection diesel system that substantially improves fuel efficiency and fuel cell processors for use in passenger vehicles. Ford expects to introduce a fuel-cell vehicle in 2004.

March 6

French petroleum company Total SA signs an agreement with the Nigerian National Petroleum Corporation to rehabilitate the Kaduna refinery in northern Nigeria. The 110,000- barrel-per-day refinery has been shut since July 1997. Currently, the 125,000-barrel-per-day unit in Warri is the only functional refinery in the country. All other refineries have been shut for safety reasons. The rehabilitation is estimated to cost around $200 million, and some units at the refinery May resume production in six months.

March 11

The United States Mineral Management Service decides to exclude an area offshore of the Arctic National Wildlife Refuge in Alaska from an oil-drilling lease sale scheduled for August 1998. The decision is a victory for environmental groups, such as the Alaska Wilderness League.

March 12

Amoco Corporation reports that it will invest $230 million to develop the Amherstia field offshore Trinidad and Tobago. The field contains 2 trillion cubic feet of natural gas reserves and first production is expected in the first quarter of 2000. Amoco also expects Amherstia to yield crude oil and condensate.

March 12

As the United States Bureau of Land Management considers proposals that would allow drilling for crude oil and natural gas on a 4.6-million-acre portion of the National Petroleum Reserve-Alaska, the American Petroleum Institute (API) announces its support for a proposal that would open it all up to drilling. However, API also wants the drilling to be conducted in an environmentally sound manner. API maintains that the oil industry has the technology to design and build facilities that minimize the industry's impact on fish, birds, and other wildlife.

March 16

State-owned China National Petroleum Corporation signs a 25-year, $325 million contract with Monde Group LLC of Houston, Texas, for an oil recovery program in China. Under the contract, Monde will take over operation of 800 wells in five sections of the Zhongyuan oil field near Puyang, Henan Province. Monde will attempt to reduce recent production declines from the field using enhanced oil recovery systems such as improved drilling techniques and 3D seismic surveys.

March 17

Following a long-time dispute over transit fees, Russia and Turkmenistan reach an agreement to transport 706 billion cubic feet of Turkmen natural gas to Ukraine via Russian territory.

March 18

The U.S. Interior Department's Minerals Management Service holds its semiannual lease sale in New Orleans, Louisiana, receiving bids on 794 of 4,180 tracts offered in the sale. The tracts cover about 4.3 million acres all of which are located in the central region of the Gulf of Mexico. Oil producers submit bids totaling $810.4 million. Conoco submits the most bids, with 122 valued at $48.7 million, and Shell Oil Company is second with 119 at $46.9 million. Other major bidders include Mobil Corporation, Chevron Corporation, Elf Aquitaine SA, and Oryx Energy Company. The next lease sale is scheduled for August 1998 covering western blocks in the Gulf.

March 19

Sergei Volchkov, deputy manager for construction for the Caspian Pipeline Consortium (CPC), reports that further delays are expected in the construction of a 838-mile oil pipeline running from Kazakhstan's Tengiz oil field to the Russian Black Sea port of Novorosiisk. Earlier this year, CPC reported that construction on the $4.5 billion pipeline would begin in the third quarter of 1998, but that date now has been pushed to the fourth quarter of 1998, and start-up has been pushed to early 2000. Securing land permits for construction has proved slower than expected. The planned capacity for the pipeline is approximately 500 million barrels per year.

March 19

Shell Exploration and Production Company announces that it plans to spend $1 billion developing the Angus, Europa, and Macaroni oil and gas discoveries located in the Gulf of Mexico. The Angus development is expected to yield 40,000 barrels of oil per day and 60 million cubic feet of gas per day by the end of 1999. The Europa development is expected to produce 60,000 barrels of oil per day and 45 million cubic feet of gas per day by early 2001. The Macaroni development is anticipated to yield 35,000 barrels of oil per day and 65 million cubic feet of gas per day by late 1999. The three discoveries will add 300 million barrels of oil equivalent to the Gulf of Mexico's deep water inventory.

March 20

In the continuing battle for market share in the United States (U.S.), Mexico and Saudi Arabia each move ahead of Venezuela as the biggest exporters of crude oil to the U.S. for January 1998, according to the U.S. Energy Information Administration. Venezuela has been the biggest foreign oil source for the U.S. on an annual basis for the past two years.

March 20

Russia's RAO Gazprom and the Bulgarian government reach a preliminary agreement providing for increased natural gas deliveries from Gazprom to Bulgaria in exchange for increased transit of Russian gas through Bulgaria to neighboring countries. Under the agreement, annual gas deliveries to Bulgaria will rise from 212 billion cubic feet in 1997 to 283 billion cubic feet in 1998. Transit of Russian gas through Bulgaria to other countries, such as Turkey, Greece, and The Former Yugoslav Republic of Macedonia, will increase to 671 billion cubic feet per year by 2010.

March 21

Russia announces plans for the sale of state-owned Rosneft, the Russian government's last major holding in the oil industry. The government is offering a 75 per cent stake in the firm at a starting price of $2.1 billion. Buyers will be required to invest $400 million in the company within three months of the sale.

March 22

In a joint statement, three major oil exporters, Saudi Arabia, Venezuela, and Mexico, pledge to reduce their own oil output in an attempt to boost sagging oil prices. Mexico and Venezuela indicate that they will cut production by 100,000 barrels per day and 200,000 barrels per day, respectively, while Saudi Arabia will cut 300,000 barrels per day. The three countries also are trying to get commitments from other Organization of Petroleum Exporting Countries (OPEC) member states and non-OPEC countries for further cuts to ultimately reduce worldwide oil production by 1.6 million to 2 million barrels per day.

March 24

After a dispute with Mobil Corporation, Elf Nigeria, a subsidiary of France's Elf Aquitaine SA, receives approval from Nigeria's Department of Petroleum Resources to develop the Amenam oil field. The field, located in Elf's OML99 area, contains estimated reserves of 400 million barrels of oil and is expected to produce 90,000 barrels per day. In a related story, France's Total SA and Brazil's Petrobras have been awarded block OPL246 in Nigeria's deep offshore zone. Some estimates put the block's reserves at one billion barrels of oil.

March 24

France's Elf Aquitaine SA acquires a 5 per cent stake for $528 million in the newly formed Russian oil company AO Yuksi. Yuksi was formed earlier this year in a merger of Russian oil producers AO Yukos and AO Sibneft. Elf will help Yuksi finance new equipment and development projects. The deal also covers joint development of the Sugmut field, located in western Siberia, which contains estimated reserves of more than 700 million barrels. Elf and Yuksi will each hold a 50 per cent stake in the field, with total development costs estimated at $1.5 billion.

March 26

In response to the recent drop in oil prices, Russia's second largest oil company, AO Lukoil, announces that it will reduce its exports of oil products by more than 7 million barrels for 1998. However, Lukoil's president, Vagit Alekperov states that the cuts will not include crude oil exports. To compensate for the cuts, Alekperov says that Lukoil will boost sales of oil products in Russia's domestic market.

March 30

The U.S. Environmental Protection Agency (EPA) sues Mobil Corporation alleging that Mobil violated the Clean Water Act by spilling more than one million gallons of oil and contaminated water into tributaries of the San Juan River. The lawsuit cites 73 spills occurring between December 1991 and January 1998 at oil and gas production fields Mobil operates on Navajo Nation lands near Aneth, Utah. EPA also alleges that Mobil violated EPA regulations by not having containment equipment in the event of a spill and by failing to report some spills. The maximum penalty under the Clean Water Act is more than $5 million.

March 31

The Organization of Petroleum Exporting Countries (OPEC) releases an official communique from its 104th (extraordinary) meeting convened in Vienna, Austria, on March 30, 1998. The communique states that member countries have agreed to voluntary cuts from each country's current production levels in an attempt to boost oil prices. OPEC has agreed to cuts totaling 1.245 million barrels per day effective April 1, 1998. The cuts, in barrels per day, break down as follows: Algeria

50,000; Indonesia 70,000; Iran 140,000; Kuwait 125, 000; Libya 80,000; Nigeria 125,000; Qatar 30,000; Saudi Arabia 300,000; United Arab Emirates 125,000; and Venezuela 200,000. In addition, non-OPEC oil-producing countries Mexico, Oman, and Yemen have agreed to cut production by 100,000, 30,000, and 20,000 barrels per day, respectively. Moreover, a third non-OPEC country, Norway, the world's third largest oil exporter, has pledged to reduce its oil production by 3 per cent, or approximately 100,000 barrels per day. However, Norway's cuts will not take effect until mid-April 1998.

April 1998

April 1

India's state-owned Gas Authority of India reaches an agreement, in principle, with U.K.-based Cairn Energy, U.S.-based Brown and Root, and Australia's Command Petroleum to build a $2.5 billion natural gas pipeline from Myanmar to India's eastern coast. The proposed deep-sea pipeline would transport nearly one billion cubic feet of natural gas annually from fields offshore Myanmar to India's Orissa state. The four companies plan to form a joint venture company by mid-1999.

April 1

Following a G-8 (G-7 plus Russia) ministers' summit, U.S. Energy Secretary Federico Peña maintains that the U.S. government will not fund the construction of a proposed oil pipeline stretching from Azerbaijan to the Turkish Mediterranean port of Ceyhan or a pipeline running from Kazakhstan's Tengiz oil field to the Russian Black Sea port of Novorossisk. However, Peña also indicates that the U.S. government does advocate multiple pipeline routes for the area.

April 2

Saudi Arabia's Petroma Refining and Marketing Company and U.S.-based Hutchison Group are planning to build a 500,000-barrel-per-day joint venture refinery in Bahrain. The refinery will be part of a larger $5 billion petrochemical complex to be located in Bahrain's new Hidd industrial zone on Muharraq Island northeast of the mainland. A joint venture company, called Bahrain Refining Company, will be set up to run the project. Construction is expected to begin in 1999, and the refinery is scheduled to come on line in 2003.

April 3

After agreeing to reduce its oil production by 200,000 barrels per day beginning April 1, 1998, Venezuela announces that it will first fill its storage tanks. Venezuelan Oil Minister Erwin Arrieta says that Venezuela will keep production at current levels, as long as it has spare storage capacity, and reduce the volume of crude oil going into the market to achieve the 200,000-barrel-per-day cut. Once the storage tanks are full, then it will cut production. Venezuela reportedly has total storage capacity of 40 million barrels.

April 5

Egypt and China sign an agreement to establish a joint drilling company with an investment capital of $73.8 million. The company will be set up by the Chinese National Oil Company and two Egyptian private concerns.

April 8

U.S. Energy Secretary Federico Peña and key members of the U.S. Congress agree to temporarily halt the sale of $207 million worth of oil from the Strategic Petroleum Reserve. The oil sale is designed to cover the Reserve's expenses so that it is self sufficient. Government officials are now looking to cover the costs from other sources. Peña has delayed the sale for one month.

April 8

In support of last month's decision by OPEC and some non-OPEC countries to cut oil production by approximately 1.5 million barrels per day, Russia and China announce that they will cut production by 61,000 and 150,000 barrels per day, respectively. In addition, Russia has pledged to reduce oil products exports by 30,000-40,000 barrels per day. The cuts by Russia amount to a 2.3 per cent reduction in crude exports and a 3.2 per cent reduction in product exports. China's cuts represent about 5 per cent of its planned 2.9-million-barrels-per-day output for 1998.

April 8

Petroleos de Venezuela SA (PDVSA) announces that due to its commitment to cut oil production by 200,000 barrels per day as part of a coordinated effort by the Organization of Petroleum Exporting Countries (OPEC) and some non-OPEC countries to boost oil prices, it will be unable to meet oil deliveries to some of its international clients and therefore, declares force majeure on affected shipments. This declaration follows recent statements from PDVSA's president and the Energy and Mines Minister that Venezuela would keep production at current levels, as long as it has spare storage capacity, and reduce the volume of crude oil going into the market to achieve the 200,000-barrel-per-day cut. PDVSA now states that the country has reduced output from 3.370 million barrels per day to 3.170 million barrels per day. No information has been released on how many cargoes or clients will be affected by the declaration.

April 8

Exxon Corporation and British Petroleum announce that they will spend $1.2 billion developing the Hoover and Diana oil and gas fields, located south of Galveston, Texas, in the Gulf of Mexico. The fields contain estimated recoverable reserves of more than 300 million oil-equivalent barrels, and production is estimated to begin by 2000. Exxon holds a 66.7 per cent operating interest in the project, with British Petroleum holding the remaining 33.3 per cent interest.

April 9

Crude oil output from the Organization of Petroleum Exporting Countries (OPEC) totals 28.86 million barrels per day for March 1998, an increase of 163,000 barrels per day from February 1998. OPEC production is now more than 1.3 million barrels per day above the 27.5- million-barrels-per-day quota level which took effect on January 1, 1998. Rising Iraqi oil production accounts for 50 per cent of the March production increase. Iraq's oil production is currently estimated to be at 1.837 million barrels per day.

April 9

Russia announces that it will cut oil exports through Latvia in a dispute over the Russian-speaking minority in the Baltic state. Andrei Pershin, spokesman for Acting First Deputy Prime Minister Boris Nemtsov, contends that Latvian laws violate the rights of Latvia's Russian-speaking population. Pershin says that Russian oil exports through the Ventspils terminal will be cut 15 per cent. The Ventspils terminal handles 11 per cent to 12 per cent of Russian oil exports.

April 13

The discovery that an existing Soviet-era pipeline between Azerbaijan and Georgia must be replaced sparks new debate over the selection of pipeline routes out of the oil-rich Caspian Sea region.

Azerbaijan's state oil company Socar now wants to lay a much bigger 42-inch line with throughput of up to 600,000 barrels per day at a cost of $590 million. The original plan was to refurbish the existing 20-inch line at a cost of $315 million. The 540-mile route runs from Azerbaijan's capital, Baku, to Supsa on Georgia's Black Sea coast. The Azerbaijan International Operating Company, a consortium led by British Petroleum and responsible for constructing the pipeline, must now decide whether the bigger line is needed and who will pay for it.

April 14

Saudi Arabia announces that it will implement crude oil output cuts by further delaying new light crude production from the Ghawar field. In March 1998, Saudi Arabia agreed to cut its crude production by 300,000 barrels per day beginning April 1, 1998. Saudi Arabia's commitment is part of an agreement between the Organization of Petroleum Exporting Countries (OPEC) and several non-OPEC producers to voluntary cuts in crude production in an attempt to boost sagging oil prices.

April 14

Approximately 1,000 pipeline workers agree to a 17.5 per cent pay increase, ending a seven-day strike against construction company Brown Root Murphy, a unit of U.S.-based Halliburton Company. The strikers are working on a $2 billion natural gas pipeline stretching between Bolivia and Brazil. The 1,863-mile pipeline will transport between 283 million cubic feet and 1.1 billion cubic feet per day of Bolivian natural gas to southern Brazil.

April 14

In the continuing battle for the biggest share of the U.S. crude oil import market, Venezuela reclaims the top spot for February 1998, while Saudi Arabia falls to fourth behind Canada and Mexico. Data from the U.S. Energy Information Administration show that U.S. crude imports from Saudi Arabia fell 31 per cent from January 1998 to February 1998 and dropped below one million barrels per day for the first time in four-and-a-half years. Meanwhile, Venezuela exported 1.32 million barrels per day of crude to the U.S., claiming 18.6 per cent of U.S. crude imports in February 1998.

April 16

A joint venture between Chevron Corporation and the Nigerian National Petroleum Corporation begins production from the Gbokoda oil field, located on the western Niger Delta. Crude oil production is expected to reach 40,000 barrels per day by the end of 1998, and by 2000, the field's output will increase the joint venture's total production by more than 85,000 barrels per day.

April 16

Kazakhstan's President Nursultan Nazarbayev suspends further privatization of the country's oil and gas sector. Analysts estimate that Kazakhstan has so far privatized less than 50 per cent of its oil and gas industries. In its short history as a newly independent state, this energy-rich nation has attracted billions of dollars in foreign investment. The country's biggest energy projects are the Tengiz field operated by Chevron and Mobil and the Karachaganak oil and gas field operated by British Gas, Agip, and AO Lukoil.

April 19

The Gulf Cooperation Council's (GCC) Assistant Secretary-General for Economic Affairs, Abdulla al-Khulaifi, reports that the GCC's natural gas reserves have risen from 343 trillion cubic feet in 1986 to 731 trillion cubic feet today. He asserts that GCC countries currently hold 15 per cent of the world's gas reserves but only account for 4.6 per cent of world output. However, current trends indicate that, in the next few years, the GCC will account for around 10 per cent of all internationally-traded gas and become a leading supplier of liquefied natural gas. In order to diversify and expand its gas industries, GCC countries also are evaluating plans to connect all GCC states by an inter-Gulf gas grid. GCC member states are Bahrain, Kuwait, Oman, Qatar, Saudi Arabia, and the United Arab Emirates.

April 19

The Kuwait News Agency reports that OPEC members countries have lost an estimated $8 billion in the first quarter of 1998 due to the global drop in oil prices.

April 22

Iran's Deputy Oil Minister Ali Hashemi announces that a new natural gas field has been discovered at Taqdis Gardan in Iran's southern Fars province. The field contains 2.1 trillion cubic feet of recoverable gas and 10.5 million barrels of natural gas liquids gas valued at $3.2 billion. The field will be developed by Iranian contracting companies and is expected to be fully operational by 2000.

April 23

Northern States Power Company (NSP) of Minneapolis, Minnesota announces that it is pulling out of a $1 billion project to build a natural gas pipeline from western Canada to the U.S. Midwest region. Other reports indicate that another partner in the project, TransCanada Pipelines, is about to withdraw as well. The 800-mile line, called Viking Voyageur, was to transport up to 1.4 billion cubic feet per day of natural gas from Alberta through Minnesota, Wisconsin and Illinois. NSP and TransCanada each hold a 40 per cent interest in the pipeline, with Nicor of Naperville, Illinois holding the remaining 20 per cent.

April 23

U.S. oil companies Mobil and Exxon sign agreements with Turkmenistan to explore for oil and natural gas in the Turkmen Caspian Sea region. Mobil and its partner Monument Oil and Gas, a British company, will explore an area in western Turkmenistan near the Caspian Sea. Mobil will hold a 52.4 per cent interest in the project, and Monument will hold a 27.6 per cent interest. Turkmenistan's national oil company will hold the remaining 20 per cent interest.

April 23

France's Total and state-owned Petroleos de Venezuela SA (PDVSA) agree to purchase Norwegian-based Norsk Hydro's 15 per cent stake in the Sincor extra-heavy crude oil upgrading joint venture project located in Venezuela's Orinoco Belt. Total's share of the project will increase from 40 per cent to 47 per cent, while PDVSA's share will increase from 30 per cent to 38 per cent. Norway's state-owned Statoil retains its 15 per cent interest. Start-up of the Sincor project is expected by the end of 2001, with production levels reaching 200,000 barrels per day of extra-heavy crude.

April 24

ARCO announces that it plans to invest at least $2 billion in Venezuela over the next 10 years. Most of the investment will be in ARCO's heavy-oil joint venture project with Venezuela's state oil company Petroleos de Venezuela (PDVSA), Phillips Petroleum Company, and Texaco, scheduled to begin production in 1999. The remainder of the investment will go toward four oil fields which ARCO and its partners are operating for PDVSA.

April 27

Bulgaria and Russia sign an agreement under which Russia will deliver between 177 billion cubic feet and 212 billion cubic feet of natural gas annually to Bulgaria for the next 12 years. The price of the gas has not been disclosed; however, Bulgaria will pay 20 per cent of the price in goods or through construction work.

April 28

Despite direct appeals for relief from Iraq's foreign minister and oil minister, the United Nations (U.N.) Security Council decides to continue economic sanctions on Iraq, imposed in 1990 following Iraq's invasion of Kuwait. The Security Council also announces that it will once again begin reviewing the sanctions every 60 days. The Council suspended 60-day reviews in October 1997 after Iraq expelled American members of U.N. weapons inspection teams.

May 1998

May 1

Iraq maintains that to meet its $2 billion oil export target, it has contracted for the export of 186 million barrels of oil under the third phase of the United Nations sponsored oil-for-food program. The third phase ends on June 3, 1998. The amount of oil for export in the third phase is higher than the previous two phases due to lower world oil prices. Iraq exported 120 million barrels during the first phase, and exported another 127 million barrels in the second phase.

May 3

Amoco Norway, a subsidiary of Amoco Corporation of the U.S., announces an oil and natural gas discovery off the coast of Norway. Amoco estimates that the find, located in the Donnatello block, holds between 200 million and 500 million barrels of oil and gas. Initial tests at the site flowed approximately 5,200 barrels per day of oil and 5.3 million cubic feet per day of gas. Company officials maintain that the find is one of the biggest off the Norwegian coast in recent years. Amoco holds a 30 per cent interest in the Donnatello block, followed by the Norwegian government's State Direct Financial Interest with 30 per cent, Britain's Enterprise Oil PLC with 25 per cent, and Mobil Corporation of the U.S. with 15 per cent.

May 4

The Atlantic Richfield Company (ARCO) announces that it will acquire Union Texas Petroleum Holdings Incorporated, an independent oil company based in Houston, Texas, for $2.47 billion. The acquisition will add 140,000 barrels per day to ARCO's oil and natural gas production and increase ARCO's total oil and gas reserves by 14 per cent. The deal also helps ARCO enter the Caspian Sea region, with ARCO gaining a 12.5 per cent interest in the Caspian Pipeline Consortium and a 5 per cent interest in Kazakhstan's Tengiz oil filed. ARCO also will gain additional interests in projects located in the United Kingdom, Indonesia, Alaska, and Venezuela.

May 6

Taiwan's Economics Ministry decides to delay the privatization of state-owned Chinese Petroleum Corporation (CPC). The decision follows a parliamentary committee's call for the delay so that legislation, detailing how the petroleum industry will operate after CPC is privatized, can be drafted. Analysts suggest that the first sale of state shares May be delayed from six months to two years. The original privatization plan called for a 15 per cent stake to be sold by the end of June 1998, and another 18 per cent by July 1999. Government ownership in CPC was scheduled to fall to less than 50 per cent by July 2000.

May 8

The United Nations (U.N.) Security Council accepts a recommendation to end a limited travel ban on some Iraqi government officials. The ban was approved on November 12, 1997 following Iraq's refusal to allow U.N. weapons inspectors access to eight "presidential palaces." In a letter to the Security Council, Richard Butler, chairman of the U.N. Special Commission, reported that Iraq has been granting weapons inspectors access to all sites and recommended termination of the travel ban.

May 11

In March 1998, the Organization of Petroleum Exporting Countries (OPEC) and several non-OPEC countries agreed to crude oil production cuts, beginning April 1, 1998, in an attempt to boost sagging world oil prices. Iraq was the only OPEC member excluded from the deal because its oil exports are conducted under the United Nations oil-for-food program. A Dow Jones poll of industry observers indicates that OPEC's April 1998 crude oil output, including Iraq, was 27.995 million barrels per day, down 865,000 barrels per day from March 1998. A second report from the Paris-based International Energy Agency (IEA) indicates that OPEC's April 1998 crude production, including Iraq, was 27.95 million barrels per day, down 800,000 barrels per day from the March 1998 rate. The Dow Jones poll shows that, since February 1998, OPEC, excluding Iraq, has cut production by 1.117 million barrels per day, while IEA reports that production, excluding Iraq, has dropped by 1.03 million barrels per day over the same period. These declines are between 83 and 90 per cent of the 1.245 million barrels per day of production cuts agreed to by OPEC at the end of March 1998.

May 11

India announces that it has conducted three underground nuclear tests, the country's first since 1974. The tests were conducted simultaneously 330 miles southwest of New Delhi, near the Pakistani border. The Indian government indicates that the three tests included a thermonuclear device, commonly known as a hydrogen bomb. Two days later, on May 13, 1998, India announces that it has conducted two more underground nuclear tests in the same desert range.

May 13

Russia postpones the privatization of Eastern Oil Company, Tyumen Oil Company, and Slavneft until at least the fall of 1998. Alexander Braverman, First Deputy Minister for State Property, says that attempts to sell government stakes in the companies have failed thus far because of a lack of demand. Braverman also indicates that if the privatization of AO Rosneft goes well, the government May reduce the stakes to be offered in Eastern, Tyumen, and Slavneft or delay the sale even further.

May 14

The United Nations Security Council issues a declaration deploring India's nuclear test explosions, urging maximum restraint and dialogue to ease tensions between India and Pakistan. The statement also calls on India, Pakistan, and other nations to sign, without conditions, international treaties banning test explosions and nuclear proliferation. Immediately following the tests, the United States and other countries imposed economic sanctions on India.

May 14

The U.S. Federal Energy Regulatory Commission approves an agreement among the seven owners of the Trans-Alaska Pipeline System (TAPS) designed to increase competition and lower tariffs. TAPS runs from Prudhoe Bay on Alaska's North Slope to the port of Valdez, Alaska.

May 18

The Caspian Pipeline Consortium (CPC) takes its first major step towards construction of a $2 billion, 940-mile pipeline route from Kazakhstan's Tengiz oil field to Russia's Black Sea port of Novorossiisk. CPC has secured approval of an investment feasibility study, and must now obtain approval for a construction feasibility study and final land allocation. The pipeline will have an initial capacity of 560,000 barrels per day of oil, eventually rising to 1.34 million barrels per day. CPC officials expect the line to be operational in 2000.

May 18

Under an agreement with the European Union and Russia, U.S. President Bill Clinton confirms that the United States will not impose sanctions on France's Total, Russia's Gazprom, and Malaysia's Petronas under the Iran-Libya Sanction Act (ILSA). Passed in 1996, ILSA imposes mandatory and discretionary sanctions on non-U.S. companies which invest more than $20 million annually (lowered in August 1997 from $40 million) in the Iranian oil and gas sectors. In 1997, the three companies signed a $2 billion deal to develop Iran's South Pars natural gas field. In exchange, the European Union and Russia agree to tighten controls on the export of weapons technology to Iran.

May 19

The U.S. Virgin Islands approves a $625 million joint venture between Hess Oil Virgin Islands and Venezuela's state-owned Petroleos de Venezuela SA (PDVSA). The approval clears the way for PDVSA to acquire a 50 per cent stake in Hess Oil, a division of New York-based Amerada Hess Corporation, over a 10-year period. The two companies will share construction costs of a new, $500 million facility, utilizing Venezuelan heavy crude oil, to make coke, a solid petroleum product used in heavy industry.

May 19

The U.S. Justice Department files suit against Texaco, contending that the oil company was involved in a nationwide effort to undervalue oil it produced on federal and tribal lands in order to pay lower royalties. The suit is separate but similar to a suit the Justice Department filed against four other major oil companies in February 1998.

May 20

U.S. oil company Chevron Corporation confirms that its has selected Australian Gas and Light Company (AGL) and Malaysia's state-owned Petronas to build, own, and operate a $938 million, 1,304-mile section of a natural gas pipeline linking Papua New Guinea with Australia. Under the deal, AGL will be primarily responsible for the Australian onshore operations of the line, with Petronas managing a 100-mile offshore section north of Cape York. Chevron says that construction of the pipeline is expected to begin in 1999, with completion scheduled for mid-to-late 2001. The pipeline will be able to deliver 185 billion cubic feet per year of natural gas to markets on the east coast of Australia.

May 21

Following two weeks of student protests, civil unrest and riots, President Suharto resigns, ending 32 years as leader of State of Indonesia. Vice President Baharuddin Jusuf Habibie is immediately sworn in as the new president.

May 21

Russia's Prime Minister Sergei Kiriyenko pledges to quickly resolve a growing miners' strike spreading across Russia. Russian miners began protests and strikes on May 6, 1998 demanding back pay and guarantees that mines will not be shut down. The protests and strikes have been spreading across the country for the past two weeks, and miners have blocked traffic along the Trans-Siberian railway. The miners maintain that wages are up to six months overdue, amounting to an estimated $1.45 billion.

May 25

Russian oil companies AO Yukos and AO Sibneft call off merger proceedings to form AO Yuksi. If the merger had been completed, Yuksi would have been one of the world's largest private oil companies. In a joint press release, the companies blame the instability of the Russian stock market and sagging world oil prices for the failure of the merger. The companies also indicate that they now will independently seek western partners. Yuksi was expected to bid for a controlling interest in Russia's state-owned oil company AO Rosneft.

May 25

Union Texas Petroleum Holdings Incorporated assumes operation of the Boqueron marginal field from Venezuela's state-owned Petroleos de Venezuela SA. In a press release, Union Texas reports that it will use high-pressure natural gas injection to boost crude oil production. Production from Boqueron, which is currently 8,100 barrels per day, will be increased to 50,000-60,000 barrels per day within three to five years.

May 26

The deadline for the long-awaited auction of Russia's state-owned oil company AO Rosneft passes without any of the Western and Russian companies expected to compete for Rosneft submitting an offer. Prime Minister Sergei Kiriyenko is expected to announce a new auction and to lower the $2.1 billion starting price the Russian government has been seeking.

May 26

China verifies that a field located in the Qaidam basin in northwest Qinghai province holds 2.7 trillion cubic feet of natural gas. This brings the total proven gas reserves in Qinghai oil field to 5.3 trillion cubic feet.

May 27

Norway's Statoil AS and Norsk Hydro AS sign an agreement with Russia's Gazprom to jointly explore the Pechora Gulf area of the Barents Sea, a region considered rich in petroleum potential.

May 27

Colombia's state-run oil company Ecopetrol signs an offshore exploration agreement with Amoco Corporation and Atlantic Richfield Company for two blocks off Colombia's Caribbean coast. The two

U.S. oil companies are expected to complete seismic studies within the next two years, and to invest $190 million drilling six test wells over the next six years. The companies will concentrate on exploration for crude oil, but also expect to find natural gas deposits.

May 28

Shell Oil Company, Texaco, and Saudi Arabia's Aramco approve final terms and conditions of a proposed U.S. refining and marketing joint venture. The alliance will be the nation's largest seller of gasoline, combining the companies' eastern and Gulf Coast refining and gas station businesses. The joint venture will be called Motiva Enterprises LLC and is expected to be operating in June 1998.

May 29

Pakistan announces that it has conducted five underground nuclear tests, the country's first ever nuclear detonations. The test were conducted at the Chagai Hills test site in western Pakistan, near the Afghan border. Pakistan follows up with two more tests on May 30, 1998. These additional tests are conducted in the remote desert hills of southwestern Baluchistan province.

May 29

United Nations (U.N.) Secretary General Kofi Annan approves Iraq's new aid distribution plan under the U.N. sponsored oil-for-food program, clearing the way for Iraq to export up to $5.26 billion worth of oil during the fourth phase of the program. Early this year, the U.N. agreed to allow Iraq to increase its oil exports from $2 billion, but Iraq maintains that it only has the capability to reach $4 billion. If oil prices remain at current levels, Iraq will need to export close to 2 million barrels per day of oil to reach $4 billion, up from current export rates of around 1.5 million barrels per day. To reach 2 million barrels per day, Iraq's production facilities are estimated to need approximately $300 million worth of repairs.

June 1998

June 1

The Russian government announces that it is reducing the price for a 75 per cent stake in AO Rosneft. Prime Minister Sergei Kiriyenko says that the minimum bid for the state-owned oil company will be set at $1.6 billion. The government was forced to cancel the sale last month after receiving no bids at the previous minimum price of $2.1 billion.

June 2

S.M. Hosseini, Iran's Deputy Oil Minister for International Affairs, reports that Iran is seeking to increase its oil production capacity and capture at least 7 per cent of the world crude oil market by 2020. Iran currently holds 5.7 per cent of the world crude oil market. Hosseini says that Iran will accomplish its goal by using enhanced oil recovery (EOR) technologies, exploring new areas, and attracting foreign partners to previously discovered fields. Iran expects to recover an additional 20-30 billion barrels of oil using EOR technologies, and another 20 billion barrels from undeveloped areas in northwest and northeast Iran.

June 2

After recent coal strikes contributed to the collapse of Russian financial markets, the Russian government begins paying back wages to the striking miners. Anatoly Chekis, a trade union official from the Kuzbass region, says that back wages have been paid for the past three months' work. However, Chekis maintains that some miners have not been paid for as long as six months.

June 3

Russian gas company Gazprom announces that it is cutting in half its natural gas supplies to Yugoslavia due to that country's failure to pay for previous supplies.

June 4

Jordan awards a $400 million contract to Amoco Corporation for the construction of a 168-mile natural gas pipeline, running from fields in Egypt's Nile Delta, across the Sinai Peninsula, and under the Red Sea to the southern Jordanian port of Aqaba. Beginning in 2001, the pipeline will transport 212 million cubic feet of gas per day, gradually increasing to 318 million cubic feet per day by 2006.

June 4

In a second attempt in recent months to raise world oil prices, Saudi Arabia, Venezuela, and Mexico agree, during a meeting in Amsterdam, to reduce their oil production by 450,000 barrels per day starting July 1, 1998. Under the agreement, Saudi Arabia will cut 225,000 barrels per day, Venezuela 125,000 barrels per day, and Mexico 100,000 barrels per day. These cuts are in addition to the reductions agreed to under the so called "Riyadh Pact" of March 1998. The three major oil exporters now will attempt to secure additional cuts from other oil exporters ahead of the June 16, 1998, meeting of the Gulf Cooperation Council and the June 24, 1998, meeting of the Organization of Petroleum Exporting Countries.

June 6

Jordan completes a feasibility study on a 650-mile oil pipeline, to run between the Iraqi and Jordanian capitals. The proposed pipeline would cost approximately $250 million, with a capacity of 100,000 barrels per day. Jordanian officials say that they expect to approach France, Italy, Germany, and Japan to help finance the line.

June 7

In support of the Amsterdam agreement reached by Saudi Arabia, Venezuela, and Mexico, Qatar pledges to cut its crude oil output by 20,000 barrels per day. This latest pledge is in addition to the 30,000-barrel-per-day cut Qatar announced in March 1998.

June 8

Nigeria's leader Gen. Sani Abacha dies after suffering a heart attack. Just hours after Gen. Abacha's death, Gen. Abdulsalam Abubakar is named the country's new head of state.

June 9

In its Oil Market Report for May 1998, the International Energy Agency (IEA) lowers its forecast for world oil demand in the second quarter of 1998 by 0.51 million barrels per day. The revision is due to the continuing Asian economic difficulties and weak U.S. oil demand in April and May 1998. At the same time, the IEA estimates that crude oil production by all members of the Organization of Petroleum Exporting Countries (OPEC) in May 1998 averaged 28.07 million barrels per day, down 250,000 barrels per day from its revised April 1998 production estimate. By contrast, a Dow Jones survey of industry participants estimates that total OPEC crude oil output in May 1998 was 28.167 million barrels per day, an increase of 172,000 barrels per day from its April 1998 estimate. The IEA also updated and revised its estimates of OPEC compliance with the 1.245-million-barrel-per-day production reduction agreed to at the OPEC meeting on March 31, 1998. OPEC crude oil production in May 1998, excluding Iraq (whose production was not included in the agreement), was estimated to be 1.0 million barrels per day below that for February 1998, the baseline month for the reduction agreement. This 80 per cent compliance in May, the IEA noted, was in contrast with their revised estimate for April, when only 560,000 barrels per day, or just over 50 per cent, of the targeted reduction was achieved.

June 10

Six days after Saudi Arabia, Venezuela, Mexico pledged additional production cuts of 450,000 barrels per day effective July 1, 1998, Iran pledges additional cuts of 100,000 barrels per day. The new cuts are in addition to the 140,000-barrel-per-day reduction Iran pledged to make in March 1998.

June 12

U.S. President Bill Clinton extends by 10 years the ban on offshore oil drilling for much of the United States. The offshore drilling ban was first imposed by President George Bush in 1990. The extended ban prevents drilling for oil and natural gas in most parts of the Atlantic and Pacific coasts until 2012. It does not include areas off the Texas and Louisiana coasts, where extensive drilling has been going on for many years.

June 13

Cairn Energy announces that it has started production at the Sangu-1 natural gas field, Bangladesh's first operational offshore gas field. The initial flow rate from the field, located 30 miles southwest of Chittagong in the Bay of Bengal, has reached 20 million cubic feet per day, and is expected to increase to 160 million cubic feet per day over the next six to nine months. The field contains proven reserves of 850 billion cubic feet. Cairn holds a 55.25 per cent stake in the field, with Halliburton holding 25 per cent and Royal Dutch/Shell holding the remaining 19.75 per cent.

June 15

Russia's gas monopoly Gazprom notifies Belarus that it will reduce natural gas exports to Belarus by 30 to 40 per cent because of unpaid bills. Belarus owes Gazprom an estimated $200 million for past natural gas supplies.

June 16

Shell Nigeria shuts in 129,000 barrels per day of Bonny crude oil output due to a combination of community unrest in the Nembe and Odeama regions of Nigeria and a pipeline leak near the Nun River flowstation. Shell has declared force majeure effective June 19, 1998, through the end of the month, affecting a total of six Bonny crude cargoes. The cargoes are expected to be delayed for one to two days.

June 16

At a meeting of oil ministers from the six-nation Gulf Cooperation Council (GCC), the United Arab Emirates (UAE) and Kuwait each pledge to cut their oil production by 75,000 barrels per day, while Oman pledges to cut its output by 20,000 barrels per day. These GCC cuts follow a second round of production cuts announced by Saudi Arabia, Venezuela, and Mexico on June 4, 1998. The latest production cuts from UAE, Kuwait, and Oman will take effect on July 1, 1998.

June 17

Boris Nemtsov, Russia's deputy prime minister, announces that Russia will cut oil exports by 63,000 barrels per day beginning July 1, 1998, to support efforts by other oil producers to boost world oil prices. Nemtsov says that the cut is equivalent to 1 per cent of Russia's daily oil production.

June 18

Shell Nigeria reopens four of five Bonny crude oil flowstations in the Nembe and Odeama Creek regions. The flowstations were shut on June 8, 1998, due to local community unrest. A spokesman for the company says that resumption of operations at the flowstations means that the force majeure announced on June 16, 1998, for the rest of June, will only cause a delay of one day to cargo loadings. Six cargoes will be affected by the delay.

June 18

U.S. President Bill Clinton nominates Bill Richardson to be the next Secretary of Energy, replacing Federico Peña, and nominates Richard Holbrooke to replace Richardson as the U.S. ambassador to the United Nations.

June 18

Belgian oil company PetroFina SA announces its seventh oil and natural gas discovery offshore Vietnam. It is the company's second gas-only discovery, yielding 36 million cubic feet per day of natural gas.

June 19

The United Nations (U.N.) Security Council unanimously approves a resolution allowing Iraq to spend $300 million on spare parts for its oil industry. The funding is intended to help Iraq increase oil exports under the fourth phase of the U.N.'s oil-for-food program. The spare parts are expected to expand Iraq's oil export capacity from 1.6 million barrels per day to 1.8 million or 1.9 million barrels per day.

June 20

Egypt joins efforts by the Organization of Petroleum Exporting Countries (OPEC) and several non-OPEC nations to support sagging world oil prices by cutting 20,000 barrels per day from its own crude oil production. Egypt says the cut will take effect on July 1, 1998. However, Egypt also announces that its oil production will decline by 20,000 barrels per day by 1999 as a result of declining output from aging fields.

June 23

Occidental Petroleum declares force majeure on the Cano Limon oil field in Colombia after five guerrilla attacks in four days on the Cano Limon pipeline, which carries crude oil out of the area. The field produces 152,000 barrels per day. Occidental last declared force majeure on the field in August 1997.

June 24

The Organization of Petroleum Exporting Countries (OPEC) agrees, at its 105th ministerial conference, to another round of oil production cuts. In recent weeks oil prices have fallen to their lowest levels in more than a decade. OPEC members have agreed to cut production by 1.355 million barrels per day, effective July 1, 1998, bringing the group's total reductions since March 1998 to 2.6

million barrels per day. Together with promises from non-OPEC nations such as Russia, Oman, and Mexico, world oil producers have pledged to cut world-wide production by approximately 3.1 million barrels per day.

June 24

Russia confirms that it will cut its oil production by 100,000 barrels per day effective July 1, 1998.

June 29

Saudi Arabia calls for a new alliance of seven to nine major oil-producing countries, establishing an ad-hoc group that would work together to raise oil prices and reduce the glut of oil in the current market. Saudi Arabia, Venezuela, and Mexico currently are in high-level negotiations to create the new alliance, which May also include Iran, Iraq, Kuwait, United Arab Emirates, and possibly Libya and Norway.

June 29

Amoco announces plans for its TransCaspian Gas Pipeline System project which will transport natural gas from Turkmenistan to markets in Turkey and Europe. The $2.4 billion, 750-mile pipeline will start near Turkmenbashi in western Turkmenistan, extending across the Caspian Sea basin, to a landing point near Baku, Azerbaijan. The line will continue across Azerbaijan and Georgia to Erzurum, Turkey, and link into the Turkish gas transportation grid. Construction is expected to take three years, and the pipeline will have an initial capacity of 350 billion cubic feet per year, eventually increasing to 1.225 trillion cubic feet per year.

June 29

Mexico's state-owned oil company Petroleos Mexicanos (Pemex) receives a $1 billion credit from the Export-Import Bank of Japan. Pemex says the money will be used to increase production at the Cantarell oil fields located in the Gulf of Mexico.

June 30

Brazil's Petroleo Brasileiro SA (Petrobras) and five other companies sign an agreement to build a 373-mile natural gas pipeline between the Brazilian cities of Uruguayana and Porto Alegre. The pipeline is expected to be operational in December 2000 with a capacity of 424 million cubic feet per day. Petrobras' partners in the project are Total Global Gas Ventures, Nova, YPF SA, Techint SA, and Compania General de Combustibles SA.

July 1998

July 1

Johnston Southern Development, a joint venture between U.S. companies Southern Energy and Johnston Development, signs an agreement with Shanxi Enhua Energy to develop a $500 million power project in China. Johnston Southern Development will build two-300-megawatt, coal-fired power plants in Shanxi province near the city of Gujiao. The project is one of the first large-scale power plants to be sponsored by China's coal industry. Site preparation is expected to begin in 1999. Shanxi Enhua will hold a 51 per cent stake in the project, with Johnston Southern holding the remaining 49 per cent stake.

July 2

Iran's National Iranian Oil Company (NIOC) announces that it is seeking foreign investors for a number of onshore and offshore development projects. The onshore projects include the Ceshmeh Khosh gas injection project and the Darquain, Masjed-e-Suleyman, Agha Jari, and the Central Zagrog oil fields, all of which would be offered on a buyback basis. Offshore development projects include the Nowrooz, Foroozan, Salaman, Sirri C and D blocks, Hendajan, Soroosh, South Pars oil fields, and a pilot development at Esfandiar. NIOC says that foreign investments May help Iran increase its offshore oil production capacity by 276,000 barrels per day by 2000 from a current rate of approximately 500,000 barrels per day.

July 2

Amoco Kazakhstan and Turkish Petroleum Overseas form an equally-held joint venture, Tepco LLC, to explore for and produce oil and natural gas onshore in western Kazakhstan. Under the terms of the production sharing agreement, Tepco will commence immediate exploration of Central Territory Block B. Turkish Petroleum is a unit of Turk Petrolleri Anonim Ortakligi, Turkey's state oil company.

July 3

The Royal Dutch/Shell Group (Royal Dutch) announces that it has withdrawn from a joint bid for Rosneft, Russia's only remaining state-owned oil company. Royal Dutch says that low oil prices, which would limit returns from the project, and financial turmoil in Russia contributed to its decision to pull out of the bid. Royal Dutch's Russian partners in the bid are Lukoil and Gazprom.

July 5

Iraq and Jordan sign an agreement for the construction of an oil pipeline between the two countries. The pipeline is expected to replace a fleet of more than 3,000 tankers that currently transport Iraqi crude oil to Jordan. Jordan depends entirely on Iraq for oil and currently incurs $50 million per year in transportation fees.

July 6

Russia and Kazakhstan sign an agreement that divides the northern part of the Caspian Sea seabed into separate Russian and Kazakh sectors, recognizing Kazakhstan's claim to the oil near its coastline. The agreement also provides both nations with equal access to the Caspian's fishing grounds. The five nations that surround the Caspian Sea–Russia, Kazakhstan, Azerbaijan, Turkmenistan, and Iran–have argued for years over who owns the oil in the Caspian seabed. Previously, Kazakhstan, Azerbaijan, and Turkmenistan, the three nations with oil near their coastlines, had called for dividing the Caspian into national sectors, while Russia and Iran, which have little or no oil near their coasts, had advocated sharing the Caspian's resources. Although Azerbaijan, Turkmenistan, and Iran are not part of the settlement, it May set a precedent for future agreements among the Caspian nations.

July 7

Following the Royal Dutch/Shell Group's action on July 3, 1998, British Petroleum (BP) becomes the second major oil company to withdraw from a possible bid for Rosneft, Russia's last state-owned oil company. BP intended to bid with its Russian partner Omeximbank, which controls Russian oil company Sidanco. The Russian government is now once again considering postponing the privatization of Rosneft until the end of 1998. The sale of Rosneft originally was scheduled for May 1998.

July 7

Hungary's new government announces that it will discontinue the privatization of the country's energy sector, leaving Hungarian Electricity Works MVM's nuclear power plant, Paks Atomeromu, under state control. This action halts a planned sale of up to 49 per cent of MVM to private investors. The previous government privatized Hungary's regional electricity and natural gas distributors and its power generators.

July 8

Following a small fire, Chevron shuts a 170,000-barrel-per-day crude unit at its El Segundo, California refinery. A date for restarting the unit is currently unknown. The El Segundo refinery has a total crude oil capacity rating of 258,000 barrels per day.

July 9

For the second time in two months, the Russian government postpones the sale of state oil company, Rosneft. The announcement comes just days after British Petroleum and the Royal Dutch/Shell Group withdrew from their respective bidding partnerships. Russian officials now say that the sale will take place in September or October 1998.

July 9

Mobil of Venezuela announces that Venezuelan engineering company Vepica and Jantesa and Japan's JGC have won a $500 million contract to develop a crude oil upgrading unit for the Cerro Negro heavy crude project located in the Orinoco Belt. Mobil is the operator of the project. The upgrading unit will transform extra-heavy 8.5° API crude into lighter 16° API crude. The unit is expected to be completed in 2001, at which time the crude output of the Cerro Negro project will double to 120,000 barrels per day from an initial level of 60,000 barrels per day in 1999.

July 9

In its June 1998 monthly oil market report, the International Energy Agency (IEA) projects that world oil demand in 1998 will be 1.1 million barrels per day higher than in 1997. This is about half the incremental demand in 1997, which was 2.1 million barrels per day. In addition, IEA reports that June 1998 Organization of Petroleum Exporting Countries (OPEC) crude production fell 330,000 barrels per day to 27.95 million barrels per day from 28.28 million barrels per day in May 1998. When compared to February 1998, OPEC production from non-Iraqi sources has been cut by 970,000 barrels per day, which is approximately 80 per cent of the 1.245 million-barrel-per-day target agreed to by OPEC in March. A Dow Jones survey of industry observers reports similar findings. The survey finds that OPEC production totaled 27.853 million barrels per day in June 1998, a decrease of 314,000 barrels per day from May 1998 output.

July 10

Monument Oil and Gas signs a production-sharing agreement with Turkmenneft, Turkmenistan's state oil company, and Mobil Exploration & Production Turkmenistan for the Garashsyzlyk concession area, located in western Turkmenistan. The agreement covers development and production of oil and natural gas over a 1,700 square mile area. Mobil holds a 52.4 per cent share in the deal, while Monument holds a 27.6 per cent share and Turkmenneft 20 per cent.

July 13

South Korea's Samsung agrees to build a $240 million oil refinery for Ghana's state-owned Tema Oil Refinery. Construction of a residual fluid catalytic cracking unit will begin later this year and is expected to be complete in 2001. The unit will refine about 93,000 barrels per day of residual fuel oil, producing various types of oil products such as gasoline and liquefied petroleum gas.

July 13

Access Industries of the U.S. suspends the extraction and shipment of coal from the Bogatyr mine located in Kazakhstan. Bogatyr is the world's largest coal mine. Sergey Karpukhovich, the company's vice-president, cites a $20 million debt for delivered coal accumulated by Kazakh consumers and the tightening of tax laws as reasons for the suspension.

July 14

Sonangol, Angola's state oil company, and Essor Exploration Angola, a subsidiary of Exxon, discover a new offshore oil field located about 224 miles northwest of Luanda, Angola. The field is called Hungo One and has an estimated production capacity of 16,000 barrels per day. It is the third discovery in Block 15 in recent months. Essor holds a 40 per cent stake in Block 15, with BP Exploration Angola holding 26.67 per cent, Italy's Agip Angola 20 per cent, and Norway's Den Norske Stats Oljeselskap 13.33 per cent.

July 15

Iraq and Syria sign an agreement to build a second oil pipeline between the two countries. No information is available on the new line's capacity or the route it will take; however, U.S. State Department spokesman James P. Rubin states that implementation of the agreement would be a violation of United Nations sanctions against Iraq. Two days earlier, Syria agreed to reopen an existing oil pipeline, linking Iraq's Kirkuk oil fields to the Mediterranean terminals of Banias in Syria and Tripoli in Lebanon. Syria closed the line in 1982 in support of Iran during the Iran/Iraq war. The pipeline has an estimated capacity of 650,000 barrels per day and is expected to be operational within a few months. However, use of the pipeline will require United Nations approval.

July 15

Australia's Boral Ltd. confirms that its Bass Strait drilling program yielded estimated reserves of between 450 billion and 600 billion cubic feet of natural gas in the Yolla Field. Partners in the Yolla Field include Boral with a 30.5 per cent interest, Premier Petroleum with 30.5 per cent, Cue Energy Resources with 14 per cent, CalEnergy Gas with 20 per cent, and Santos with 5 per cent.

July 16

The U.S. blocks the proposed August price formula for Iraqi crude oil over concerns that the Kirkuk grade is priced too low for the European market. It is the first time Iraq's crude oil price formula has been questioned since the United Nations' oil-for-food program began in 1996.

July 17

Mobil Oil and the Royal Dutch/Shell Group announce that they are pulling out of the $3 billion Camisea natural gas project in Peru. The companies indicate that demands from the Peruvian government for local gas supplies at reduced rates made the project unprofitable. Camisea holds an estimated 11 trillion cubic feet of natural gas, and the development project was considered to be a key to Peru's economic development.

July 20

Abida tribesmen blow up part of Yemen's Marib-Ras Issa oil pipeline in protest of government price hikes on gasoline, kerosene, and cooking gas. This marks the eighth attack on Yemeni oil and gas pipelines since June 19, 1998. The 272-mile pipeline carries 220,000 barrels per day of oil and runs from Marib to the Red Sea province of al-Hudaydah, 140 miles west of Yemen's capital Sanaa.

July 21

British Petroleum, Amoco, and Ramco Energy sign contracts to develop Azerbaijan's Caspian Sea oil resources. In three separate contracts, the companies have pledged to invest as much as $13 billion over the next 15 years. The Caspian Sea region is said to contain approximately 30 billion barrels of oil reserves, comparable to the North Sea region.

July 21

Oil analysts calculate that oil export revenues of the Organization of Petroleum Exporting Countries (OPEC) have fallen by a third in 1998 due to the decline in oil prices. According to Petrostrategies of Paris, OPEC's collective export revenues have fallen from $89.6 billion in the first six months of 1997 to $60.7 billion in the first six months of 1998. Analysts predict that for all of 1998, OPEC oil export revenues will fall 30 per cent from a 1997 total of $121.5 billion. Export revenues have declined for almost all OPEC members, with the exception of Iraq, whose revenue has increased 20 per cent in the first six months of 1998 under the United Nations= oil-for-food program.

July 22

Unocal agrees to pay fines of $43.8 million to California for contaminating the state's Central Coast with millions of gallons of petroleum over four decades. Unocal allowed a refined oil product, called diluent, to gradually leak from an underground pipeline at its Guadalupe Oil Field in San Luis Obispo County. As part of the settlement, $9 million will be used to restore natural resources, $15 million will fund water quality improvement projects, and $11.1 million will create emergency response programs for future oil spills and other pollution incidents on the Central Coast.

July 22

In response to poor market conditions, Singapore Refining Company (SRC) cuts crude oil throughput by 45,600 barrels per day, or 16 per cent, at its 285,000-barrel-per-day refinery. SRC will continue to run the refinery at this lower throughput level until mid-August 1998, when the company will decide on an optimal operating level. SRC is a joint venture made up of Singapore Petroleum, British Petroleum, and Caltex Petroleum.

July 22

Rep. Thomas Bliley (R-VA), Chairman of the U.S. House Commerce Committee, abandons efforts to pass electric industry restructuring legislation in 1998. The legislation is designed to end utilities protected monopoly service territories and institute retail electric competition nationally.

July 23

In response to U.S. opposition to the August pricing plan for Iraqi crude oil, Iraq offers the United Nations a revised pricing mechanism for Iraq's August oil exports. The revised pricing formula adjusts the price of Kirkuk and Basrah Light for the European, U.S. and Asian markets. The U.S. had been concerned that, under the original pricing plan, prices for crude oil headed for European markets were too low.

July 27

Russia announces plans to sell a five per cent stake in RAO Gazprom, the country's largest company and the world's largest natural gas company, with the Russian government retaining a 35 per cent stake in the company. The five per cent stake is valued at $500 million, although the government May get more if the shares are sold to foreign investors. The Royal Dutch/Shell Group and Italy's ENI SpA are considered to be probable contenders for the shares. Gazprom holds nearly 40 per cent of the world's proven natural gas reserves.

July 27

Nigeria's leader General Abdulsalam Abubakar cancels oil concessions granted to Nigerian companies at the time of previous leader General Sani Abacha's death on June 8, 1998. General Abubakar also reportedly is no longer awarding new oil concessions or prospecting leases. Companies that are affected by the decision include Malabor Petroleum, DICS Petroleum, Feuimarc Petroleum, and Continental Petroleum.

July 29

British Petroleum (BP) begins oil production at its Schiehallion field at an initial rate of 30,000 barrels per day. The field is located offshore northern Scotland. The development is expected to produce 340 million barrels from the main field and an additional 85 million barrels from the nearby Loyal reservoir. Peak production is expected to reach 154,000 barrels per day. BP is the operator of the field and holds a 33.35 per cent stake. BP's partners include Royal Dutch/Shell Group (33.35 per cent), Amerada Hess (15.66 per cent) and Murphy Oil (5.88 per cent) of the U.S., Norway's Statoil (5.88 per cent), and Australia=s OMV (5.88 per cent).

July 29

Anadarko Petroleum announces that it has discovered oil and natural gas at the Eugene Island South Addition Block 346 in the Gulf of Mexico, located approximately 75 miles offshore Louisiana. Anadarko estimates that the Tanzanite field contains reserves of at least 140 million barrels of oil equivalent.

July 29

The Indonesian government postpones exploration of the giant Natuna natural gas field due to the economic crisis in Thailand and the oversupply of liquefied natural gas in the world market. Thailand was expected to be the main buyer of Natuna gas. Production from the field was expected to begin between 2003 and 2007, but Indonesia government now says exploration May not restart until 2007. The Natuna field is located in the Natuna Sea about 700 miles north of Jakarta.

July 29

The Russian nuclear safety federation approves the construction of two 640-megawatt nuclear power reactors on the Kola peninsula, located in northwest Russia near Finland. The reactors will be VVE640 type units, considered to be the most advanced reactors in the world. However, Heikki Aulamo, a spokesman at the Finnish Radiation and Nuclear Safety Authority, says that the reactors May never be built due to Russia's current financial situation,.

July 30

Dutch energy and water company Nuon NV opens Asia's largest and first commercial wind farm, located on the island of Nan'ao near the southern Chinese city of Shantou. Nuon developed the farm with three Chinese partners, Shantou Electric Power Board, China Fulin Wind Power Development and Zhen Neng Wind Power. The wind farm has a capacity of 24 megawatts, and the electricity generated by the farm will be fed directly into the Chinese grid. It is expected to service approximately 60,000 households.

July 31

The U.S. Senate confirms the nomination of Bill Richardson as Secretary of Energy. Richardson replaces Federico Peña, who resigned in June 1998.

August 1998

August 2

Oman LNG signs an initial 20-year agreement with MetGas of India, an affiliate of Enron Corporation, to supply MetGas with 1.2 million metric tons (58.44 billion cubic feet) per year of liquefied natural gas (LNG). Deliveries are scheduled to begin in late 2001, and the LNG will be used to generate electricity at a 2,450-megawatt power plant currently under construction on the Indian coast, 150 miles south of Bombay.

August 3

The Caspian Pipeline Consortium (CPC) submits the final draft of a construction feasibility study for a Caspian Sea oil pipeline to Russia's regional authorities for approval. The $2 billion, 981-mile pipeline will transport 1.3 million barrels per day of oil from Kazakhstan's Tengiz oil field to the Russian port city of Novorossiisk. Construction is scheduled to begin in late 1998 or early 1999, and the line is expected to be operational in 2001.

August 3

The Mexican and Venezuelan governments renew the San Jose Accord, under which these two countries provide crude oil and refined products at favorable terms to other countries in Central America and the Caribbean. Under the annually-renewed accord, participating countries pay for oil at the market price when the international price is below $15 per barrel. When the price of oil is above $15 per barrel, a portion of the oil price is converted into credits that are used to finance development projects in participating countries and trade between participating nations and the two supplier countries. Countries participating in the accord, in addition to Mexico and Venezuela, include: Barbados, Belize, Costa Rica, the Dominican Republic, El Salvador, Guatemala, Haiti, Honduras, Jamaica, Nicaragua, and Panama.

August 4

Japan's Nuclear Safety Commission approves plans for the construction of a commercial nuclear power plant, the country's first nuclear plant in 10 years. The light-water reactor will be located in the village of Higashidori, 392 miles northeast of Tokyo, and will have a generation capacity of 1.1 million kilowatts. Construction is set to begin in December 1998, with operations expected to begin in July 2005. The plant will be operated by Tohoku Electric Power.

August 5

In protest against economic sanctions imposed on Iraq following the Persian Gulf War, President Saddam Hussein freezes cooperation with United Nations (UN) weapons inspectors. UN weapons inspectors have been searching suspected weapons sites in Iraq since 1991 in an attempt to eliminate Iraq's weapons of mass destruction.

August 5

Lone Star Gas International, a unit of Texas Utilities, wins a 30-year concession to distribute natural gas throughout Mexico City, Mexico. Over the first five years of the concession, Lone Star and its two local partners, Grupo Diavas and Controladora Comercial & Industrial, will invest $125 million expanding Mexico City's natural gas distribution network.

August 5

Chevron U.S.A. and Arco Alaska agree to equal ownership of leases on the Alaskan North Slope. Each company will hold a 50 per cent stake in 35 tracts consisting of 200,000 acres southwest of the Kuparuk and Alpine oil fields. In addition, the companies will each hold a 50 per cent stake in 16 tracts in the McCovey/Salmon area, including 196,000 offshore acres of the Prudhoe Bay oil field in the Beaufort Sea.

August 6

The U.S. Interior Department announces that it will open a new area of the Alaska North Slope wilderness to oil and gas leasing. Interior Secretary Bruce Babbitt says exploration will be allowed on a four million acre parcel in the northeastern quarter of the National Petroleum Reserve, west of the Prudhoe Bay oil field. However, an area north and east of Teshekpuk Lake near the Beaufort Sea, will remain off limits. The National Petroleum Reserve is estimated to contain 2.2 billion barrels of oil.

August 6

Abdulillah al-Tikriti, director-general of the economy department at Iraq's Foreign Ministry, reports that Iraq has discovered a new oil field and a new natural gas field. The oil field is located in the northern part of Iraq, with estimated reserves of 500 million barrels. The natural gas field, located in the desert west of Baghdad, contains estimated reserves of 21.2 trillion cubic feet.

August 7

In its July 1998 report, the International Energy Agency (IEA) estimates that the Organization of Petroleum Exporting Countries (OPEC) produced 27.81 million barrels per day of oil in July 1998, down 355,000 barrels per day from 28.17 million barrels per day in June 1998. Excluding Iraqi production, OPEC produced 25.53 million barrels per day in July. The figures show OPEC to be 56 per cent compliant with production cuts agreed to earlier this year. In addition, the IEA reduces its projection of world oil demand in 1998 by 200,000 barrels per day to 74.7 million barrels per day, a rate just 900,000 barrels per day higher than demand in 1997.

August 7

A Dow Jones survey of oil market observers shows Organization of Petroleum Exporting Countries (OPEC) production for July at 27.67 million barrels per day, down from 27.853 million barrels per day in June. Excluding Iraq, OPEC produced 25.41 million barrels per day, representing 61 per cent compliance with agreed production cuts.

August 10

The Middle East Economic Survey (MEES) reports that U.S. imports of Iraqi crude oil have jumped to 100,474 barrels per day in the first four months of 1998, compared to an average of 85,989 barrels per day for 1997. MEES also notes that the number of U.S. companies importing Iraqi crude oil has fallen from 10 in 1997 to 3 thus far in 1998–Chevron, Clark, and Exxon.

August 11

British Petroleum announces that it will acquire Amoco for $48.2 billion in stock. If the merger is approved by regulators and shareholders of both companies, it will be the largest oil industry merger and the largest foreign take-over of a U.S. company to date. The company will be known as BP Amoco, and it will be the world's third-largest multinational oil company in terms of net income behind Exxon and Royal Dutch/Shell Group.

August 11

Bolivia announces the discovery of a new gas field in the southern part of the country near its border with Argentina. The San Alberto X field contains natural gas reserves of 1 trillion cubic feet and is expected to produce 35.5 million cubic feet per day. Production from the field will feed into a 1,900-mile pipeline currently under construction between Santa Cruz, Bolivia and the Brazilian state of Rio Grande do Sul. Interests in the field are held by Petrobras Bolivia, Petrolera Andina, and Total Exploration Bolivie.

August 13

As part of Russia's continuing privatization program, the government announces that it will sell a five per cent stake in Gazprom as a single lot, at a starting price of $1.65 billion. The sale will leave the government with a 35 per cent stake in the natural gas monopoly. In other privatization news, Gazprom President Rem Vyakhirev announced on August 6, 1998, that Gazprom will not bid for a controlling stake in Rosneft, Russia's only remaining state oil company.

August 13

Saudi Arabia's state oil company Aramco notifies U.S., European, and Japanese crude oil lifters that it will cut September 1998 exports by 18 per cent. The cut is much higher than traders anticipated, causing a jump in September crude oil prices.

August 13

Data from the U.S. Energy Information Administration show that Saudi Arabia was the leading exporter of crude oil to the highly competitive U.S. market in June 1998. U.S. imports of Saudi crude oil averaged 1.585 million barrels per day in June, followed by Venezuela at 1.424 million barrels per day, and Mexico at 1.379 million barrels per day.

August 13

Petroecuador, Ecuador's state-run oil company, reports that it has increased its crude oil transport capacity through the country's two main oil pipelines by nearly 30,000 barrels per day, to more than 400,000 barrels per day. The company says the capacity expansion work cost $787,000 and was completed by the Ecuadorean unit of Argentine oil company.

August 16

The National Iranian Drilling Company signs a contract worth $25 million with Turkmenistan to drill four wells in the Turkmen-Bashi region. The drilling work will be carried out over the next 18 months.

August 17

Shell Philippines Exploration, a unit of Royal Dutch/Shell Group, awards a $500 million contract to Allseas Marine Contractors for the construction of a 314-mile natural gas pipeline in the Philippines.

The new line will link the Malampaya gas fields, located near Palawan Island, with a landing terminal in Tabangao, Batangas.

August 18

Phillips Petroleum begins oil and gas production from Ekofisk II, the company's $2.5 billion North Sea redevelopment project. As part of the project, Phillips phased out or modified 14 existing platforms and installed 31 miles of new pipeline. Production capacity is expected to be 306,000 barrels per day of oil and 789 million cubic feet per day of natural gas, approximately 20 per cent above initial expectations.

August 18

BG PLC and Atlantic Richfield (Arco) report that proven natural gas reserves contained within their production sharing contract blocks in Irian Jaya, Indonesia have more than doubled to 14.4 trillion cubic feet. The reserves are located in the BG-operated Muturi block and the Arco-operated Wiriagar and Berau blocks in Berau Bay. Natural gas from these blocks will supply the Tangguh liquefied natural gas project of Pertamina, the state-run oil company of Indonesia. The project is expected to come online between 2003 and 2004.

August 20

Iraq and Syria sign an agreement to reopen an oil pipeline linking Iraq's Kirkuk oil fields to Mediterranean terminals in Banias, Syria and Tripoli, Lebanon. Syria closed the pipeline in 1982 in support of Iran during the eight-year Iran-Iraq War. The pipeline has an estimated capacity of 650,000 barrels per day. Since Iraqi oil exports currently operate under the United Nations (UN) oil-for-food program, Iraq will need UN approval before the pipeline can be reopened.

August 20

Following its regular 60-day review of trade sanctions against Iraq, the United Nations (UN) Security Council decides to extend the sanctions and expresses concern over Iraq's continuing refusal to cooperate with arms inspectors. On August 5, 1998, Iraq suspended cooperation with UN weapons inspectors, stating that it has destroyed its banned weapons of mass destruction and calling on the UN to lift the trade sanctions.

August 21

Due to Afghanistan's continuing civil war and recent U.S. attacks on suspected terrorist training camps inside Afghanistan, Unocal suspends plans for the construction of a $2 billion natural gas pipeline that would run from Turkmenistan to Pakistan via Afghanistan. Unocal has stressed that the gas pipeline project will not proceed until an internationally- recognized government is in place. To date, only three countries–Saudi Arabia, Pakistan and the United Arab Emirates–have recognized Afghanistan's Taliban government.

August 22

A senior official from Aramco, Saudi Arabia's state oil company, announces that production has started at the company's new light crude Shaybah oil field. Production capacity is expected to be 500,000 barrels per day. Shaybah is located in an area of the empty quarter in southeast Saudi Arabia, near the border with the United Arab Emirates. The field contains an estimated seven billion barrels of 40EB 42E API gravity extra light crude oil.

August 25

French oil company Elf Aquitaine abandons its plan to invest more than $500 million and acquire a 12 per cent stake in Russian oil company Sibneft. This latest actions follows Elf's announcement on August 21, 1998 to delay its plan to invest in the Russian oil firm.

August 25

Unocal reports that its West Seno #2 discovery well offshore Indonesia tested at a rate of 10,069 barrels per day of 39E API gravity oil and 9.5 million cubic feet per day of natural gas. The West Seno #2 well is located on the Makassar Strait production-sharing contract area. Unocal is the operator of Makassar Strait and holds a 50 per cent interest, with Mobil holding the remaining 50 per cent.

August 27

Belgian petroleum firm Petrofina announces that production from the Badami oil field in Alaska commenced on August 24, 1998. This field's reserves are estimated at 120 million barrels and production capacity is expected to be 20,000 barrels per day. Petrofina holds a 30 per cent stake in the field, with British Petroleum holding the remaining 70 per cent. Badami is located 35 miles east of Prudhoe Bay.

August 28

At the Daqing oil field, China has closed 543 wells, 1,104 wells are inundated, and daily production is down approximately 1 per cent due to flooding from the Nenjiang River in northeast China's Heilongjiang Province. Daqing is China's largest oil field and accounted for about 50 per cent of the country's total onshore production in 1997. Flooding along the Nenjiang and Songhua Rivers has reduced crude oil output at the Jilin, Liaohe, and Daqing oil fields by around 5 per cent.

August 29

Mobil Corporation agrees to pay $56.5 million to settle lawsuits with the U.S. Federal Government and the State of Texas over royalties for oil extracted from public and Indian lands. Both suits claimed that Mobil undervalued the oil taken from these lands. Mobil will pay $45 million to settle with the Federal Government and the Navajo and Jicarilla Apache tribes and $11.5 million to settle with Texas.

August 31

Japan's Mitsubishi Oil reports that production has started at its Rang Dong oil field, located in Block 15-2 offshore southern Vietnam. Peak production is expected to reach 45,000 barrels per day. Mitsubishi holds a 82.5 per cent interest in Block 15-2, with PetroVietnam, Vietnam's state-run oil and gas company, holding the remaining 17.5 per cent.

August 31

Indonesia's President B.J. Habibie approves Atlantic Richfield's plan to build a $4 billion liquefied natural gas (LNG) complex in Irian Jaya, Indonesia. LNG from the complex will be used to meet future demand for natural gas from power utilities in Japan, South Korea, and Taiwan.

September 1998

September 2

China discovers its first large coalbed methane gas field, located in northern China's Shanxi province. The field's proven reserves are estimated to be 1.9 trillion cubic feet, with annual production expected to reach 35 billion cubic feet by the end of 1998.

September 3

France's Elf Aquitaine discovers its fourth oil field in Block 17 offshore Angola. The discovery is located at the Lirio 1 well which tested at an initial rate of 11,000 barrels per day. Elf holds a 35 per cent stake in Block 17 and is the block's operator for a consortium consisting of Esso Exploration, a unit of Exxon (20 per cent); British Petroleum (16.67 per cent); Den Norske Stats Ojeseskap (13.33 per cent); Norsk Hydro (10 per cent); and Petrofina (5 per cent).

September 3

The U.S. Senate approves a spending bill providing the U.S. Department of Energy with $420 million at current prices to buy oil for the Strategic Petroleum Reserve (SPR). If a joint House-Senate conference committee preserves the measure, it could add 29 million barrels of crude oil to the SPR, which currently contains 563 million barrels.

September 4

Royal Dutch/Shell and Texaco announce that they will combine their European refining and marketing activities. Shell and Texaco expect to save approximately $200 million by joining operations. Shell will hold 88 per cent of the venture, with Texaco holding 12 per cent. Together, the two companies will control 19 refineries and process 2.1 million barrels per day of oil, or about 15 per cent of European capacity. The companies say that the deal could be finalized by mid-1999.

September 7

Norway's Saga Petroleum signs a joint study agreement with the National Iranian Oil Company to develop an onshore/offshore area in the northern part of the Persian Gulf. The area to be studied includes the Dara Block and an undeveloped section of the Hendijan field. The agreement provides for a 12-month exploration study.

September 9

The United Nations (U.N.) Security Council unanimously agrees to a resolution suspending periodic reviews of economic sanctions imposed on Iraq. The resolution is in response to Iraq's August 5, 1998, decision to break off cooperation with U.N. arms inspectors. The Security Council has been conducting regular 60-day reviews and more comprehensive six-month reviews of the sanctions.

September 10

The U.S. Energy Information Administration (EIA) reports that in 1997, for the first time in 10 years, U.S. proved reserves of crude oil increased. Crude oil reserves reached 22.546 billion barrels, up 2.4 per cent from 22.017 in 1996. In addition, EIA reports that natural gas reserves increased for the fourth year in a row.

September 11

The Paris-based International Energy Agency estimates that in August 1998, the Organization of Petroleum Exporting Countries (OPEC) implemented 83 per cent of their pledged oil production cutbacks. OPEC production, excluding Iraq, fell to 24.83 million barrels per day. Total OPEC production for August 1998 declined to 27.26 million barrels per day from 27.65 million barrels per day in July 1998. Meanwhile a Dow Jones survey of oil market observers indicates that OPEC was 86 per cent compliant with its oil production cutbacks. According to Dow Jones, OPEC's August production, excluding Iraq, fell to 24.757 million barrels per day. Total August output declined to 27.058 million barrels per day from 27.67 million barrels per day in July 1998.

September 14

Phillips Petroleum and Japanese-owned Indonesian Petroleum Company sign an exploration agreement with Kazakhstan to acquire stakes in 10 blocks totaling 2,000 square miles in the Kazakhstan sector of the Caspian Sea. The exploration area is said to contain an estimated 6 billion barrels of recoverable oil reserves. Other stakeholders include Royal Dutch/Shell, Mobil, British Petroleum, Statoil, Agip, Total, and British Gas. Over the next six years, the consortium will drill six wells and conduct seismic surveys.

September 14

Anadarko Petroleum reports that its Tanzanite No. 1 well tested 21,917 barrels per day of oil and 29.7 million cubic feet per day of natural gas. The well is located in the Gulf of Mexico approximately 75 miles offshore Louisiana.

September 17

Lasmo announces that peak oil production from its Dacion development in eastern Venezuela will hit 120,000 barrels per day by the end of 2002. This is a 650 per cent increase compared to the 16,000 barrels per day forecast for the end of 1998. The Dacion development contains estimated reserves of 550 million barrels of oil. Lasmo holds a 90 per cent stake in the project, with Venezuelan state oil company Petroleos de Venezuela holding the remaining 10 per cent.

September 18

Mobil's Canadian unit receives six oil and gas exploration licenses totaling more than one million acres on the Grand Banks, offshore Newfoundland. Two of the licenses are within the eastern edge of the Flemish Pas Basin, while the remaining four are within the Jeanne d'Arc basin. Mobil plans to invest $100 million exploring these areas over the next five years.

September 18

Arco China (a unit of Atlantic Richfield), Texaco China, and Chinese National Offshore Oil agree to develop the Qinhuangdao 32-6 offshore oil field, located in China's Bohai Bay. Oil production is expected to begin in mid-2001 and to reach 60,000 barrels per day by 2002. Chinese National Offshore Oil will be the operator of the field with a 51 per cent interest and Arco and Texaco will each hold a 24.5 per cent interest.

September 23

Texaco acquires a 65 per cent stake in the North Buzachi oil field in west Kazakhstan from Nimir Petroleum, a privately-owned Saudi Arabian oil company. Nimir will continue to hold the remaining 35 per cent stake. North Buzachi is an undeveloped onshore oil field, located 120 miles north of the Caspian Sea port of Aktau. The field contains estimated reserves of 1 billion to 1.5 billion barrels of oil. Texaco reports that production could begin within two to four years.

September 25

Egypt and Jordan announce the testing of the linkage of their power systems as the first step toward a proposed electricity grid extending east to the Persian Gulf and west to Morocco and even

Europe. The Egypt-Jordan link should begin commercial operation in October. The power is carried by cables that are lifted over the Suez Canal by 740 foot towers and then run along the floor of the Gulf of Aqaba between the two nations.

September 27

Chile switches to summer time three weeks earlier than in previous years, a move ordered by the government to save energy in the wake of a long drought.

September 27

Turkey restores full diplomatic ties with Iraq after nearly six years of all-but-frozen relations. Iraq's ambassador to Ankara left last year. It is not known whether Iraq's ambassador will now return.

September 28

Japan's Ministry of International Trade and Industry decides to liquidate 27 oil exploration companies affiliated with the debt-ridden, state-run Japan National Oil Corporation (JNOC) as part of the Ministry's restructuring scheme. The Ministry estimates JNOC's losses for fiscal year 1998-1999 will amount to approximately $3 billion, some of which can be covered by proceeds from the sale of some of the subsidiaries.

September 28

India's Foreign Investment Promotion Board approves 23 project proposals, including a liquefied natural gas (LNG) terminal in Kakinada on the eastern coast. The board also gave the go-ahead for a disinvestment proposal by state-owned Bharat Petroleum in relation to its liquefied petroleum gas storing and marketing joint venture with Caltex Petroleum.

September 29

Finance Minister Mikhail Zadornov says that Russian oil companies' tax payments were 10 per cent less than expected in September. Zardonov also indicates that natural gas monopoly Gazprom has not paid any taxes so far during September.

September 30

Saudi Arabia asks several U.S. oil companies for proposals to develop its oil reserves. The topic is raised in a meeting between Saudi Crown Prince Abdullah and top executives from Atlantic Richfield, Chevron, Conoco, Exxon, Mobil, Phillips, and Texaco.

October 1998

October 1

South Korea's oil refining sector fully deregulates, allowing for 100 per cent foreign investment. Originally, South Korea had expected to fully deregulate its refining industry by January 1999, but it decided to move up the date in order to help reform its economy.

October 2

An underwater pipeline leaks an estimated 155,000 gallons of oil into the Gulf of Mexico, forming a slick about 100 miles south of the mouth of the Mississippi River.

October 4

Mexican Energy Minister Luis Tellez announces that his country will maintain oil export cuts for an additional six months, or until June 1999. As part of agreements earlier this year with producers from the Organization of Petroleum Exporting Countries, Mexico had already agreed to trim its exports by 200,000 barrels a day through the end of 1998.

October 5

China begins exploration at one of its largest gas fields in Chengdu, capital of Sichuan province. The field is estimated to contain approximately 3.5 trillion cubic feet of proved reserves.

October 7

European Union (EU) nations approve an accord in which European car makers will voluntarily agree to cut carbon dioxide emissions 25 per cent by 2008. EU officials say they will seek similar deals with automakers in Asia and North America.

October 7

China National Star Petroleum Corporation, the smallest of the country's three exploration companies, discovers 730 million barrels of crude oil reserves in the Tarim Basin. The oil field is located in the remote northwest province of Xinjiang.

October 7

Oil production at Shell Nigeria is cut by 269,000 barrels per day as a result of protests against Shell's activities in the region.

October 9

In a report to Congress, the U.S. Energy Information Administration (EIA) assessed the macroeconomic and energy market impacts of six energy-related, carbon-reduction scenarios in relation to a reference case. EIA undertook this study in response to a request from the Chairman and Ranking Minority Members of the House Committee on Science.

October 9

Phillips Petroleum Company and Ultramar Diamond Shamrock Corporation (UDS) agree to combine their gasoline stations and refineries into a joint-venture company to be called Diamond 66. The joint venture will be owned 55 per cent by UDS and 45 per cent by Phillips. Moody's Investor Services says Diamond 66 will be the largest independent refining and marketing company in North America.

October 9

French petroleum company Total SA starts oil production at the Sirri A field in the Persian Gulf near the coast of Iran. Total says production in the start-up phase will reach 7,000 barrels per day from three wells, with a goal of 20,000 barrels per day from 11 wells. Total heads the development of the two Sirri fields, along with Malaysia's Petronas.

October 11

Pakistan Prime Minister Nawaz Sharif announces a 50 per cent cut in power tariff surcharges to domestic consumers. The decision is made in the context of savings realized after fraud is unearthed in payments to the Hub Power Co., Ltd., or Hubco, an independent power producer. Recent moves against fraudulent activity are helping to increase foreign investment in the nearly bankrupt power sector.

October 12

The Paris-based International Energy Agency (IEA) estimates that the Organization of Petroleum Exporting Countries (OPEC) achieved 98 per cent of its targeted production reduction in September. The IEA reports that OPEC's September crude oil output, excluding Iraq, was 24.44 million barrels a day, compared with its target of 24.387 million barrels per day. Meanwhile, Dow Jones reports that crude oil output from OPEC members, excluding Iraq, totaled 24.319 million barrels per day in September, down from 24.757 million barrels per day in August, and 68,000 barrels per day below the production level agreed to at the June OPEC meeting.

October 13

Russian crude oil exports soar to an average of 2.54 million barrels per day for the first eight months of 1998, up 10 per cent from a rate of 2.3 million barrels per day during the same period last year.

October 13

Chad's Minister of Oil and Mines Abdoulaye says the 621-mile Chad-Cameroon oil pipeline will definitely go ahead as planned. The pipeline will transport 225,000 barrels of crude oil per day from the Doba oil fields in eastern Chad to the port of Kribi in Cameroon. Work on the pipeline is expected to start in early 1999.

October 14

China says it will give North Korea almost 600,000 barrels of crude oil. China cites "valuing the North Korea-China friendship and preserving the excellent traditions of helping each other in times of adversity" as reasons for the donation. Cash shortages have forced North Korea to cut its crude oil imports.

October 14

China's Xinhua News Agency reports that three new medium-sized oilfields have been discovered at Shiijiuto off Bohai Bay in northern China, with estimated proved reserves of 440 million barrels of oil and 300 billion cubic feet of natural gas.

October 15

Japan's Ministry of International Trade and Industry changes its forecast for domestic oil consumption in Fiscal Year 1998 from a growth of 0.3 per cent to a decline of 3.7 per cent.

October 15

The ITAR-Tass news agency reports that Russia's financial crisis has not affected foreign investment in Sakhalin Island's oil and gas projects. Sakhalin Island has attracted several joint ventures aiming to develop Sakhalin's eight onshore oil and gas fields.

October 18

Villagers trying to tap into a trans-national oil products pipeline in Nigeria touch off a blazing inferno which razes several villages and destroys a swath of surrounding cropland. Hundreds of people are killed.

October 18

Leftist rebels blow up an oil pipeline in Machuca, Colombia, a village 180 miles from Bogota, killing 45 villagers. President Andres Pastrana pleads for "gestures of peace" from the rebel group following the attack. The blast shuts down the Colombia Central pipeline, which transports 100,000 barrels per day from oil fields in east-central Colombia to the Caribbean coast.

October 19

The Middle East Economic Survey (MEES) reports that Iraq's methods for producing crude oil are causing permanent damage to the country's oil reservoirs. MEES reports this damage has been underscored by a deterioration in the quality of Iraqi oil exports.

October 19

The threat of confrontation between Iran and Afghanistan's ruling Taliban militia is defused after a series of diplomatic efforts by United Nations special envoys. Iran had sent more than 200,000 troops to its border with Afghanistan after Taliban troops killed eight Iranian diplomats and a journalist in northern Afghanistan in August 1998.

October 20

Norway's state oil company Statoil says its natural gas commitments to Europe are being met by increased output from its North Sea Troll and Statfjord fields, after production in the Sleipner area was shut down as a result of fire damage.

October 20

Turkish and Syrian officials sign a security agreement after several weeks of escalating tensions between the two countries. Turkey had threatened military action against Syria, claiming Syria was supporting Kurdish rebels.

October 20

Turkey invites Russia to cooperate in the construction of the Ceyhan oil pipeline route, which would transport Caspian Sea oil from the Azerbaijani capital of Baku to the Turkish Mediterranean port of Ceyhan.

October 20

China rejects Taiwan's offer to join an international consortium struggling to fund nuclear reactors for North Korea. Chinese Foreign Ministry spokesman Tang Guoqiang says the consortium was founded under a U.S.-North Korean agreement and "should be composed of sovereign states. As a province of China, Taiwan is not qualified to participate."

October 23

The seven biggest makers of diesel engines in the U.S. agree to pay over $1 billion in fines and corrective actions for allegedly cheating on antipollution tests. The Environmental Protection Agency (EPA) charges that the manufacturers used a computerized timing device to make engines conform to EPA emissions standards in laboratory tests, but that once the engines were put on the highway, their timing changed, spewing 1.3 million tons of nitrogren oxide above what the standards allow. This amount is equivalent to 65 million cars' total pollutants. Although they signed consent agreements, the manufacturers say they didn't cheat.

October 23

Canada announces tough new rules aimed at cutting sulfur levels in gasoline by about 90 per cent by the year 2005.

October 23

Israeli Prime Minister Benjamin Netanyahu and Palestinian leader Yasser Arafat sign a land-for-peace West Bank accord at a White House ceremony.

October 25

Turkey says it will limit oil tanker traffic through the Bosporus Straits due to environmental concerns. According to Turkish Foreign Minister Ismail Cem, about 440 million barrels of crude oil pass through the Bosporus each year on 4,500 oil tankers. Turkey is concerned about the potential for oil tanker accidents.

October 26

Peru and Ecuador sign a historic peace treaty, ending a 57-year dispute over a part of the Amazon jungle that had caused three wars between the two countries. The peace accord also outlines cooperation on several large projects, including the interconnection of oil pipelines. Peru and Ecuador are in the process of selecting an investment bank to work as an advisor for the project.

October 26

Georgia negotiates a $200 million loan from the European Bank for Reconstruction and Development to help build an oil pipeline that would carry some of the early oil produced by an international consortium in Azerbaijan's Caspian Sea shelf through Georgia to the Black Sea.

October 27

Indonesia's state-owned electricity company PT Perusahaan Listrik Negara says it will not buy electricity from independent power producers (IPPs) next year due to decreased electricity demand. This decision will adversely affect the IPP market in Indonesia.

October 27

The U.S.-led Korean Peninsula Energy Development Organization (KEDO) purchases 360,000 barrels of 2 per cent sulfur fuel oil for delivery into North Korea in November 1998.

October 28

South Korean Hyundai Oil Refining Company, a unit of the Hyundai Group, plans to open its first gasoline station in North Korea this year.

October 28

Japan's Nippon Oil Company, the country's second largest petroleum distributor and Mitsubishi Oil Company, the sixth-ranking company in the industry, agree to merge as of April 1, 1999. The combined company will be the largest oil distributor in Japan.

October 29

Turkey, Azerbaijan, Georgia, Kazakhstan, and Uzbekistan sign a declaration that declares the Baku-Tblisi-Ceyhan pipeline project as the main oil export pipeline for Azerbaijan. U.S. Secretary of Energy Bill Richardson serves as an official witness at the ceremony.

October 30

The presidents of Turkmenistan and Turkey sign an agreement under which Turkey would buy natural gas from Turkmenistan for 30 years. The gas would be shipped through a proposed trans-Caspian pipeline. Turkey would buy 565 billion cubic feet of gas annually, with the amount eventually ramping up to 1 trillion cubic feet per year.

November 1998

November 2

Russian President Boris Yeltsin approves the sale of an additional 5 per cent stake in Gazprom, the national natural gas company. The decree will allow foreign ownership of Gazprom stock to increase from its former limit of 9 per cent to 14 per cent.

November 3

The fourth conference on global climate change is held in Buenos Aires to discuss implementation of the Kyoto Protocol, which aims to reduce greenhouse gas emissions. Argentina and Kazakhstan push to include the topic of voluntary greenhouse-gas-emission reductions by developing nations on the agenda, but are defeated by other developing countries.

November 4

The president of the China National Petroleum Company says that natural gas will supply 8 per cent of China's energy demand by 2020, up from its current share of 2 per cent. China proposes building a pipeline grid to transport natural gas from the resource-rich western part of the country to the economically-more-developed east. China also aims to import Russian and Central Asian gas by 2020.

November 4

Russian Prime Minister Yevgeny Primakov cancels the sale of a controlling stake in state oil company Rosneft. The Russian media reports that the country's financial crisis has made it unlikely that the government would be able to find a buyer. The auction, which was scheduled to take place October 30, was the third attempt to sell off the company.

November 5

The South Korean Ministry of Trade and Industry reports that South Korean energy consumption fell 3.6 per cent in September 1998 compared to a year ago. However, this 3.6 per cent drop is smaller than the 10.6 per cent drop in August 1998 compared to August 1997. South Korean energy consumption during the first nine months of this year was down 6.7 per cent compared to a year ago. The energy demand decrease is a result of the country's economic slowdown.

November 6

The Norweigian government approves development of the North Sea Huldra Field, which contains approximately 6.9 trillion cubic feet of dry natural gas. The field will be operated by Statoil in conjunction with Total, Conoco, Petro-Canada, and Svenska Petroleum Exploration. Production is expected to begin in 2001.

November 8

China begins operations at the Shinxi oil field, the biggest desert oil field in the Xinjiang province, according to Chinese media reports. The field is expected to produce 1 million barrels of crude oil by the end of 1998. The field is believed to hold 840 million barrels of oil and 1 trillion cubic feet of natural gas.

November 9

Turkish Prime Minister Mesut Yilmaz breaks ground for part of a pipeline that will carry Iranian natural gas to Turkey and the rest of Europe. Turkey is scheduled to receive Iranian gas by mid-1999 through the pipeline under construction between the Iranian border and the Turkish city of Erzurum.

November 9

Interfax news agency reports that Russian crude oil exports rose by 9.9 per cent in the first nine months of the year, despite a government pledge to the Organization of Petroleum Exporting Countries to cut back production to help combat the oil price slump on the world market. Despite the rise in Russian crude exports, Russian oil revenues dropped by 28 per cent compared with the same period last year due to the lower world oil prices.

November 9

President Clinton lifts most of the economic sanctions that were imposed on India and Pakistan following nuclear tests each country conducted in May 1998. According to a senior U.S. official, the decision is a "recognition that some real progress has been made on the nuclear issueBbut not enough". The decision lifts the punitive measures hitting U.S. companies the hardest, and also will restore the availability of assistance from the U.S. Export-Import Bank, the Overseas Private Investment Corporation, and the Trade and Development Agency.

November 11

In a visit to Taiwan, U.S. Secretary of Energy Bill Richardson praises Taiwan's efforts to privatize the island's electricity sector.

November 12

The United States signs the Kyoto Protocol, the global climate change agreement aimed at reducing greenhouse gas emissions. The United States is the 60th nation to sign the protocol. Under the protocol, during the period 2008 to 2012 the United States is required to reduce emissions of carbon dioxide and other specified heat-trapping gases to 7 per cent below 1990 levels. The agreement still must be submitted to the U.S. Senate for ratification.

November 12

Nigerian youths abduct seven foreign oil workers, including three Americans, from an offshore Nigerian drilling rig that is under contract by Texaco. Oil rigs in Nigeria have been the subject of attack and protest on several occasions this year by Nigerian youths alleging that the targeted oil companies are destroying their villages.

November 13

Several major oil and gas companies close their Jakarta offices during the day due to civil unrest in the capital of Indonesia. Companies affected included Esso Indonesia, Limited, PT Caltex Pacific Indonesia, and Shell Company. At least 16 people, including eight students, were killed during the week in a wave of protests and rioting in Jakarta. The violence was the worst Jakarta has seen since May, when hundreds died in violence that led to the resignation of President B.J. Habibie's predecessor, ex-President Suharto.

November 13

Texaco and China launch a $500 million joint initiative to exploit coal-bed methane in eastern China. According to the news agency Xinhua, the Huaibei mine in Anhui province contains over 2.1 trillion cubic feet of methane. Some 300 wells with a combined annual output capacity of 18 billion cubic feet will be drilled in the joint effort.

November 14

President Clinton aborts airstrikes against Iraq after receiving a last minute letter from Iraq pledging unconditional cooperation with United Nations weapons inspectors.

November 16

The Organization of the Petroleum Exporting Countries (OPEC) crude oil production fell by 420,000 barrels a day in October to 26.85 million barrels per day from 27.27 million barrels per day in September, according to the Middle East Economic Survey. The Paris-based International Energy Agency estimates that OPEC October output was 26.99 million barrels per day including Iraq, and 24.56 million barrels per day excluding Iraq.

November 17

The Israeli Knesset approves the Wye Pact with the Palestinian Authority, setting the stage for the beginning of a phased Israeli troop pullback from parts of the West Bank.

November 17

Crude oil and gasoline prices are low and should stay that way for the foreseeable future, according to the Annual Energy Outlook released by the Energy Information Administration (EIA). EIA says the Asian economic crisis will continue to depress crude oil prices for much of the next decade, keeping them under $20 per barrel until at least 2007.

November 20

Entergy Corporation, a major utility in the Southern United States, emerges as the winning buyer in the first multiple-bid auction for a U.S. nuclear power plant. Entergy purchases the Pilgrim power plant, a 26-year-old reactor on the Massachusetts coast, for $80 million cash.

November 24

North Korea denies it is building an underground nuclear facility and disallows U.S. weapons inspectors to view the site. North Korea also threatens to abrogate the 1994 Korean Peninsula Energy Development Organization (KEDO) agreement requiring it to abandon its suspect nuclear weapons program in exchange for fuel oil and the construction of two light-water nuclear reactors by consortium members. North Korea is demanding a cash compensation of $300 million for the inspection, a demand the United States has rejected.

November 25

Russia, Kazakhstan, and major oil companies agree to build a multi-billion dollar pipeline linking the oil-rich Caspian Sea region with the Black Sea. The project is scheduled to be completed in 2001 and would be the first major new pipeline in the region. It would connect the Tengiz oilfield in Kazakhstan to the Russian Black Sea port of Novorosiisk. The Tengiz pipeline has had a tangled history, being under discussion for the better part of a decade.

November 26

The International Monetary Fund (IMF) agrees to resume assistance to Pakistan, which had been cut off earlier this year after it detonated five nuclear devices. The agreement follows a decision by the United States to support the IMF's resumption of aid to Pakistan, which has been on the verge of defaulting on its international debts.

November 26

Organization of Petroleum Exporting Countries (OPEC) member countries are unable to agree on additional production reductions beyond those agreed to in June 1998 or even on extending current reductions beyond June 1999. OPEC members do agree to reschedule their twice-yearly ministerial meetings to March and September from June and November.

November 27

Russian Prime Minister Yevgeny Primakov issues a resolution approving a draft agreement between the Russian and Iranian governments on cooperation in the oil and gas industry. The document stipulates the aim of the agreement is to promote economic, scientific, and technical cooperation in the areas of exploration, development, and storage construction.

November 30

A United Nations human rights envoy calls for an independent inquiry into claims of widespread environmental damage in Nigeria caused by international oil companies.

December 1998

December 2

Exxon Corporation agrees to buy Mobil Corporation for approximately $75.4 billion, which will make the company the largest corporation in the U.S. The companies say they expect to cut about 9,000 jobs from their combined worldwide workforce of 122,700 and to close offices, saving $730 million. The merger comes in the context of low oil prices, which have hurt profits at many oil companies.

December 3

President Clinton waives economic sanctions against India and Pakistan six months after imposing them as punishment for a series of nuclear weapons tests.

December 4

The director of Brazil's state-owned oil company Petroleo Brasileiro SA (Petrobras) says the company will not be privatized. Instead, the government plans to sell more Petrobras shares, but will remain the majority shareholder. The statement puts to rest speculation that Petrobras will be privatized as part of the country's privatization program.

December 5

Mobil Oil Qatar Inc., a unit of Mobil Corporation, signs a final agreement with state-owned Qatar General Petroleum Corporation to develop part of Qatar's natural gas reserves. The production sharing contract calls for the development of reserves in the North Field with a design capacity of 1 billion cubic feet per day, beginning in 1999. The gas will be supplied to power, chemical, and petrochemical industries at Qatar's Ras Laffan Industrial City.

December 5

Omani Oil Minister Mohammed bin Hamed Al-Rumhy says Oman has lost nearly 50 per cent of its oil revenues because of the sharp fall in oil prices this year. Al-Rumhy warns that the current lack of capital for oil and natural gas investment likely will cause shortages in these commodities down the line.

December 7

Unocal Corporation drops out of a consortium that has been trying to build a $4.4 billion Central Asian oil pipeline passing through war-torn Afghanistan. A company spokesman attributes the decision to continuing political instability in the region and low oil prices.

December 8

A massive power outage lasting most of the day brings much of San Francisco to a standstill. The outage is blamed on human error.

December 9

Kazakhstan signs an agreement with three international oil companies to conduct a feasibility study on building parallel pipelines for shipping the country's oil and natural gas to Western markets. The study will look at the commercial viability of shipping oil and natural gas from western Kazakhstan across the Caspian Sea to Baku, Azerbaijan, and then on to the Turkish port of Ceyhan on the Mediterranean. Signing of the agreement takes place in a ceremony at the U.S. Department of Energy.

December 11

Enron Corporation forms a $450 million joint venture with a leading South Korean natural gas distribution company, marking the company's first significant investment in the region. Enron says it will invest more than $220 million in a new natural gas distribution company in partnership with SK Corporation, a unit of global conglomerate SK Group. SK Group, which has about 20 per cent of South Korea's natural gas market, will contribute the assets of five city distribution companies valued at approximately $225 million. The new company will be called SK/Enron Company. The deal awaits South Korean government approval.

December 12

The United Nations approves 11 contracts for the export of 60.1 million barrels of Iraqi crude oil through the oil for food program. Iraq is allowed to export up to $5.26 billion of oil over 180 days in the fifth phase of the program, which began November 26, 1998.

December 12

New England Electric System (NEES) of Westborough, Massachusetts, and National Grid Group of Coventry, England, agree to merge. Under the proposal, which still requires regulatory approval, NEES will become a wholly-owned subsidiary of National Grid. Earlier in the week, Scottish Power announced it would take over U.S. PacificCorp for $7.9 billion in what would be the first purchase of an entire U.S. electric utility by a foreign company. In March 1998, a Massachusetts law took effect that encouraged electric companies to sell off their generating plants.

December 14

India's Petronet LNG Limited signs an agreement with Ras Laffan Liquefied Natural Gas Company Limited (RasGas) of Qatar and Mobil LNG and Power Incorporated of the United States for the supply of liquefied natural gas (LNG). The agreement provides for the supply of five million metric tons a year of LNG to India's Dahej terminal, and 2.5 million metric tons a year to the Cochin terminal, for a period of 25 years. The Dahej terminal is expected to meet the growing demand for gas by consumers in the state of Gujarat, while the Cochin terminal will meet gas requirements in the state of Kerala.

December 15

Royal/Dutch Shell announces it will sell 40 per cent of its chemicals' businesses and record changes of $4.5 billion, as part of a sweeping reorganization. The cutbacks, which will include job losses, are expected to affect Shell operations in Texas, Venezuela, Britain, Norway, and France.

December 16

Chevron Corporation, the third-largest oil company in the U.S., says it plans to cut costs by $500 million in 1999, primarily by cutting older projects in North America. Chevron Chairman Ken Derr said that significant spending would continue in promising areas such as Kazakstan, West Africa, and the Gulf of Mexico. The cuts announced by Chevron follow spending cuts announced recently by the Royal Dutch/Shell Group, and an announcement from Exxon and Mobil, which agreed to merge, to cut costs by $2.8 billion a year.

December 16

U.S. Energy Secretary Bill Richardson and Bangladesh Secretary of Energy and Mineral Resources Tawfiq-e-Elahi Chowdhury sign an agreement on U.S.-Bangladeshi cooperation in the development of Bangladesh's natural gas resources.

December 16

The United States begins a four-day air strike campaign against Iraq hours after United Nations (U.N.) military inspectors declare that their work in Iraq has been blocked. President Clinton says he ordered the attacked after a "stark, sobering, and profoundly disturbing report" from the chief U.N. arms inspector that described Iraq's actions to thwart weapons inspections.

December 18

A team of international scientists conclude that the period from December 1997 to December 1998 was the warmest year in recorded history. The United Nations Meteorological Organization reports that, altogether, the record of warming over the past 25 years clearly indicates a "greenhouse" effect warming of the climate from accumulating greenhouse gases, particularly carbon dioxide from the burning of coal, natural gas, and oil.

December 21

Venezuelan President-elect Hugo Chavez says he is waiting for additional information from state-owned company Petroleos de Venezuela SA before deciding when the country will comply fully with its commitment to cut crude oil output by 525,000 barrels a day.

December 21

Ruhrgas AG, a major German natural-gas distributor, buys a 2.5 per cent stake in giant Russian gas monopoly RAO Gazprom for $660 million, strengthening the two companies' cooperation. The successful tender of the Gazprom stake brings relief to the cash-strapped Russian government, which is battling a deep economic crisis and has $17.5 billion in foreign debt payments due in 1999. According to German press reports, Ruhrgas will eventually hold a 4 per cent stake in Gazprom once both companies have set up a joint venture.

December 21

European Union countries approve rules ensuring that consumers will be informed through labels, guides, and posters about the fuel efficiency of new cars. The rules are expected to come into force after approval in early 1999. They will require fuel economy labels to be attached to cars in showrooms, fuel efficiency posters to be hung in dealerships, fuel economy guidebooks to be made available for free in booklet, and electronic media form and fuel consumption data to be provided in promotional literature.

December 23

The Colombian government says it will allow gasoline and diesel prices to float with international oil prices starting January 1, 1999. The move will end a system of artificial price fixing which has cost the government more than $3.2 billion in subsidies over the past five years.

December 25

A Japanese consortium, including Japan Petroleum Development Corporation, Itochu Corporation, Indonesia Petroleum Limited, and Teikoku Oil Company, signs a production sharing contract with Azerbaijan's State Oil Company, also known as SOCAR, for an oil and gas concession in the Caspian Sea. The contract involves the development of an offshore Caspian Sea petroleum exploration block encompassing the Yanan Tava, Atashkah, and Mugan Deniz fields of Azerbaijan. The contract marks the first time Japanese oil developers have undertaken a project in the Caspian without an alliance with foreign partners.

December 28

A natural gas pipeline explodes in a rural hamlet in northern Colombia, killing at least 12 people, bringing down power lines, and setting homes on fire. Another 80 people are injured. Authorities have not determined the cause of the explosion, but say it appears accidental.

December 28

A consortium led by CMS Energy Corporation wins a bid to build a $1.6 billion natural-gas-fired power plant in India, beating out four other major energy companies, including a group headed by Enron Corporation. The 1,886-megawatt power plant will be built on India's eastern coast. The construction is scheduled to begin in April 2001 and commercial operation is expected to start by 2003.

December 30

U.S. federal regulators approve the proposed purchase of Amoco by British Petroleum P.L.C. after extracting a handful of antitrust concessions from the new company. With a market capitalization of about $140 billion, BP Amoco will rank third in the global oil industry behind the proposed Exxon-Mobil merger and Royal Dutch/Shell.